Riverscape Dynamics and Contemporary Hazards

This book provides a comprehensive exploration of morpho-hydro-sediment interactive processes to explain channel planform dynamics, integrating advanced quantitative techniques and assessing flood risks, including model-based mitigation strategies. It delves into the temporal evolution of riverscape dynamics driven by autogenic and allogenic factors, while evaluating sediment generation within catchments. Additionally, it features comparative case studies of riverscapes across diverse climatic regimes at comparable scales.

Key features of the book include:

- **Quantitative analysis of hydro-morphological indices and sediment components**: Offers detailed insights into sediment dynamics and their interactions with hydrological processes.
- **Land cover and land use linkages**: Explores the relationship between land use patterns and hydrologic regimes, highlighting their interconnected impacts.
- **Hydrology, ecology, and environmental linkages**: Discusses vulnerabilities and interactions between hydrological systems, ecological processes, and environmental factors.
- **Application of HEC-RAS flood modeling**: Examines spatial floodwater distribution scenarios with and without bridges, as well as strategies for reframing and relocating bridge structures.
- **Sediment generation and transportation**: Reviews sediment generation processes, transportation mechanisms, and physio-chemical characteristics in depositional environments.

This book is designed for graduate/postgraduate students and researchers in civil and environmental engineering, hydrology, and related fields, offering valuable insights into the complex interplay of hydrological, morphology, and sediment aggradation processes.

Earth, Environment, and Sustainability

Series Editors: Narendra Kumar and Vertika Shukla

Understanding of earth and environment includes earth's natural environment and reasons of alteration in natural environment, role of environmental dynamism, environmental degradation, growing concern about health and ecological damage caused by elevated levels of certain chemicals has always attracted attention. This series on "Earth, Environment and Sustainability" aims to cover different aspect of earth, environment, and sustainability to provide a connecting link between environment and earth with emphasis on importance of sustainability. The major topics focussed are earth composition and natural/anthropogenic phenomenon, Environmental dynamism, Earth's water profile, Environmental Degradation, Climate change, Environmental Health and so forth.

Climate Change and Sustainable Developments: Current Perspective and Technological Solutions
Edited by Chitralekha Nag Dasgupta, Rachana Singh and Ram Boojh

Adaptive Agro-Management for Environmental Sustainability
Edited by Sanjeev Kumar, Susmita Karmakar, Anita Rani and Ritu Singh

G20 and Energy Security: Roadmap for Sustainable Development (Argentina to Italy)
Edited by Richa Kothari, Neeta Rani and Anita Singh

Riverscape Dynamics and Contemporary Hazards
Mery Biswas, Adrija Raha and Paramita Haldar

For more information about this series, please visit: www.routledge.com/Earth-Environment-and-Sustainability/book-series/CRCEES

Riverscape Dynamics and Contemporary Hazards

Mery Biswas
Adrija Raha
Paramita Haldar

CRC Press
Taylor & Francis Group
Boca Raton London New York

CRC Press is an imprint of the
Taylor & Francis Group, an **informa** business

Designed cover image: © 2026 Mery Biswas

First edition published 2026
by CRC Press
2385 NW Executive Center Drive, Suite 320, Boca Raton FL 33431

and by CRC Press
4 Park Square, Milton Park, Abingdon, Oxon, OX14 4RN

CRC Press is an imprint of Taylor & Francis Group, LLC

ISBN: 978-1-032-89867-4 (hbk)
ISBN: 978-1-032-91381-0 (pbk)
ISBN: 978-1-003-56297-9 (ebk)

DOI: 10.1201/9781003562979

Typeset in Times
by SPi Technologies India Pvt Ltd (Straive)

Contents

About the Authors

Mery Biswas is Associate Professor in the Department of Geography at Presidency University, Kolkata, India. She is specifically concentrated on several aspects/methods of morphotectonic, riverscape geomorphology, and related geo-hazards (field-based, instrumental survey techniques, analytical models, and micro landforms).

Adrija Raha is Junior Research Fellow in the Department of Geography at Presidency University, Kolkata, India. She is a Doctoral Research Fellow in Geography at Presidency University, specializing in Fluvial Geomorphology and Tectonics. Her interests lie in understanding Earth's surface processes and tectonic dynamics, contributing to advancements in geoscience research.

Paramita Haldar is Assistant Professor in BITS Pilani at the K. K. Birla Campus, Goa. Haldar offers expertise in Molecular Modeling.

Preface

This is a great pleasure that this preface comprises the riverscape dynamics and the contemporary hazards with a special emphasis on the Himalayan foothill section of the North Bengal foreland area. This book embarks on a comprehensive exploration of riverscape dynamics, a field that integrates fluvial geomorphology, hydrology, ecology, and increasingly, human influences, to understand the complex and ever-changing nature of river systems. The treatise is the result of intimate personal knowledge—accompanied by modern methodology, modeling, statistical analysis, and intensive field works in last few years. Riverscapes, as a dynamic interface between terrestrial and aquatic realms, are characterized by a mosaic of landforms, habitats, and ecological communities that are linked by the flow of water, sediment, and patterns of recent hazards associated with engineering constructions. Fluvial processes, including erosion, transport, and deposition, are the main forces driving the evolution of landscapes and are also the natural disturbance mechanisms that uphold significant landscape diversity in river corridors. While specific landscape elements may change frequently and largely due to the interplay between fluvial dynamics and successional processes, a significant but little-understood factor in maintaining the wide range of riverine landscapes is hydrological connectedness or the interchange of materials, energy, and biota across the fluid medium in respect to contemporary hazards such as floods. The construction of bridges across the channel controls the riverine morphology and hazardous interceptions in both up and down stream sections resulting in the landscape dynamism over time. Thorough studies of connectedness in various river systems ought to shed a great deal of light on functioning endeavors at the landscape level.

Riverscape dynamics encompass the interplay between hydrological, ecological, and geomorphological processes that influence river environments. Grasping these dynamics is essential for managing risks such as flooding and sediment movement, which can greatly affect ecosystems and human endeavors. Riverscape dynamics include hydrological processes: Variability in water flow due to seasonal changes, precipitation, and human interventions. Influence on river morphology and sediment transport. Ecological Interactions: Biodiversity within riverscapes, including flora and fauna adapted to specific hydrological conditions. Role of riparian vegetation in stabilizing banks and providing habitat. Geomorphological Changes: Erosion and sedimentation processes that shape river channels and floodplains. Formation of features such as bars, pools, and riffles that contribute to habitat diversity and hazards associated with riverscapes. Flooding: Natural floods can rejuvenate ecosystems but pose risks to human settlements. Management strategies must balance flood risk reduction with ecological health. Sediment Transport: Excessive sediment can lead to channel blockage and habitat degradation. Understanding sediment dynamics is essential for effective river management. Anthropogenic Impacts: Dams and levees disrupt natural flow and sediment processes, leading to ecological consequences. Urbanization increases runoff and alters natural hydrology, exacerbating flood risks. Management Approaches: Emphasizing the importance of working with natural processes rather than against them. Strategies include restoring floodplains and enhancing riparian buffers to mitigate hazards. Integrated River Management: A holistic approach that considers the interconnectedness of hydrological, ecological, and social systems. Involves stakeholder engagement and adaptive management to address changing conditions. Monitoring and Assessment: Continuous evaluation of river health and dynamics to inform management decisions. Use of geomorphological assessments to understand river behavior and predict responses to disturbances.

Beyond a river's main course, aquatic and terrestrial animals and their habitats are impacted by hydrological and hydraulic processes like sediment transfer. This region, known as the "riverscape," consists of a high diversity of riparian habitats, which can change over time and space based on sediment dynamics and water discharge. The shifting environment benefits the plants and animals that have adapted to live in riverscapes. Sediment dynamics, in particular, can produce and momentarily change habitats, encourage reproduction, and supply nutrients. Contemporary hazards, such as flooding, are becoming increasingly critical due to factors such as climate change, urbanization, and deforestation. Flood modeling is a modern, scientific, and technological approach used to understand, predict, and mitigate flood risks. Here's an overview of how contemporary flood modeling works and why it matters. Flood modeling and bridge construction technologies have advanced significantly, enhancing resilience against flooding. The integration of advanced modeling techniques and materials in bridge construction has led to improved resilience against flood events. This includes the use of sophisticated hydraulic models that simulate flood scenarios and their impacts on infrastructure. Recent studies propose frameworks that categorize research on flood management systems, highlighting the need for comprehensive assessments of flood risks associated with bridge structures. Technological advancements, such as bridge constructions for rail and road across the river, improve flood risk assessments, while real-time management frameworks help protect critical infrastructure during extreme weather events. This book discloses the role of bridges in assessing flood coverage area both in up and downstream of the riverine section. The increasing or decreasing extent of flood-inundated areas has been modeled for scenarios with bridges, without bridges, and with the relocation or redesign of the bridges across the Rivers Lish, Gish, and Chel, which are the three tributaries of the River Teesta that join the floodplain. The changing riverscape dynamics of the River Teesta after the coronation bridge is also enhanced by the hydro-geomorphology and sediment dynamics of these three tributaries. Slope failure, earthquake events, and landslide occurrences in the catchment area of these tributaries are the fundamental factors influencing the variation in topographic slopes. This undulating terrain includes hill side, valley slopes, and cliffs formed by direct uplift along faults, as well as deposits from landslide-generated sediments or debris flows of different grain sizes. These features are connected with streamlines and deposit near bridges through aggradation processes, while scouring around bridge piers acts as a degradation mechanism over time.

The outcome of the study is a unique one and portrays a spatio-temporal hydromorphology and constructional effect on riverscape dynamics with an in-depth study of contemporary hazards of the area. A number of research papers have been published by the authors on various aspects of geomorphology and related hazards with a scientific approach in the domain of north-eastern Himalayan drainage basins. It is a definite contribution to the applied geomorphology that opens a new avenue of research in contemporary times, with critical significance for recent constructional developments affecting the natural riverine layout of the foreland area, including alluvial fans and floodplains.

Acknowledgments

The corresponding and first author of the book Dr. Biswas is grateful to DST-SERB (File No. CRG/2020/000523) for funding the project (January 2021–March 2024). I offer my sincere honor and respect to the late Prof. S.C. Mukhopadhyay, my supervisor and former Head of the Geography Department at Calcutta University, for his direction and mentorship of my research perspective during my PhD, which allowed me to create such a theme in a book format. It is my pleasure to acknowledge my mother, Mrs. Krishna Biswas, whose mental support inherited and channelized my academic achievements in every step of my life. I am also deeply indebted to the Department of Geography and Geology Presidency University, Kolkata.

Acknowledgements

[illegible]

1 Riverscape Dynamics
A Global Perspective

1.1 INTRODUCTION

Riverscapes are complex, landscape-scale mosaics of connected river and stream habitats embedded in diverse ecological and socio-economic settings (Hand et al., 2018). Changes in riverscapes have occurred through both natural and anthropogenic ways, and the dynamism incorporates all ecological, economic, and social aspects as well. Globally, over 3.2 billion people were impacted by 4713 floods that occurred in 168 countries between 1990 and 2022, resulting in 218,353 fatalities and more than 1.3 trillion US dollars in economic losses. More than 2.0 billion people were impacted by floods in the WHO Western Pacific Region, which accounted for 63.19% (2,024,599,380/3,203,944,965 of all impacted populations). Deaths were the highest in the South-East Asia Region (71,713; 32.84%). In 2022, the highest number of flood-related deaths and injuries per 100,000 inhabitants occurred in the Eastern Mediterranean and African regions. Compared to high-income nations, low-income countries had a considerably greater chance of floods, resulting in more than 50 fatalities (adjusted odds ratio: 14.34; 95% CI: 7.46 to 30.04). India is extremely susceptible to flooding. Over 40 million hectares (mha) of the 329 million hectares (mha) of the total land area are vulnerable to flooding. Floods are a frequent occurrence that destroy properties, infrastructure, public utilities, and livelihood systems in addition to causing a significant number of fatalities (Rowland et.al; 2021). Damages connected to flooding are showing a rising tendency, which is concerning. In the last ten years, from 1996 to 2005, the average annual flood damage was Rs. 4745 crore, while the equivalent average for the previous 53 years was Rs. 1805 crore. Numerous factors, such as a sharp rise in population, fast urbanization, an increase in economic and developmental activity in floodplains, and global warming, contribute to this. Flood risk is the first and foremost effect of the dynamic riverscape. To reduce the susceptibility of flood risk because of the changes in riverscapes, all flood mitigation project designs should be multidisciplinary from early stages to compromise among human health, the environment, cultural heritage, and economic activity (Poulard et al., 2011). These riverscape dynamics mostly have an adverse effect on bridge construction as highlighted by many research studies. According to the records from the Papua New Guinea Department of Works, it is estimated that over (285 bridges, fords (causeways), and major culverts were damaged by flood action alone in the last five years between 2013 and 2017 due to the flooding rivers in Papua New Guinea (Holemba & Matsumoto, 2019). Using a two-dimensional mathematical model, a numerical study of the flood effect of a bridge at tail of the Ganjiang River, entrance of Poyang Lake, showed the construction of a river-crossing bridge would have an impact on the flow movement of the river section, which would bring an adverse impact on flood control (Ge & Deng, 2020). A study in Japan reveals that in July 2018, a heavy rainfall that occurred in western Japan caused massive and widespread damage to social infrastructures and lifelines, and the structural analysis reveals the flood hydrodynamic force on the bridge girder during the flood that caused the collapse (Ario et al., 2022). Changes in riverscapes due to massive flooding (Hall, 2022) and their adverse effects on bridges are now at their peak worldwide. In the Western Black Sea region, several investigations in Turkey in which river bridges are severely damaged or collapsed showed that these events are usually experienced during flooding and most of the bridges collapsed due to the excessive scouring around bridge piers resulting in river bridges being at greater risk (Koçyiğit et al., 2016). In central Chile, at the Rapel Bridge, over the Rapel River, the local scour and sediment deposition play a pivotal role and are very sensitive to excess sediment supply, producing important refilling of scour holes (Link et al., 2020). In between 2010 and 2011, in the Lockyer valley of Queensland in Australia, 43 out 46 bridges were damaged during the 2013 flood due to scouring of abutments and piers, as well as damage to bridge decks from urban debris impact (Setunge et al., 2014). In a city named Carlisle in the United Kingdom, the hydrodynamic force impacting inundated bridges has been studied and the preliminary observation suggests that the riverine bridges are particularly prone to damages during flood events (Pregnolato et al., 2022). In the Darjeeling–Jalpaiguri district of West Bengal, the scenario is quite different from the worldwide view. The Jaldhaka River, along with its tributaries, carries a huge discharge during the monsoon period, resulting in floods that affect the entire riverscape (Raha et al., 2023).

1.2 RIVERSCAPES AND THEIR DYNAMICS

Changes in riverscapes from the very past represent a dynamic process. Riverscapes seem to be a great way to communicate conceptual thoughts and exchange information. Riverscapes may also be helpful in interdisciplinary studies (Poulard et al., 2011). River ecosystems are increasingly understood as dynamic riverscapes, changing in size, extent, and arrangement in response to discharge pulses (Malard et al., 2006). In the case of medieval Vienna on the Danube River, major channel changes occurred in 1565–1566, with repeated attempts to force the main arm into the old riverbed, and the first organized, successful flood control measures were put in place between 1775 and 1792. In certain areas of the former floodplain, this allowed for urban growth and had a significant impact on river dynamics. It took until 1875 for the dynamic riverscape to completely change into stabilized regions that allowed for the expansion of cities and safe canals (Hohensinner et al., 2013a).

An urban riverscape combines human effects and benefits with natural forms and processes in a holistic way. Specialized tools for technical design, strategic planning, and transdisciplinary communication can help the management in this regard. Although sediment movement is a vital link between channel shapes and processes and is essential for physical equilibrium, it is one of the least understood natural elements of an urban stream system. This design and assessment tool is particularly helpful for places restricted by constructed infrastructure and damaged streams that lack time and space for self-recovery; however, its potential uses are not confined to urban riverscapes. Via the creation of pipework, bridges, dams, culverts, buildings in urban settings, and other engineering feats, humans have continuously attempted to alter their surroundings for better adaption. These activities notably reframed the physical environment as well as stream geometry, resulting in changes in riverscapes due to flooding and their adverse effects on bridge construction, which is an alarming issue nowadays.

In recent years, there has been a significant increase in flood damage caused by natural disasters, such as torrential rains which alters the riverscape, and many reports have been published all over the world. Bridge damage caused by the flooding of rivers has also been reported in various places. Local scour around bridge pier is the main factor for the failure of the bridge structure, such as bridge piers and abutments, in the alluvial channel. The main concept comes from excessive flooding events and destruction of bridges happening all over the world. The picture is more or less the same throughout the world as several studies reveal. In the Tigris River in Turkey, the HEC-RAS tool was utilized in this work to determine the Tigris River bridges, the cross-section changes brought forth by the hydraulic structures, and the existing natural section changes and floods brought on by siltation issues in reservoirs, and natural and artificial waterways filled with sediment should not be undervalued: the Sadi Bridge and the historical Ten-Eyed Bridge are insufficient to pass the flood flow which results in bridge damage (Ogras & Onen, 2020). In the case of the Lockyer valley in Queensland, Pritchard notes that the primary factors causing bridge damage during the 2011–2012 Queensland floods were urban debris, such as automobiles, and inadequate bridge spans to allow debris to pass through. Lokuge and Setunge came to a conclusion that it is essential to look at the failure patterns and building methods used during bridge construction and rehabilitation in the years following the 2013 flood incident in the Lockyer valley. The subject of what failure mechanisms and contributing factors should be taken into account when constructing bridges to withstand extreme flood events was highlighted by the findings of Setunge et al. (2014). The flood-prone city of Carlisle (UK) is used as a case study where scour is recognized as the greatest threat to bridges over water. Hydrodynamic forces can be critical for bridge piers on bedrock (where scour is unlikely) and for the decks of all flooded bridges. In particular, the impacts of flooding have been exacerbated in recent years by urbanization. Results for the city of Carlisle (UK) and a flooding event occurring once every 500 years indicated that the impact of the flood on the highway bridge resulted in damages exceeding GBP 500,000, of which 93.5% were indirect damages (Pregnolato et al., 2022). In Cumbria, UK, the 2012 floods caused 131 bridges to be destroyed overall, mostly as a result of scour. The majority of the documented bridge failures are hydraulically caused, specifically scour (S), which is brought on by debris buildup (D), hydraulic forces (F), and floods. These factors may become more severe as a result of climate changes (Mitoulis et al., 2021). At the Western Black Sea region in Turkey, the occurrence of hydroelectric power plants or dams, the removal of sand and gravel from streams, changes in the rainfall-runoff relationship, changes in the amount of runoff resulting from changes in land cover (LC) and land use (LU) in the basin, flooding caused by a brief but extremely severe precipitation or sudden snowmelt due to climatic conditions, all have an impact on stream discharges. Therefore, these elements have an impact on bridges in river environments, either directly or indirectly (Koçyiğit et al., 2016). Another case is in Nanchang City, the capital of Jiangxi Province, located at the end of the Ganjiang River, which flows into China's largest freshwater lake, Poyang Lake. The study found that the proposed bridge has minimal effect on the general flow field of this river section, and changes in flow velocity following the project's execution are mostly concentrated in the vicinity, both upstream and downstream of the proposed bridge (Ge & Deng, 2020). In the case of Papua New Guinea, the country has been experiencing frequent bridge failures and collapses due to flooding rivers in the recent past. It has been noted that bridge foundations sustain the majority of damage. More precisely, the floodwaters impair the bridge's ability to withstand the weight of the flood and ultimately destroy it by eroding the bridge abutments and scouring the bridge piers. Furthermore, it has been confirmed that erosion of road approach embankments and riverbanks caused by flooding rivers is one of the main reasons for bridge collapses in Papua New Guinea. The superstructure and substructure that make up a bridge are damaged by flooding. Hydrostatic load failure under the substructure category, bed aggradation, contraction scour at the riverbed and riverbank, and local scour at bridge piers and abutments (Holemba & Matsumoto, 2019). Severe flooding is brought on by an increase in atmospheric precipitation. Globally, scouring—primarily brought on by flooding—is the primary cause of bridge collapse. Between 1846 and 2013, the effect of scouring caused the failure of 140 railroad bridges in the United Kingdom and about 6000 bridges in the United States of America. Removal of material from the riverbed by water leads to a phenomenon known as scouring, which clears the channel of debris (Abrar & Farooqi, 2021).

The notion of the Himalayan foothill portion and the effect of bridge construction on flooding are alike with the worldview. Engineering constructions such as bridges, dams, culverts, pipe, and building in urban landscapes—stream geometry, hydro-sediment flow dynamics, and the physical environment were all significantly altered by these actions. Such actions cause a channel's deterioration processes to worsen over time. Rivers such as the Lish, Gish, Chel, Neora, Mal, and others have caused congestion, with a broad transverse upstream riverbed and a downstream cross-sectional area that has undergone accentuated alteration. Due to the region's heavy rainfall, which amounts to around 1582 mm during the rainy season, there is a propensity for flooding every two to five years. This is extremely disastrous and severely impairs people's means of subsistence. As one moved farther downstream of the bridge, the speed rose. The flood propagancy of River Chel after the bridge increase in monsoon (Banerjee & Biswas, 2021). Teesta, Jaldhaka, and Torsa originate in the Himalayan region, flow in a south-west direction, and have their own hydro-sediment discharge before debouching into the floodplains. Massive amounts of water flow through the small opening during the monsoon, impacting both the upstream and downstream sections and raising the risk and vulnerability to flooding (Biswas & Raha, 2023; Biswas and Raha 2024). Because of the regular change in gradient, River Chel has deposited all of the carried sediments, maintaining the sediment regime. However, the river near the bridge crossing exhibits a bottleneck condition because it must pass through a narrow, constricted section that compressed the channel. As a result, the river's velocity is slowed down upstream of the crossing area because of the obstruction that has been placed in its path, and sediments have begun to accumulate. This also has an afflux impact upstream, which increases the consequences of flooding in this area and due to scouring of the bridge piers and bed degradation processes that occur close to the bridge site, the average velocity at the downstream increases relative to the upstream (Biswas & Banerjee, 2018); Sarma & Talukdar, 2019. The rapid shift in the streamflow and sediment discharge mechanism has also been caused by the frequency of extreme weather events including tropical cyclones and torrential rain in the Himalayan foothill rivers, and numerous research studies have determined that the building of the rail and road bridge has resulted in an increase in the area covered by the food hazard zone downstream. River Teesta and its tributaries drain the Darjeeling-Jalpaiguri region's Himalayan foothill zone. The rivers' velocity profile indicates a decrease in velocity from the foothills to the confluence or river mouth. Nevertheless, the building of bridges over the rivers has changed this tendency (Biswas & Raha, 2023).

So, it has been evident that the concept emerges from the overall changes in dynamics of riverscape due to construction of bridges, dams etc. throughout the world resulting in severe flooding and the overall changes in flow dynamics. This concept belongs to one of the most predominant and prevalent issues in recent times.

1.2.1 Consequences of Riverscape Dynamics

As discussed earlier, the change in riverscape dynamics is mainly driven by some anthropogenic activities, e.g., bridge construction, dam construction, and some natural causes like change in flow dynamics, torrential rains, change in velocity which results in severe flooding and damage of bridges that can threat lives, inundate properties and businesses, destroy belongings, damage vital infrastructure, and prevent access to essential public services.

The devastation and damage to homes and personal items is one of the biggest effects of flooding. This frequently leads to sorrow, anxiety, and anguish due to the loss of priceless items and equity that has been accrued. There are also additional, long-term challenges for some people locating resources for rebuilding, moving, or repairing after the flood. Infrastructure is frequently damaged as a result of flooding. This may involve harm to crucial infrastructure and services like telecommunications, energy, wastewater, and sewerage. When these services fail, individuals may be forced to flee, often for an extended length of time, as their dwellings become hazardous and unlivable. In severe floods, roads, railroads, tramlines, and bus services may be impacted or shut down entirely. Frequently, portions of roads and quite frequently, portions of the road as well as the infrastructure that supports them, such traffic lights and signage, can be damaged or swept away. Infrastructure repairs can take a long time to complete, which is quite expensive and inconvenient for the general people.

The consequences of bridge failures are diverse, naming a few: life loss, injury, critical or emergency routes, economic loss, environmental concerns, and historical significance. Millions of people and a great amount of property had been under the water throughout the world because of it.

A study on local scour and sediment deposition at bridge piers during floods estimates predicted scour at a bridge's foundation—its piers and abutments—as part of hydraulic design entails taking the maximum equilibrium scour depth into account and they said scour is responsible for more than 50% of the bridge failures over the world (Link et al., 2020). Queensland's floods caused significant damage to state-owned properties, including 4748 rail lines, 9170 roads, 89 seriously damaged bridges and culverts, 411 schools, and 138 national parks. The devastation was concentrated in central and southern Queensland. During this period, almost 18,000 residential and business properties in Brisbane and Ipswich suffered serious damage. While small enterprises, primary producers, and non-profit organizations received grants totaling more than $121 million, individual, family, and household support exceeded $42 million. Additionally, concessional loans totaling over $12 million have been given to primary producers and small enterprises. The governments of Queensland and Australia have pledged $6.8 billion to the state's reconstruction and a rough estimate can be made considering the case studies of failure. For example, for a given structural configuration and a given intensity of flood, if 36 out of 72 bridge piers have failed due to scour, probability of failure of piers due to scour can be crudely estimated as 50% (Setunge et al., 2014). In a city of the United Kingdom, Carlisle, in terms of consequences, in addition to inflicting direct physical damage, natural disasters can also contribute to functional failures in bridges, which can result in rerouting and travel time delays that create indirect costs. Any bridge failure, whether structural or functional, might have severe repercussions for the owners or accountable authorities, along with exorbitant costs and the impact analysis showed that indirect damages covered 93.5% of the total cost of damages to the bridge. In the city of Carlisle (UK) and a 1-in-500-years flooding, results showed that it caused 93.5% of indirect damages which include rerouting and delays (Pregnolato et al., 2022). For instance, at least 20 bridges were damaged during the 2009 floods in Cumbria, UK, resulting in £34 million in repair expenditures and a significant impact on society (Cumbria County Council, 2010). In the same region, the 2012 flood events caused 131 bridges to sustain damage, mostly from scour. An estimated $7 billion was spent on repairs to Queensland, Australia's road network following a string of severe weather events in 2011 and 2012. Of those repairs, 5% went toward bridges. Thirty-seven bridges in Serbia were damaged during the 2014 floods; the projected total damages for transportation infrastructure were €166.5 million, with restoration costing €128 million (Mitoulis et al., 2021). Worldwide, riverine flooding results in enormous losses—roughly $314 million annually. For example, the Schoharie Creek Bridge in New York City collapsed in 1987, the Hatchie Bridge in Tennessee collapsed in April 1989, and the Portuguese Hintze Ribeiro Bridge collapsed in 2001. Floods were the cause of these bridges' collapse. In Europe, over the four years from 1998 to 2002, there were over 100 significant floods. There were more than 30 floods in a year. The flood that occurred in 2017 during the spring season in Quebec and Ontario was ranked as the third most noticeable outcome in the Canadian Meteorological and Oceanographic Society's list of the top ten weather stories in Canada. Floods nearly always cause fatalities, agricultural losses, infrastructural damage, and harm to other assets like livestock in Pakistan. In 2010, Pakistan saw the greatest flood in a century, brought on by a four-day period of rain from 27 to 30 July, in the country's northwest province. More than 1900 people died as a result, and the economy suffered losses of billions of US dollars. The Zijin Bridge experienced the same situation. Previous research indicated that scouring over 47 years removed bed material, which in turn led to the bridge's destruction (Abrar & Farooqi, 2021). Gökçeler in Muğla in January 2010, Curidere in Ordu in 2011, Yarpuz I in Osmaniye, Güzelceçay 2 and Abdest in Sinop, Gökdere in Adana and Kadıköy in Mersin in 2012, Dolbazlar in Antalya in January 2014, and two river bridges—one new and the other old—in Malatya Doğanşehir in 2014. These are some of the severely damaged or collapsed river bridges. In April 2012, the Çaycuma River Bridge in Zenguldak collapsed, resulting in the deaths of 15 individuals and damage to property (Koçyiğit et al., 2016). Rivers spanned embankments in several areas of Germany, and flash floods seriously damaged bridge infrastructure. Unexpected rainfall causes rivers to flow faster, and the force of the water causes bridges to collapse. Japan is also prone to periodic natural calamities. For instance, there was a lot of rain in July 2018, Typhoon No. 19 hit the Boso Peninsula in September 2019, and in July 2020, there was a lot of rain, which damaged roads and bridges, raising the possibility of damage to the road traffic networks. The year 2018's intense rains in western Japan devastated a large region after the Japan Meteorological Agency sent out special warnings to 18 western prefectures. This intense rain caused extensive damage to a variety of physical assets (bridges over the Misasa River), including roads and railroad bridges, as well as disruptions to logistics across the board. Road and rail infrastructure, including the JR Sanyo Main Line, Kure Line, and Geibi Line, suffered bridge damage even in Hiroshima Prefecture, isolating numerous districts. Reports indicated that two prefectural bridges and 70 municipal bridges were destroyed, according to a statement released by the Hiroshima Prefectural Disaster Countermeasures Headquarters (Ario et al., 2022).

A numerical study of the flood effect of a bridge at tail of the Ganjiang River, entrance of Poyang Lake, shows that the proposed bridge has minimal effect on the study river section's general flow field, and changes in flow velocity following the project's execution are mostly concentrated in the vicinity, both upstream and downstream of the proposed bridge (Ge & Deng, 2020). Rivers overflowing in the previous several years have caused numerous bridge collapses and failures in Papua New Guinea. Records from the Department of Works in Papua New Guinea indicate that in only the last five years, flooding has caused damage to over 285 bridges, causeways (also known as causeways), and large culverts. That translates to 57 bridges annually on average (Holemba & Matsumoto, 2019). Another study on bridge construction and river channel morphology—a comprehensive study of flow behavior and sediment size alteration of the Chel River, India, concludes that the construction of the Odlabari road bridge (after 1970) changed the river's natural hydrological behavior, increasing the likelihood of floods upstream and bank erosion (Lovric & Tosic, 2016) downstream, creating a situation where a bottleneck exists in the river (Biswas & Banerjee, 2018; Chakraborty and Mukhopadhyay, 2015).

1.3 DYNAMIC NATURE OF WORLD'S MAJOR RIVERS, THEIR SURROUNDINGS, AND HUMAN INTERFERENCES

The concept of a "riverscape" is dynamic in nature itself from ancient times to recent times. In Vienna, north of the city, the Danube branched out, creating a sizable floodplain. Documentary evidence from the fourteenth century indicates that the Danube was progressively displacing the old city center. From that point on, the river was prevented from being further displaced by the authorities in Vienna, and the waterway was maintained as an essential supply line for the city for more than five centuries and the study focuses on natural and human-induced transformation processes of the Viennese Danube riverscape from the late fifteenth century to the present. Approximately between 1455 and 1565, a new, extremely sinuous arm formed. It extended southward over several kilometers, becoming nearer to the city. A period of natural channel straightening as a result of a river bend cut-off came next. A very winding path was also built by the new main channel till 1663/1700. After that, the river's primary stream would move even farther north. Up until about 1805, river bend evolution was in its latter stages, during which time a characteristic double-meander loop formed. Over the long time, the average channel migration rate was around 20 m per year, and each stage of the evolution of river bends and meanders as well as the channel shift lasted *100–130 years (Hohensinner et al., 2013b). A project on co-conception of integrated flood management solutions: riverscapes to facilitate dialog on the rivers of Gard, Meuse, Roanne, Nivelle in France, River Bóbr in Japan and Rivers Miami and Santa Ana in the USA reveals how the construction of dams can bring effect on riverscapes (O'Connor et al., 2003). According to their research, changes in riverscapes are multidisciplinary in nature and it includes both the themes of biodiversity and water management strategy. Nutrients from soil drainage and upland runoff are carried by surface waterways. Generally speaking, groundwater has less nutrient content, unless there is heavy fertilizer use. Water exchanges between surface and groundwater are part of the aquatic metabolism as a whole. Biofilms, which need a solid surface to adhere to, and nutrients carried by water flowing over and through river substrates, are both involved in this metabolism. However, water head gradients and channel bottom permeability are both necessary for an efficient water circulation system. River depletion happens occasionally during floods when the river's water level rises over the level of the surrounding groundwater table. A biological reset can be induced by significant floods that remove, mix, and dilute the silt that has held pollution. Often, high-quality fluids that transport nutrients also improve the hyporheic system's capacity for self-purification during flood recessions. Even though the dam allows water to flow freely, during flood events, there may be some fluctuations in the export of silt. The geomorphic result of this could significantly diminish the shallow hyporheic exchanges in the close downstream reaches, together with a lowering of the associated water table because of dam foundations (Poulard et al., 2011). A comparison between the Missouri River in eastern Montana, the United States, and the Elbe River in Saxony-Anhalt, Germany, shows that the Elbe River groynes have changed a number of floodplain ecosystems, cut off floodplain wetlands from channel flow, and caused erosion of the former sand and gravel banks, which has encouraged the development of revetments as a means of artificial stabilization. Many river floodplains are essentially left stranded when dams are built to control river flow. The natural complexity and integrity of floodplain networks were significantly diminished by the construction of revetments for traffic along river corridors, gravel mining, harvesting, channelization, and agricultural and urban growth. To a greater or lesser extent, this pattern of human settlement along floodplains has happened globally. Global trends in human development and migration are reflected in the size and scope of floodplain simplification. Because of this, the average floodplain in Europe is simpler than those in North America, which is indicative of the fact that human expansion—and, in particular, the practice of agriculture on river floodplains—started far earlier in history (Peipoch et al., 2015). Roseg River in Switzerland shows seasonal fluctuations in the density of benthic invertebrates in the main channel were partially explained by dilution–concentration processes, as indicated by the high correlation between the total aquatic area and the local zoobenthic density. The intricate replenishment of the channel mosaic was caused by the yearly pulsing of river discharge. In 1997 and 1999, the average relative change between two consecutive months was 30% and 38%, respectively. This difference was significantly greater during the phases of channel network expansion and contraction (April–June, 41%; October–January, 41%) than during the minimum and maximum extension phases (February–March, 2%; July–September, 12%). As river discharge increased and decreased, changes in the make-up and arrangement of the mosaic of clear and muddy channels functioned in a particular order. A rise in river discharge caused new clear channels to arise one after the other, flood murky water into dry channels, and flush turbid water into clear channels. Certain turbid channels dried out during the flow recession, whereas other channels turned clear. Eventually, a few obvious channels dried up. Riverscapes are highly dynamic across a range of temporal scales due to variations in the flux of various water sources and hydrological pathways within a catchment. Seasonal change in the inflow of melting water and surface and subsurface flow routes produced a predictable pattern in the composition and shape of the riverscape in Roseg, a braided glacier river (Malard et al., 2006). In Columbia River basin, since the 1930s, 56 significant hydroelectric dams have been built on various mainstem rivers in the Crater of Refuge Basin (CRB), causing significant fragmentation of the riverscape and severing of links between vital habitats for migratory salmonids (Hand et al., 2018). A study on dynamic slowdown: a flood mitigation strategy complying with the integrated management concept—implementation in a small mountainous catchment reveals that many river works, and dams in particular, modify river dynamics, sediment balance, and ecosystems. Also, it takes into account that sediments are carried by the Isepnica down to a reservoir on the Soła River, hillslope erosion is an issue. It is true that some places are vulnerable to landslides, which endanger infrastructure and personal belongings (Christine et al., 2005).

The change in riverscape dynamics is broadly brought by the construction of river bridges above the river. Scouring at bridge piers significantly results for the change. Research on local scour and sediment deposition at bridge piers during floods analyses the behavior of the Rapel Bridge, over Rapel River in central Chile, local scour and sediment deposition during a two-year period in two different scenarios: pre-dam (controlled flow regime) and post-dam (natural flow regime), taking into account various extra sediment supplies. The complex interactions between flow and sediment over time are illustrated by the fact that the highest scour depths in the pre-dam scenario with excess sediment supply were only briefly around the peak discharges before sediment deposition, and they are of comparable magnitude to those computed for a post-dam scenario (Link et al., 2020). A study with HEC-RAS on the Tigris River indicates that due to the presence of social and productive (fruit, fisheries, and nursery) facilities, lush agricultural grounds, hospitals, and partial settlements along the study path, significant damage is caused during floods. This part of the road narrows since the Sadi Bridge is not wide enough. Due to the variations in the gap and beam heights of the bridges constructed side by side at different eras, significant alterations in the surface profiles of the bridges are noticed. The floods had an impact on the Hevsel Gardens, which are located between the historic Ten-Eyed Bridge and the University Bridge. In 2015, UNESCO inducted the gardens into the list of world cultural heritage. Private and commercial business sectors were noted to have stayed below the flood levels, both upstream and downstream of the iconic Ten-Eyed Bridge (Ogras & Onen, 2020). Assessing flooding impact to riverine bridges: an integrated analysis shows in Carlisle's Eden River, at the deck's upstream edge. Since the incoming flood was on the upstream side of the deck, it carried the greatest positive horizontal forces at the upstream edge of the deck, where it has its principal impact at its highest velocity. The positive horizontal forces were at the deck's downstream edge because the flow was affecting its bottom flange and the lower part of its web. However, for flow heights above 14.0 m, this force decreased until it reached a negative value at a flow height of 16.0 m. Differences in the vertical surface face areas and flow velocities on its upstream and downstream sides may be the cause for the gradual decrease in the horizontal force. This is because the fluid pressures acting on the downstream faces were greater than those acting on the upstream faces. There is more space on which the fluid pressure could act upstream on the downstream faces because their total vertical surface area was larger than the upstream faces by an amount equal to the deck section. The bridge was sufficiently submerged by a flow height of 17.0 m that buoyancy allowed the vertical forces on the upstream portion of the bridge to grow once more. Uplift caused by buoyancy increased for the bridge's downstream half up to a flow height of 15.0 m. At this moment, the flow passed over the bridge's midway superstructure crest. The vertical stresses on the downstream portion of the bridge decreased as a result of this shift in flow behavior, and the bridge was eventually fully buried at a flow height of 18.0 m (Pregnolato et al., 2022). Climate change is also predicted to affect the frequency and/or intensity of flood events, which will affect bridge restoration in Cumbria, Serbia, and Zurich. Regionally, significant rainfall events might intensify even while yearly mean precipitation patterns may eventually decline. There might not be a clear correlation between the rise in flood flows (anticipated as a result of climate change) and the change in scouring danger. The latter results from the fact that, once a threshold is reached, scour depth does not always rise in tandem with flow. The majority of bridge failures that have been documented are caused by hydraulic forces, namely scour (S), which is caused by two floods, debris buildup (D) and hydraulic forces (F). These forces may become more intense as a result of climate change. With an average yearly restoration cost of over $50 million, hydraulic activities (SDF) have been identified as the most catastrophic causes of bridge failures in the USA, accounting for over 50% of instances (Mitoulis et al., 2021). The parts of bridge structures most vulnerable to flooding and its impacts are the piers. For instance, erosion of the bed near piers due to increased flow velocities and water pressure may induce instability in the bridge construction. The total stability of the bridge foundation decreases as a result of scouring. The risk of a bridge collision grows as the scour depth rises because the foundation becomes less stable. While scour is acknowledged as the primary hazard, bridge piers on bedrock and the decks of all flooded bridges may also be severely affected by hydrodynamic pressures brought on by floods owing to excessive force in addition to scouring. The parts of bridge structures most vulnerable to flooding and its impacts are the piers. For instance, erosion of the bed near piers due to increased flow velocities and water pressure may induce instability in the bridge construction. The total stability of the bridge foundation decreases as a result of scouring. The risk of a bridge collision grows as the scour depth rises because the foundation becomes less stable. While scour is acknowledged as the primary hazard, bridge piers on bedrock and the decks of all flooded bridges may also be severely affected by hydrodynamic pressures brought on by floods owing to excessive force in addition to scouring. As for the outcome, the main causes of flood-generating processes are precipitation from above and moisture beneath the surface, which results in rain. These elements have varying effects in different parts of the Earth. The effects of backwater mixed with river pressure erode piers, causing damage to the bridge. Even while scouring is a major factor in bridge collapse, hydrodynamic pressures brought on by floods can also contribute to bridge failure, whether or not the riverbed is rocky. Because they obstruct and reduce the area of flow for floods, bridge piers can have the worst effects on the actual floods. Higher flow velocities occur close to the piers, and these velocities remove bed material, which eventually causes the bridge structure to settle and fail. Increased water flow rates cause the bed material to be removed, which leads to scouring. Bridge piers are severely damaged by the backwater effect of flooding, and as a result, the piers need to be built to withstand the weight of the flood. The bridge's seismic behavior is demonstrated by scouring. The bridge is destroyed as a result of piers being shifted by seismic pressures brought on by scouring. A typical bottleneck stretch is the Yangtze River's Tianjiazhen reach, a channel border that rises from the hills. Water from the Yangtze River can be practically confined upstream for two to three days due to tight rock-controlled places, causing floods larger than 50,000 m^3/s. In the reach and upstream toward Wuhan, high water periods lasting three to five days are common because the Yangtze River's water covers narrow rock-controlled spots. Three to five-day high-water spells are common in the reach and upstream into Wuhan. In conclusion, the new bridge in the bottleneck reach expands the backwater effect's region, and depending on how the flood level changes for various plans, the backwater scope may surpass 500 m (Han et al., 2018). It has been noted that bridge foundations sustain the majority of bridge damage. To be more precise, the inundating floods degrade the bridge's resistance to the flood load, scour the bridge piers, and eventually demolish the bridge due to erosion of the abutments. Furthermore, it has been confirmed that erosion of road approach embankments and riverbanks caused by flooding rivers is one of the main reasons for bridge collapses in Papua New Guinea. The two types of resistance to flow are shear resistance and resistance caused by the drag force created by the pressure differential between an object's upstream and downstream sides. The fluid experiences a local acceleration and streamline deflection due to the geometry of the boundary, which is a bridge pier. This results in what is known as a normal stress, or a shift in pressure from the boundary's upstream to downstream side. A drag force acts on the boundary and a pressure resistance resists the fluid as a result of the forces accumulated over the surface. All of the bridges under investigation were river bridges, and these rivers had their source in steep valleys and mountains. As a result of erosion in the upstream section, high flow velocity carry large amounts of logs and flood debris that alter the hydro-geomorphology of the downstream part of the river. Few rivers have braided or swamp systems; most have meandering patterns. Rivers close to the coast swelled and witnessed tidal surges. Thus, this study establishes that flooding, which resulted in debris and log impact, overtopping, sedimentation, scour, embankment erosion, and channel migration, was the primary cause of these bridge failures in Papua New Guinea (Holemba & Matsumoto, 2019). A numerical study of flood effect of a bridge at the tail of the Ganjiang River, the entrance of Poyang Lake, reveals that the planned bridge's upstream area would see an increase in water levels due to the impact of flooding, whereas the local area downstream of the piers would see a decrease in water levels. The squeezing impact of piers causes an increase in flow velocity between them, whereas the water blocking effect of piers reduces flow velocity in the area upstream and downstream of piers. The influence of a bridge on water flow increases with discharge. The research area is situated in Nanchang City's main metropolitan region, with the river embankment on the left bank and a 100-year return period flood as the dike's flood management standard. The East Jiangxi Dykes is the right bank. Following recent construction, the non-major urban area's flood control standard is once every 50 years, while the main urban area's flood control standard is once every 100 years (Ge & Deng, 2020). Study on bridge collapse phenomena due to heavy rain floods indicates that the structural and hydrological changes which includes the condition of exceeded stress-resistance limit at the bridge. The increased hydrodynamic force, resulting from higher flow velocity, pushed the substructure that supported the superstructure, to fall and migrate downstream section of the river. The fallen pier was encircled by a tangle of intertwined branches and a few floating items, like big trees. These branches did not harm the water, but the bridge gave way. Due to the concentration of stress at the upstream side edge (water cutting tip), the hydrodynamic force brought about by an increase in the water level at this stone masonry pier produced the maximum bending moment (Ario et al., 2022). Flooding brought on by brief but intense precipitation or abrupt snowmelt as a result of climatic changes, adjustments to the rainfall-runoff relationship, and the quantity of runoff resulting from modifications to the basin's land use and cover (LU and LC), the presence of hydroelectric power plants or dams, and the removal of sand and gravel from streams all have an impact on stream discharges. Therefore, these elements have an impact on bridges in river environments, either directly or indirectly. Degradation of the riverbed, severe scouring near bridge piers and abutments, a strong backwatering effect, and the potential for hydraulic jump development because of excessive constriction at bridge opening are some of the hydraulic variables affecting river bridges. The water level rises during floods as a result of an increase in the flow output. As a result, in the event of significant flooding, the water level may rise to the bridge deck's lower girder. In these situations, the flow is referred to as pressurized flow rather than free flow. The water level will rise in tandem with the additional stream discharge, rising well past the lower girder level and potentially surpassing the bridge deck. The flow in this scenario is referred to as a weir type of flow, and the bridge will function as a broad-crested weir. These circumstances may have a major impact on the transfer of silt on the riverbed beneath the bridge deck. Given that the majority of river bridge failures are found to be caused by excessive scouring around bridge piers, further research is necessary to determine how sediment transport occurs in these kinds of flows and how it affects scouring in bridge openings. It's important to keep an eye out for any obvious deterioration and aggravation of the riverbed, bank failure, approach embankments, and other like changes near bridges. In order to take the appropriate measures before these impacts accumulate and reach a critical point where the river bridge is at risk, assessment of these changes is important. The stream meanders and is encircled by fast growing commercial and residential areas. Shrubs and the trunks of thin trees cover both banks of the stream. More fine sediment was being carried by the excited and intense flow beneath the bridge. This indicates that the basin has a wide drainage area and that precipitation transforms into runoff quite quickly. Thus, it can be said that there is a considerable chance that a heavy downpour will result in floods that will damage the area's bridges. In this instance, bridges located in such areas may be subject to more intense and frequent flooding, raising the possibility of flooding and excessive scour at piers and abutments. Moreover, a skewed bridge would also have an impact on the flow's velocity field, creating a more intricate three-dimensional flow structure and raising the risk of scouring near the piers. The rainfall-runoff connection can be negatively impacted by declining vegetation and forest cover as well as by the development of residential areas. This can lead to more precipitation continuing as runoff, which raises the danger of floods in the area. Streams supply the building industry's requirement for sand and gravel, especially in the Western Black Sea Region. These pits must be located at least a kilometer downstream of the bridge, according to Turkish rules. In actuality, due to site-specific changes in flow, bed composition, river geometry, and bridge characteristics, this distance must to be established for each bridge individually. The site and method of sand and gravel withdrawal are not in compliance with rules, leading to significant degradation of the river bottom and severe scouring around bridge piers. The flow structure becomes fully three-dimensional when a river meanders close to a bridge, which intensifies scouring around piers and abutments.

Tree trunks, branches, and other similar wastes of all sizes are seen to be lodged and deposited around the bridge piers of every bridge that is studied during field research. When these materials build up around bridge piers, the flow structure in those regions is impacted and becomes complicated and three-dimensional, which raises the possibility of scouring near piers and abutments. Moreover, objects that build up around piers over time or during flooding may obstruct the bridge spans, decreasing the flow area. This in turn causes an increase in flow velocity in this area, which intensifies or initiates scouring.

It was observed that large rocks are placed around the bridge piers as a precaution against scouring during field investigations at bridges when observations and bathymetric measurements are made or other bridges in the area are evaluated. But in the spaces between and around these big rocks and boulders, three-dimensional turbulent flow will happen, which raises the risk of scouring on the riverbed beneath these rocks. Moreover, scouring issues would worsen if these boulders and pebbles were randomly moved as a result of strong flows or flooding (Koçyiğit et al., 2016). Pressure also rises when floods cause water to flow more quickly. As a result of this increased pressure, pressure is spread over the whole height of the pier when water flows along the bridge's piers. In the water channel's flow area, scouring delays the deposition of sediments, and vice versa. As a result, it leads to flow characteristics and foundation instability. The water channel experiences greater flow depth and velocity during damaging floods. Because of the high velocity, new processes were developed, such as the removal of the bed material from the flow channel. This process weakens the bridge's abutments, jeopardizing the structure's overall stability. The building is completely destroyed by the flooding forces (Abrar & Farooqi, 2021).

An overview of the open channel monsoon hydraulics of Himalayan foothill rivers, West Bengal, India (mainly in the rivers: Lish, Gish, and Chel) shows a bottleneck condition. Bridge construction on alluvial rivers results in modifications to the shape of the riverbed, as well as changes to the flow of sediments, aggradation, and degradation mechanisms. Its geographical properties are significantly correlated by the rail and road bridges over the Rivers Lish, Gish, and Chel, particularly during the monsoon. These anthropogenic causes frequently affect the hydraulics and dynamics of the channel above and downstream. The hydro-sediment interaction has been significantly impacted and changed by road and rail bridges. Beginning in 1936, there has been a discernible increase in the channel width both upstream and downstream of the rail bridge. Significant increases in the channel width of River Lish have been seen downstream of the rail bridge, indicating a potential acceleration of the sedimentation process. It has been determined that Lish is 807 m wide downstream. The Gish River has seen the largest increase in channel width both upstream and downstream from the train bridge. The average widths downstream and upstream in 1943 were measured to be 1370.35 m and 687.06 m, respectively. The breadth of the Gish River at the bridge has significantly decreased from 396.63 m in 1913 to only 162 m in 2021, resulting in a bottleneck situation for the river. Conversely, Chel has been identified with reduced amplitude in the channel width variations. It has been observed that the construction of the rail bridge has almost no impact on the downstream reaches. The 1960s road bridge construction significantly changed the hydro-sediment interaction and changed the way the channels formed. The bottleneck situation is caused by the River Gish's maximal growth in the upstream stretch directly below the mountain front and just downstream from the road bridge. The condition remains the same for River Lish also. The increased sheer stress value in the foothill area and 300 m upstream from the rail bridge indicates significant degradation of the sediments, which are being carried downstream and depositing in areas with lower sheer stress. The degraded sediments from these reaches are carried downstream and deposited as the stream power decreases downstream. The easy flow of river discharge and sediment transfer are impeded by the decreasing stream energy close to the bridge piers, which speeds up the aggradation process. The general stream energy distribution in Gish reduces the potential for erosion, which further promotes sediment deposition and, as a result, increases the channel width (Biswas & Raha, 2023). There have been noticeable changes to the River Chel's planform as well (1913–2014). The Rail Bridge was the only structure along its path from 1913 to 1942, but after 1986, the Road Bridge of NH31 was constructed, which had an enormous impact on the river at downstream reaches. These changes began to occur when the rail bridge was constructed. River Chel has a total channel width of

400–1100 m when it meets the Menzingkhola at the Pathorjhora area. This width decreases to 600–800 m further downstream and narrows to less than 100 m between the two bridge areas. Once again, the river widens to nearly 1500 m width further downstream of the bridges, giving an overview of the river's changing characteristics. The following are the first observations that were made: shortly after the rail bridge was built in 1913, the river began to deposit its weight at the upstream section. Due to the characteristics of the river, a massive load is brought downstream during heavy flows, increasing the riverbed extension and starting bed scour, and is deposited at the upstream side and at the base of the bridge piers. This alteration of the river is still going on today. Large rocks or other silt were carried downstream by the river's increasing kinetic energy during high flows, and these materials are now being deposited upstream as a result of the barrier imposed by the bridge piers lowering the river's velocity. The number of debris deposited has decreased toward downstream from the debris pile at Pathorjhora, but it has increased around the bridge area once more. Pathorjhora has the largest grain sizes, both upstream and downstream of the bridges. Pathorjhora, a fan area where the river empties larger rocks from the rocky landscape, is obstructed by the bridge piers, resulting in an accumulation of gigantic boulders close to the bridge. The spherical sediments are carried downstream and deposited during high flow (Biswas & Banerjee, 2018). A study on the Jaldhaka River basin concludes that, according to the flood susceptibility map, all riverine areas are vulnerable to flooding in their lowest reaches. In the last 20 years, there has been a significant change in the LU/LC pattern. Large parcels of land have been set aside for seasonal cultivation on the majority of the flood-affected regions. The populated area has expanded. It is spread in a random pattern instead of being concentrated or clustered in one area. Though precipitation patterns have been steady over the previous 30 years, with the exception of a few exceptional years like the floods in 1991, 1993, 1996, 2000, and 2013, the amount of natural vegetation coverage has decreased, potentially leading to soil erosion (Raha et al., 2023). Massive amounts of water flow through the small opening during the monsoon, impacting both the upstream and downstream sections and raising the risk and vulnerability to flooding. Three tributaries, Lish, Gish, and Chel, enter the Teesta River on its left bank. The Rivers Lish, Gish, and Chel join the Teesta, respectively, upstream and downstream of the Gajoldoba barrage. The rising water level of the Teesta River and the inundation of the nearby plains in North Bengal are caused by the monsoon season's increased water volume in these rivers (Biswas & Raha, 2023).

REFERENCES

Abrar, M., and M.U. Farooqi. 2021. Impacts of floods on bridges—a review. In *1st International Conference on Applied Engineering and Natural Sciences*. Konya, Turkey.

Ario, I., T. Yamashita, R. Tsubaki, S. Kawamura, T. Uchida, G. Watanabe, and A. Fujiwara. 2022. Investigation of bridge collapse phenomena due to heavy rain floods: structural, hydraulic, and hydrological analysis. *Journal of Bridge Engineering*, 27(9). https://doi.org/10.1061/(asce)be.1943-5592.0001905

Banerjee, P., and M. Biswas. 2021. Application of habitat modification score and fluvial functioning index in discussion of eco-hydrological behavior and flood risk zonation of Himalayan foothill rivers, West Bengal, India. *Acta Geophysica*, 69: 877–893.

Biswas, M., and P. Banerjee. 2018. Bridge construction and river channel morphology—a comprehensive study of flow behavior and sediment size alteration of the River Chel, India. *Arabian Journal of Geosciences*, 11: 1–23.

Biswas, M., and A. Raha. 2023. An overview of open channel monsoon hydraulics of Himalayan foothill rivers, West Bengal, India. *Modeling Earth Systems and Environment*, 9(2): 2969–2979.

Biswas, M., and A. Raha. 2024. Assessment of flood hazard along the NS section of North Bengal plains, India. *Natural Hazards*, 120: 2333–2348.

Chakraborty, S., and S. Mukhopadhyay. 2015. An assessment on the nature of channel migration of River Diana of the sub-Himalayan West Bengal using field and GIS techniques. *Arabian Journal of Geosciences*, 8(8): 5649–5661.

Christine, P., J. Jerzy, H. Wrona, and K. Rzepa. 2005. Dynamic slow-down: a flood mitigation strategy complying with the integrated management concept–implementation in a small mountainous catchment. *International Journal of River Basin Management*, 3(2): 75–85.

Cumbria County Council. 2010. https://elibrary.cumbria.gov.uk/elibrary/Content/Internet/536/6357/42347104432.PDF

Ge, H., and C. Deng. 2020. A numerical study of flood effect of a bridge at tail of Ganjiang River, entrance of Poyang Lake. *E3S Web of Conferences*, 198: 04006.

Hall, H.R.Y. 2022. Innovative approaches to urban riverscape planning and design. Doctoral dissertation, University of Georgia.

Han, J., H. Wang, H. Luo, and W. Duan. 2018. Numerical simulation on the influence of bridge construction on river flood control in a bottleneck reach. *Journal of Engineering & Technological Sciences*, 50(4): 451–460.

Hand, B.K., C.G. Flint, C.A. Frissell, C.C. Muhlfeld, S.P. Devlin, B.P. Kennedy, and J.A. Stanford. 2018. A social–ecological perspective for riverscape management in the Columbia River Basin. *Frontiers in Ecology and the Environment*, 16(S1): S23–S33.

Hohensinner, S., B. Lager, C. Sonnlechner, G. Haidvogl, S. Gierlinger, M. Schmid, and V. Winiwarter. 2013a. Changes in water and land: the reconstructed Viennese riverscape from 1500 to the present. *Water History*, 5: 145–172.

Hohensinner, S., C. Sonnlechner, M. Schmid, and V. Winiwarter. 2013b. Two steps back, one step forward: reconstructing the dynamic Danube riverscape under human influence in Vienna. *Water History*, 5: 121–143.

Holemba, G.A., and T. Matsumoto. 2019. Flood-induced bridge failures in Papua New Guinea. *MATEC Web of Conferences*, 258: 03014.

Koçyiğit, M.B., H. Akay, and A.M. Yanmaz. 2016. Flooding and its effects on river bridges in Western Black Sea Region. *Disaster Science and Engineering*, 2(1): 29–35.

Link, O., M. García, A. Pizarro, H. Alcayaga, and S. Palma. 2020. Local scour and sediment deposition at bridge piers during floods. *Journal of Hydraulic Engineering*, 146(3). https://doi.org/10.1061/(asce)hy.1943-7900.0001696

Lovric, N., and R. Tosic. 2016. Assessment of bank erosion, accretion and channel shifting using remote sensing and GIS: case study–lower course of the Bosna River. *Quaestiones Geographicae*, 35(1): 81–92.

Malard, F., U. Uehlinger, R. Zah, and K. Tockner. 2006. Flood-pulse and riverscape dynamics in a braided glacial river. *Ecology*, 87(3): 704–716.

Mitoulis, S.A., S.A. Argyroudis, M. Loli, and B. Imam. 2021. Restoration models for quantifying flood resilience of bridges. *Engineering Structures*, 238: 112180.

O'Connor, J.E., M.A. Jones, and T.L. Haluska. 2003. Flood plain and channel dynamics of the Quinault and Queets Rivers, Washington, USA. *Geomorphology*, 51(1–3): 31–59.

Ogras, S., and F. Onen. 2020. Flood analysis with HEC-RAS: a case study of Tigris river. *Advances in Civil Engineering*, 2020: 6131982.

Peipoch, M., M. Brauns, F.R. Hauer, M. Weitere, and H.M. Valett. 2015. Ecological simplification: human influences on riverscape complexity. *BioScience*, 65(11): 1057–1065.

Poulard, C., A. Lenar-Matyas, M. Łapuszek, M. Lafont, L. Schmitt, C. Jézéquel, and P. Breil. 2011. Co-conception of integrated flood management solutions: riverscapes to facilitate dialog. *Ecohydrology & Hydrobiology*, 11(1–2): 107–119.

Pregnolato, M., A.O. Winter, D. Mascarenas, A.D. Sen, P. Bates, and M.R Motley. 2022. Assessing flooding impact to riverine bridges: an integrated analysis. *Natural Hazards and Earth System Sciences Discussions*, 22(5): 1559–1576.

Raha, A., S. Gupta, and M. Biswas. 2023. Flood risk assessment of Himalayan Foothill Rivers: A study of Jaldhaka River, India. In *Floods in the Ganga–Brahmaputra–Meghna Delta*. Springer International Publishing, 63–90.

Rowland, J.M., G. Lucarini, and G.J. Tassie. 2021. *Revolutions: The Neolithisation of the Mediterranean Basin: The Transition to Food Producing Economies in North Africa, Southern Europe and the Levant.*

Sarma, S., and B.P. Talukdar. 2019. Effect of channel width contraction at Bogibeel bridge site on the morphology of the River Brahmaputra. *Bulletin Pure and Applied Science-Geology*, 38f(2): 128. https://doi.org/10.5958/2320-3234.2019.00009.x

Setunge, S., W. Lokuge, H. Mohseni, and W. Karunasena. 2014. Vulnerability of road bridge infrastructure under extreme flood events. In *AFAC & Bushfire & Natural Hazards CRC Conference 2014*. University of Southern Queensland.

2 Channel Dynamics in Spatio-Temporal Scale

2.1 CHANNEL SHIFTING, MIGRATION, CAUSES, AND GROUND VERIFICATION

The natural process that explains how a stream or river channel changes over time is called channel migration. For example, when a stream erodes away one bank and dumps sediments along the opposing side, channel migration might happen gradually. During floods or other high water events, it can also happen very quickly. Numerous elements, including gradient, geology, sediment supply, stream flow, vegetation, natural instability, and human activity, influence the flow path of the rivers.

The morphology of alluvial river channels continuously changes over time in response to variations in both external and internal factors, eventually reaching a new state of equilibrium (Zheng et al., 2017). External factors such as tectonic activity, geomorphic characteristics, and channel boundary materials influence fluvial system evolution over long geomorphological timescales (Varrani et al., 2019). In contrast, internal factors, including hydrodynamic forces, water discharge, and sediment concentration, affect river morphology on shorter hydrological timescales. The evolution of river channels at hydrological timescales is primarily driven by interactions between incoming water sediment flow and channel boundary conditions, responding swiftly to changes in sediment deposition and erosion patterns (Burroughs et al., 2009; Poeppl et al., 2019; Gu et al., 2020; Kong et al., 2020).

A hierarchy of channels, with minor channels nestled inside larger, major channels, may exist in a fluvial system. Minor channels include those that have been partially or entirely abandoned and chute channels that cross point bars etc. Three distinct methods of bar formation were hypothesized by Leddy et al. (1993) after modeling braid bar formation in a laboratory setting. In diverse natural environments over spatio-temporal scales, river shape is dynamic and subject to change.

An avulsion is a mechanism that redirects flow away from the parent canal into a fresh floodplain course. According to Heller and Paola (1996), a local avulsion creates a new channel that joins its parent channel downstream. The floodplain area is where the channel diversion mostly takes place, and lateral course shifting holds back bank erosion. Although it is often a natural process, human intervention has made it semi-natural throughout time. The socioeconomic well-being of the local population is severely impacted by shifting channel routes (Sonowal et al., 2022). Many people occasionally lose their houses, farms, and infrastructure as a result of channel migration. River discharge (Q), water velocity (m/sec), channel width (m), and channel depth (m) are some of the interrelated parameters that determine a channel pattern (Matsuda, 2004). Human involvement has a tremendous impact on channel shifting. High rates of sedimentation, unfavorable channel gradients, and

DOI: 10.1201/9781003562979-2

human-induced activities lead to river levee degradation and avulsion (Sharma et al. 2010; Heyvaert et al. 2012).

Avulsion, or the movement of channels into new locations in the floodplain, is the most significant floodplain process. Usually, this is explained and explored in terms of an abrupt process. River avulsion is the relatively quick movement of river flow from an established portion of the channel belt onto a new flow pathway somewhere on the floodplain, according to Mohrig et al. (2000). Avulsions dominate the long-term distribution of sediment and water throughout the alluvial surfaces of river systems where they occur. Therefore, understanding and characterizing the controls on avulsion is essential for geomorphologists studying the evolution of depositional landscapes, sedimentologists recreating the history of ancient river systems, and civil engineers interested in regulating the location of waterways.

The migration of channel can also be seen within the river valley. It is typically noticed in multi-thread channel setting. The multi-thread river migration occurred along a continuum from low energy to high energy. Lateral adjustment, hydrological regime, and sediment transport regime are associated with secondary channel dynamics (Table 2.1).

The design of bridge crossings and other transportation facilities in impacted areas must take into account channel migration (alluvial river meander, planform deformation), which deteriorates the alignment and approach conditions during construction as the upstream channel location shifts. Bank failure is often preceded by toe erosion and commonly occurs upstream of diversion dams and hydraulic structures. Such failures pose a threat to dams and nearby infrastructure, making prevention essential. A numerical model's capability to predict the location, extent, and depth of potential erosion is crucial for designing effective protection measures (Lai & Greimann, 2008). Excessive scour on bridge piers and abutments, hazards to bridge approaches and other highway infrastructure, exacerbated debris issues, and blocked passage through bridge openings are all possible outcomes of channel migration. For highway engineering purposes, a stream channel can be deemed unstable if the rate or magnitude of change is such that the planning, location, design, or maintenance considerations for a highway crossing are significantly affected during the facility's life, even though geomorphologists may consider channel stability from the perspective of hundreds or thousands of years. The following types of changes are concerning: (1) erosion of the lateral bank, including erosion from meander migration; (2) gradual deterioration or aggravation of the stream bed, and (3) transient variations in streambed elevation that are typically linked to floodplains (scour and fill). The specific focus of this study is meander-induced lateral channel instability, which includes down valley migration.

The dynamic nature of sedimentation along with flow behavior is responsible for this channel migration. It is important to note that the aforementioned two factors are also shaped by human intervention, i.e., construction of dam or bridges on rivers.

The Himalayan foothill rivers in the North Bengal plain portray significant signature of channel shifting both within their courses and channels. Riverine chars, bars, and inland areas exist in stable and unstable mode in the lower part of all the major rivers. Sediment aggradation process occurred in terms of landforms like alluvial fans and floodplains. It discloses the river planform dynamics and morphological

TABLE 2.1
Details of Secondary Channel Types and Behavior

Lateral Adjustment	Hydrologic Regime	Sediment Transport Regime: Primary Channel Transport	Sediment Transport Regime: Secondary Channel Supply	Secondary Channel Type	Characteristics
Laterally Inactive	Peak-flow and/or Base-flow	Low to moderate fine and coarse material bedload transport	Suspended Bed material and wash load	**Dam Distributed**	• Flow distributed laterally by beaver dam(s) • Multi-thread backwater channels of variable width • More than one outlet channel at various elevations • Dense riparian vegetation and abundant in stream woody material • Multiple individual stable channels that persist over time in the same location
	Base-flow	Low to moderate coarse material bedload transport	Suspended bed material and wash load	**Valley-fill Sub-parallel**	• Channels separated by vegetated floodplain, upland terraces, or stable islands • Dense riparian vegetation and abundant instream woody material • Associated with bedload deposition and channel aggradation
Laterally Active	Peak-flow	Moderate coarse material bedload transport	Primarily suspended bed material and wash load; moderate coarse bedload	**Valley-fill Distributed**	• Multiple small-scale avulsion channels along outside of meander bend carving new channels • Dense riparian vegetation limits side channel expansion • Beaver dam development following side channel formation
		Moderate to high coarse material bedload transport	Bedload, suspended bed material, and wash load	**Meander-Relict**	• Associated with point-bars and lateral channel migration • Small-scale avulsion into relict channel scar along outside of meander bend

signatures of the area. The main three left-hand tributaries of River Teesta—Rivers Lish, Gish, and Chel—have shown the shifting of channel over time. It is important to note that the Chel River, which is joined by River Neora, then flows as Dharala that has changed its confluence with River Teesta over time due to shifting near Domohani. The channels of the three rivers are seen to be shifting in between the banklines too. The dynamic nature of water and sediment discharge is responsible for the shifting nature of channels.

2.2 EROSION-DEPOSITION AND MEANDERING PATTERN, SPATIAL CHANGES IN CROSS-SECTIONS

Rivers with gravel beds that meander actively are dynamic environments that are extremely sensitive to changes in the surrounding environmental circumstances (Hooke, 2008). Key topics in fluvial geomorphology include our understanding of the dynamics of channel development, the significance of varying flood occurrences, and the sensitivity or responsiveness of rivers. These understandings have a significant impact on the effects of changing climate and altering hydrology as a result of human activity. Numerous studies have been conducted on the specific causes and processes of bank erosion, the dynamics of sedimentation, and the amounts and patterns of channel deposition on these types of rivers (Hooke, 2008). Meandering rivers experience longitudinal and lateral migration due to erosion at concave (convex) banks. Erosion happens sporadically when a bank collapses during a flood and when smaller particles are removed from the surface of the bank. This is dependent on a number of variables (scour at the bank toe, cohesiveness of the bank, wetting and drying of the banks, etc.), and the stream's capacity to remove sediments deposited at the bank foot ultimately determines how quickly this happens (Seminara, 2006).

The erosion and deposition sequence is an outcome of the changes in the volume of river discharge. Seasonal variation of rainfall is a key controlling factor of river discharge. The changing nature of erosion and deposition is also influenced by the sediment load of the channel. If the sediment load exceeds the river's carrying capacity, accretion will take place. The increase of sediment is further influenced by the catchment sediment generation. It is to be mentioned that if the geo-lithological composition of the catchment facilitates erosion, huge sediments are added to the river downstream. Human intervention in the catchment areas also contribute to the sediment generation process. Research indicates that human activities, such as deforestation and agriculture, can greatly influence erosion rates on hillslopes and affect the sediment yield at the catchment level. Various studies (e.g., Montgomery, 2007; Maetens et al., 2012) have highlighted the impact of land use changes on erosion, while others (e.g., Walling, 1999, 2006; Vanacker et al., 2007, 2014; Vanmaercke et al., 2015) have examined how these changes alter sediment transport within watersheds. This huge sediment deposit in downstream sections alters the existing channel pattern and river morphology. The river often loses its competency to carry this bulk sediment load due to certain hydraulic constraints and thus accretion of deposition takes place. Constantine et al. (2014) identified a strong positive relationship between sediment supply, morphological activity, and variations in channel width in meandering rivers of the Amazon basin. Their findings align with Brice's (1975) classification, which suggests that highly dynamic meandering rivers tend to have the widest sections near bend apexes. In contrast, when width maxima are closer to meander inflections, it may indicate lower morphological activity (Brice, 1975; Luchi et al., 2011), though this pattern does not appear to be universally applicable (Monegaglia et al., 2019). Hydraulic conditions such as river velocity and stream power are two of the leading controlling factors in this regard (Biswas & Raha, 2022). The slope of the channel is also responsible for controlling the flow of water and the carrying capacity of river. It is interesting to note that the deposition of sediments on the river bed helps to elevate the river bed, thus reducing the carrying capacity of the river. Another important aspect of channel aggradation is the role of riparian vegetation which creates obstacle and helps in the diversion of channel. This in turn helps in the trapping of sediments, especially suspended and finer bed load sediments, to get deposited. Gurnell et al. (2012) mentioned the role of riparian and aquatic vegetation as physical ecosystem engineering in shaping the riverine morphology.

Meandering rivers undergo lateral shifting and longitudinal migration due to erosion at concave banks and deposition at convex banks. Erosion occurs both gradually, through the removal of fine particles, and sporadically, through bank collapses during flood events. This process is influenced by various factors, such as bank toe scour, cohesion, and cycles of wetting and drying, while its overall rate depends on the stream's capacity to transport accumulated sediments. While detailed studies can analyze individual erosion events, long-term assessments benefit from a more integrated approach. This method identifies eroding outer bank regions and represents the process as a continuous spatial and temporal phenomenon, effectively capturing its average impact (Seminara, 2006). The sinuous pattern of channel with prominent bed load and suspended load help in the formation of alternate bars. The stability of channel is reflected in the cross-section of the rivers. However, frequent migration and change in the channel course along the dynamic nature of bar morphology is manifested through cross-sections as dynamic channel. The volume as well as the flow of hydro-sediment discharge is responsible for this dynamic nature of the channel morphology. It is interesting to note that the shifting of channels also occurs within the bed which is confined between the two banks. These mechanisms are a response to the dynamic sediment load of the river which changes over season. The sediment deposition and aggradation in response to the river channel hydraulics largely shape the position of the river bars—longitudinal, point etc., which are not stationary in nature. The river channels change their path and oscillate within the banks in response to the shifting nature of these mid-channel deposits. In case of multi-thread rivers such as Teesta, Jaldhaka, and Brahmaputra, this mechanism is more complex and harder to comprehend. The detailed analysis of the hydro-sediment interaction might help to understand this complex mechanism. It is important to remember that the sediment generation and its aggradation-degradation process are also modified by allogenic factors. The construction of roads, bridges, dams, etc. on the rivers largely impact the sedimentation characteristics.

Various methods have been developed to estimate erosion and deposition in meandering bends. Analytical models, such as those proposed by Engelund (1973) and Odgaard (1981), are applicable primarily to relatively simple flow conditions. Additionally, tools for calculating near-bank velocity and shear stress have been introduced to aid in designing and sizing riprap protection (e.g., Bridge 1977; Thorne et al. 1995). However, these methods, while suitable for stability control in engineering design, often tend to be overly conservative (Thorne et al. 1995).

The physical, ecological, and socioeconomic aspects within the fluvial environment area are affected by the fluvial dynamics, bank erosion, and accretion (Neal and Anders, 2015). The morphological aspects of erosion are necessary to understand the dynamic nature of alluvial rivers (Hasanuzzaman et al. 2024).

A study by Gogoi and Goswami (2014) has pointed out the nature and type of channel shifting of River Subansiri, an important right-hand tributary of River Brahmaputra. The authors have noted five types of channel shift from 1828 to 2021: (a) chute cut off, (b) shifting of meander, (c) neck cut off, (d) channel avulsion, and (e) alteration bar-induced shifting. Huge rainfall and consequent flooding make River Subansiri highly unstable with a huge sediment load. The embankment of river initially restricts its lateral flow to some extent. Thus, bad aggradation is noticed. At some point, the bank loses its competency in order to accommodate this huge hydro-sediment discharge and overbank flow occurs leading to channel shifting and changes in channel morphology.

The erosion and accretion of a river are often controlled and altered by anthropogenic factors which include the construction of engineering practices such as dams, bridges, and barrages. It is to be noted that the sediment aggradation in response to hydraulic and channel slope parameters is also influenced by the anthropogenic factors. This includes engineering practices. The constructions of bridges with several piers with the narrow passage of water flow often create obstacles to the easy flow of the hydro-sediment discharge of the river. This in turn facilitates the deposition of sediments behind the piers toward the upstream section of the rivers as well as the boulders, pebbles, and cobbles often increase the scouring of the piers with the increase of stream power, velocity, and discharge near the bridges. This has a direct influence on the changes in the channel width and depositional character of the channels over time. The huge sediments are accommodated by laterally expanding the banks in the upstream reaches specially in River Gish.

Gaeumen et al. (2005) in their research depicted the Duchesne River's channel reaction to several silt kinds. The lower Duchesne River's gravel-bed and sand-bed portions demonstrated a unique adaptation to the sediment supply and outflow. While gravel-bed reaches mostly adjusted through width changes, sand-bed reaches tended to adapt to all disturbances with quick changes in the bed level. Only when gravel was added to the river by local bank erosion did the gravel-bed sections deteriorate, and only during significant floods would they erode. The Duchesne River's avulsion is brought on by bed aggradation.

The construction of Farakka Barrage in 1975 has a significant impact in the alteration of the hydro-sedimentological mechanism of River Ganga. This has resulted in accelerated degradation, avulsion, channel shifting, and bar formation (Hasanuzzaman et al. 2024, Chakraborty 2017). It has been noted by Das and Samanta (2022) that a land area of about 42.74 km^2 got eroded due to the lateral shifting of River Ganga between 1979 and 2024. Fluctuating discharge and sediment deposition disrupt the river's morphology, resulting in instability and entropy maximization (Hasanuzzaman et al. 2021). Saleem et al. (2019) calculated the erosion and accretion rate of River Padma (Bangladesh), which suggests that erosion is predominant on the left bank while accretion was high along the right bank, over the study period. The volumes of morphological change for two epochs correspond to a net volume gain of 338.75 million m^3 sediment between 1984 and 1992 but a net loss of 295.20 million m^3 during the period 1992–2008. The area of large mid-channel bars increased over time, which may have had a role in shaping the erosion and accretion processes of the river. Suryawanshi et al. (2015) have demonstrated the impact of K. T. Weirs on the erosion and accretion nature of River Urmodi. Their study has revealed how aggradation has taken place in the upstream reaches of K. T. Weirs and degradation is dominant at the downstream section (Figure 2.1).

The shifting nature of the channel with positive and negative changes has been studied on a temporal scale (1913, 1936, 1943, 1965, 1984, 2008, and 2021) using topographic maps and 2.5 × 2.5 m resolution DEMs from the Cartosat satellite, Toposheet (LISS 3, Landsat ETM, and OLI for 2008 and 2021), and satellite picture (for 1913, 1936, 1943, 1965, and Survey of India: Topographical Map for 1984), respectively.

It is evident from the figure that accretion is predominant for River Lish during 1913–1936 (3.097 km^2) and 1965–1984 (3.915 km^2). Negligible amount of accretion is noticed from 1936 to 1943 (0.300 km^2). The erosion process seems to be predominant during 1965–1984 (1.531 km^2) and 1984–2008 (1.27 km^2). The unchanged areas are maximum during 1984–2008 (5.366 km^2) and 2008–2021 (5.097 km^2) (Figure 2.2).

Gish has shown a massive accretion rate during 1913–1936 (17.567 km^2). Subsequently, in the next considered tenure, i.e., 1936–1943, the erosion process (10.953 km^2) was predominant. It is important to note that the unchanged area is high throughout the period from 1936 to 2021. There was a decrease in the rate of the

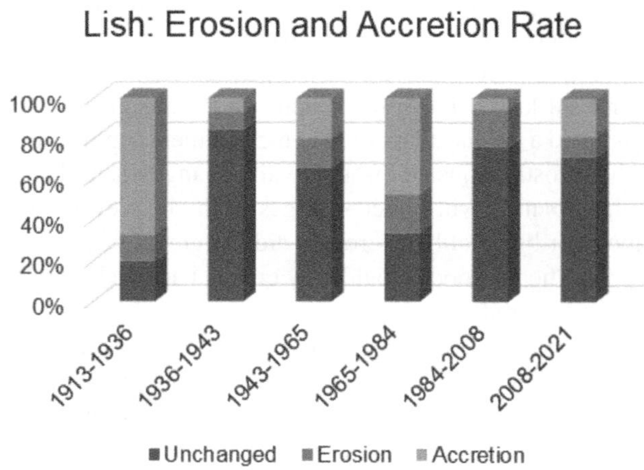

FIGURE 2.1 Erosion and accretion rate of River Lish over time.

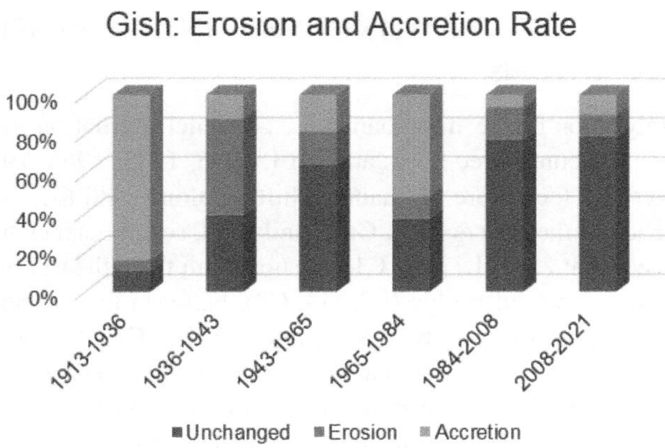

FIGURE 2.2 Erosion and accretion rate of River Gish over time.

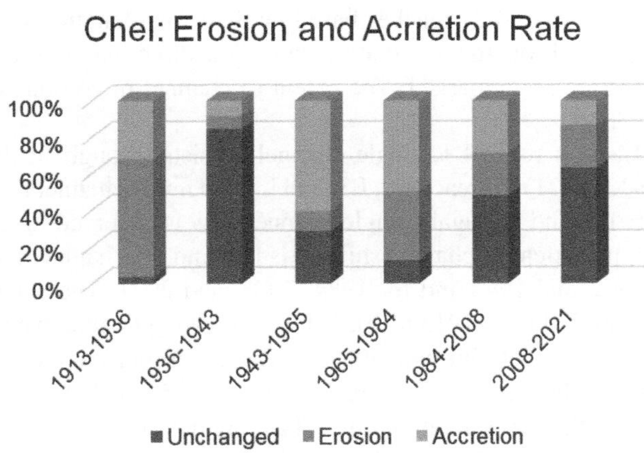

FIGURE 2.3 Erosion and accretion rate of River Chel over time.

erosion process from 1943, but the rate remained the same to some extent at an average of 2.789 km^2 from 1943 to 2021 (Figure 2.3).

The scenario of Chel is somewhat different from Rivers Lish and Gish. The tenure from 1913 to 1936 shows a remarkable erosion rate (12.677 km^2). The rate of erosion further increased drastically during 1965–1984 (12.408 km^2). The process of accretion gained a momentum from 1943 to 1965 when an area of 11.280 km^2 got aggraded. It further increased (16.690 km^2) during 1965–1984. Thereafter, the accretion rate gradually decreased during 1984–2008 (8.504 km^2) and 2008–2021 (3.537 km^2). Subsequently, the erosion rate during 1984–2008 (6.731 km^2) and 2008–2021 (6.155 km^2) decreased (Figure 2.3). This period thus portrays the maximum unchanged area (14.168 km^2 and 16.518 km^2 during 1984–2008 and 2008–2021, respectively) (Biswas et al., 2024).

2.3 TIME-DISTANCE ANALYSIS OF SHIFTING CHANNEL ALONG CROSS-SECTIONS

The channel migration between the banks, i.e., channel oscillation, is mapped on a temporal scale. The considered years are 1913, 1936, 1943, 1965, 1984, 2008, and 2021. To understand the nature of channel shifting along with the change in width at various reaches of the Rivers Lish, Gish, and Chel, seven reaches have been considered, i.e., piedmont zone (L1, G1, C1), far upstream from the rail bridge (L2, G2, C2), upstream from the rail bridge (L3, G3, C3), between the rail and road bridge (L4, G4, C4), downstream from the road bridge (L5, G5, C5), far downstream from the road bridge (L6, G6, C6), and near the confluence (L7, G7, C7). The changes were more predominant after the construction of the road bridge (Biswas et al., 2024). The evolution of channel from single channel flow to multiple channel flow is noted. To analyze the extent and direction of migration, channel migration analysis toolbox of ArcGIS 10.4 is used. The calculation is done for 1913 and 2021, i.e., the extent of migration is measured for 2021 from 1913. Rivers Lish, Gish, and Chel are divided into 8, 18, and 21 reaches based on their total lengths and required gap between the reaches. From the observation points, the direction of the channel migration from 1913 to 2021 is marked. The extent of channel migration is measured in meters (Figure 2.4).

For River Lish, at reach L1, single channel persists throughout the considered time frame (1913–2021). At reach L2, L3, and L4, the multi-channel flow is observed in 2021. At L5, L6, and L7, again single channel flow is observed. It is important to note that at L4, the extent of channel shifting is less and the channel flow is the same for the years 1936 and 1943 and for 1984, 2008, and 2021. The extent of channel migration is quantified from 1913 to 2021. It is interesting to note that the channel of Lish has shifted toward left throughout its upstream section from the rail bridge. The rate of migration in the three upstream reaches are 297.058 m, 202.61 m, and 340.564 m. The channel in the downstream sections has migrated in both the directions. The maximum extent is observed for the downstream reach L5 with a value of 327.88 m. The far downstream sections show a negligible degree of migration toward both left and right (Figure 2.5).

River Gish has the most prominent migration pattern. The G1 reach shows no change in terms of channel migration and throughout the considered tenure (1913–2021), a single channel flow is observed. At G2 reach, the flow of channel is the same for 1936, 1943, 1965, and 2021. A double channel flow is observed for 2021 with an intermediate distance of 483 m suggesting the bifurcation of channel flow with the deposition of sediments in the formation of a mid-channel bar. The most diverse channel pattern is observed for G6 reach. The multi-channel flow is marked for 1936, 1943, 1984, 2008, and 2021. The braiding pattern of the channel is thus prominent suggesting the aggradation of sediments along with the formation of several mid-channel bars.

The direction and extent of migration suggest that in the upstream reaches from the foothill, the channel has shifted toward right from 1913 to 2021. At the 3rd and 5th reaches, the extent of migration is notable (638.178 m and 507.884 m).

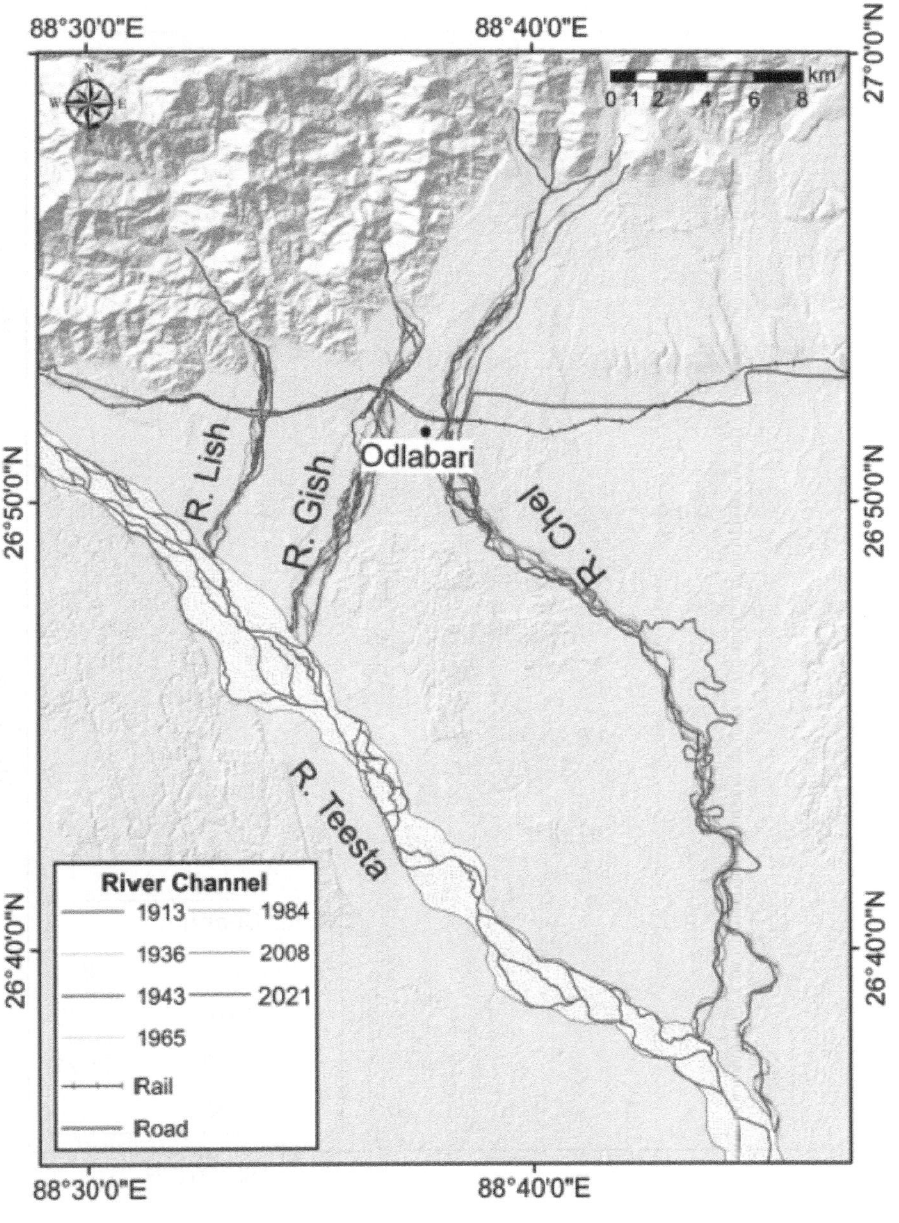

FIGURE 2.4 River channel migration from 1913 to 2021 of all three respective rivers.

The immediate upstream reach of the rail bridge seems to have a channel migration toward left (62.035 m) (Biswas et al., 2024) (Figure 2.6).

The downstream reaches from the road bridge shows a leftward migration from 1913 to 2021. The 8th, 9th, and 10th reaches have shifted 502.149 m, 783.033 m, and 579.463 m, respectively. Far downstream reaches exhibit a negligible extent of

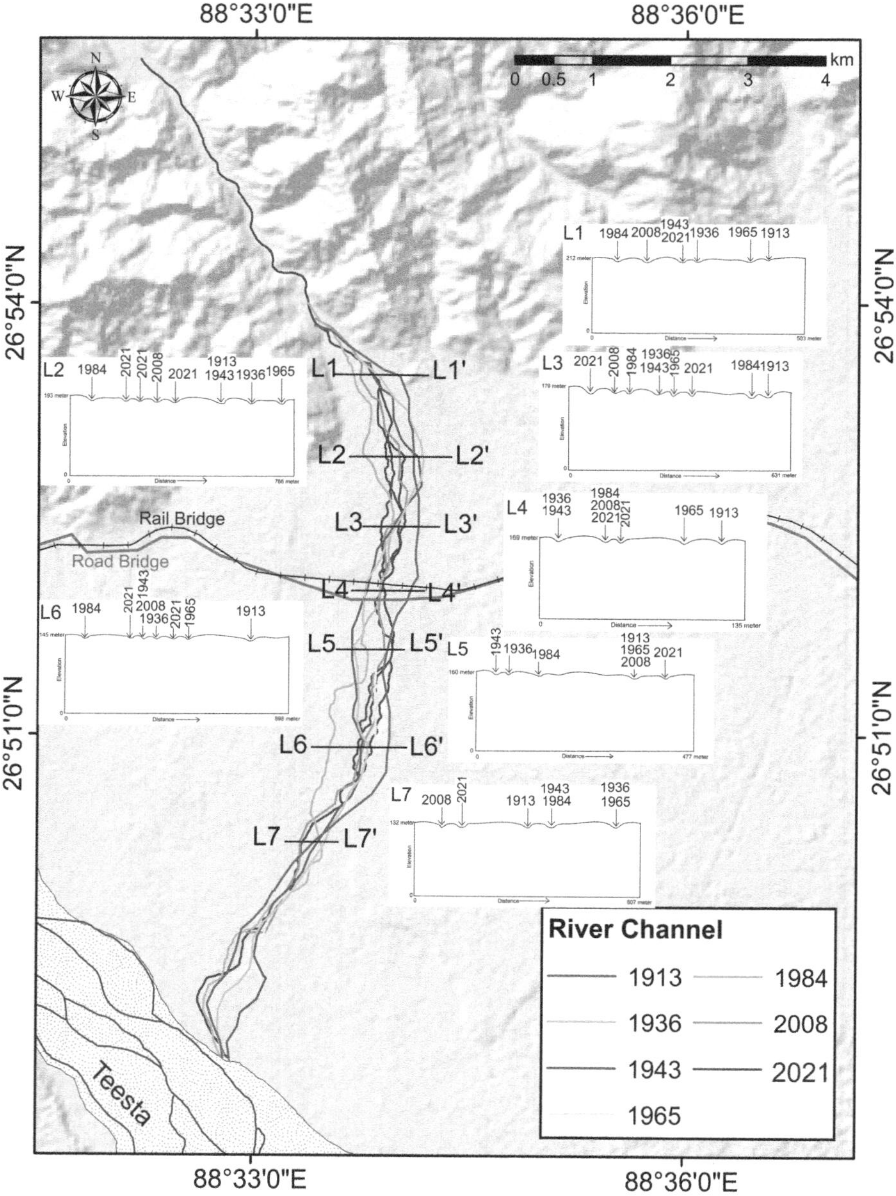

FIGURE 2.5 River channel migration of River Lish from 1913 to 2021.

migration with an average of 51.54 m. The reaches near the confluence again exhibit noteworthy migration toward right (444.88 m and 662.849 m).

River Chel shows an important degree of channel shifting on temporal scale. At C1 reach, the channels in 1936, 1943, and 1965 were flowing through the same thalweg points after their migration toward left from the year 1913. The channels further

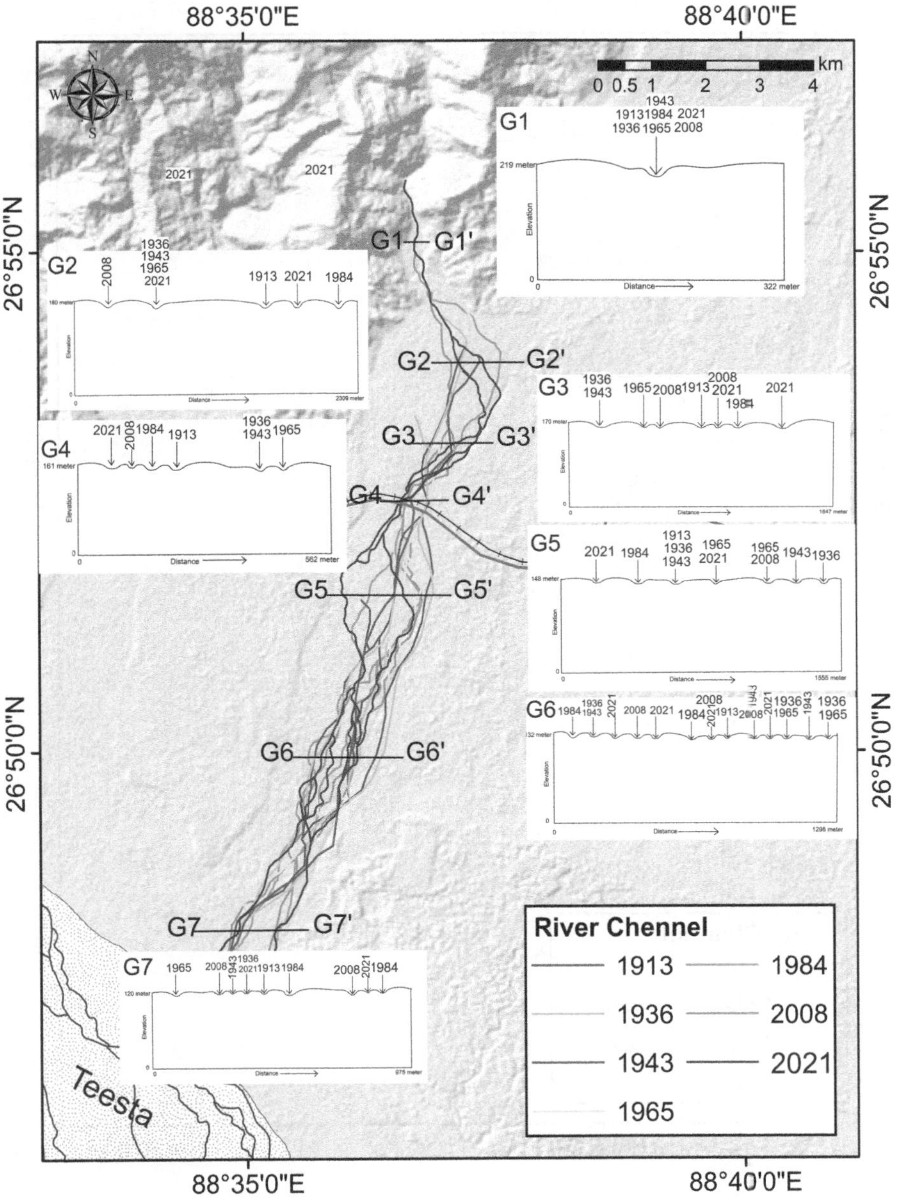

FIGURE 2.6 River channel migration of River Lish from 1913 to 2021.

migrated toward left and were flowing through the same thalweg points in 1984, 2008, and 2021. The multi-channel flow for reach C2 was observed for 1984 only. At reach C3, the multi-channel flow is noted for the years 1943 and 1984, suggesting the dynamic nature of channel along with the dynamic sediment aggradation and degradation mechanism. At C4, C5, C6, and C7 reaches, the same channel is observed for 1984, 2008, and 2021 (Biswas et al., 2024) (Figure 2.7).

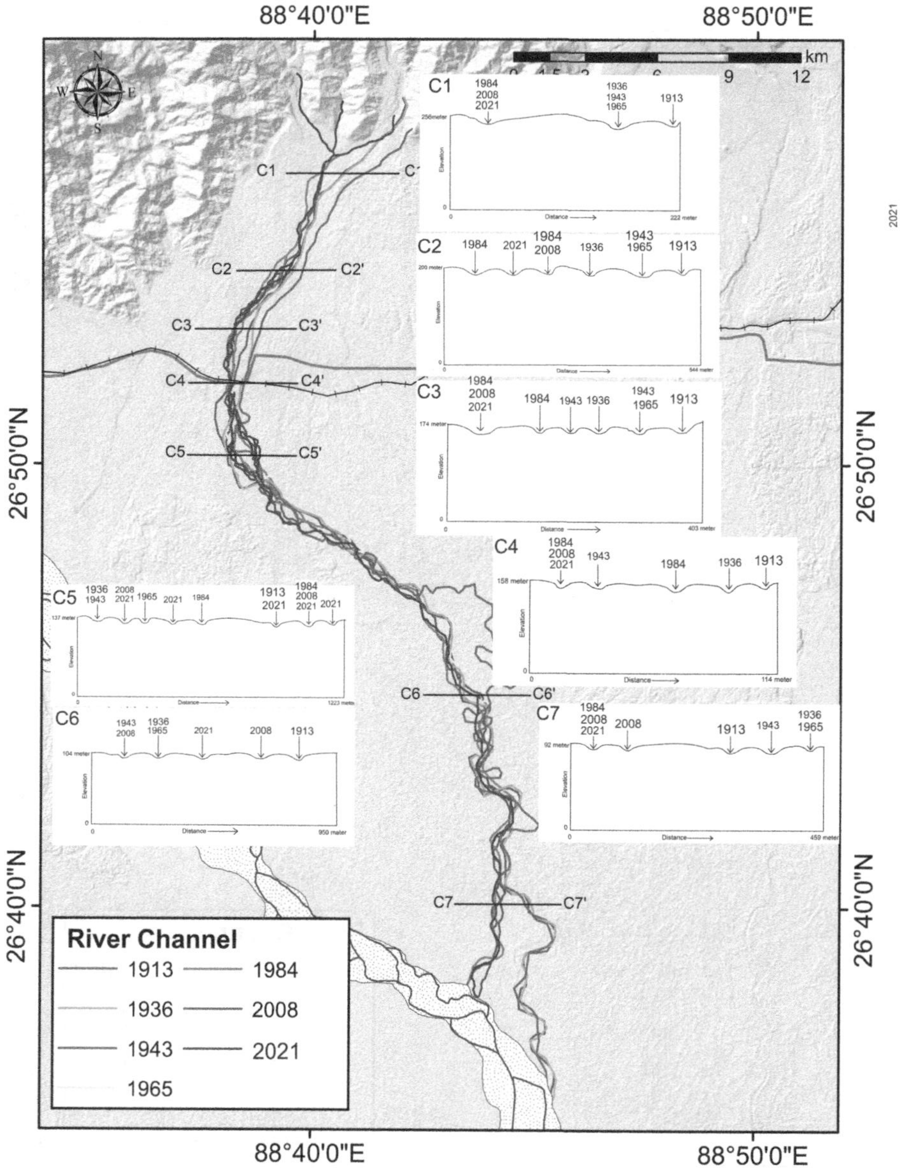

FIGURE 2.7 River channel migration of River Lish from 1913 to 2021.

The direction and extent of migration as portrayed in Figure 2.4c suggest a leftward migration of channel at the three upstream reaches from the rail bridge. The 1st and 3rd reaches seem to have a channel migration of 891.125 m and 982.824 m, respectively. A rightward shifting is noted for 5th reach of about 810.521 m. Significant migration is noted for reaches 9, 10, and 11 (2165.44 m, 1498.65 m, and 1754.51 m, respectively) toward left (Biswas et al., 2024). The confluence of the River Chel is also marked to be shifted toward left (1159.35 m and 1191.89 m).

2.4 SPATIO-TEMPORAL CROSS-SECTIONS ANALYSIS

The period under consideration (1913–2021) witnessed a noteworthy change in the channel morphology. The considered three rivers, i.e., Lish, Gish, and Chel, had a single channel during 1913. The width of the rivers was less. Gradually, in 1936, the width of the three rivers increased considerably. The reach L2 of River Lish marks an increase of 176.29 m in the width. A considerable increase in the width is marked for reaches L4 (497.88 m), L5 (382.13 m), and L6 (669.29 m) (Biswas et al., 2024).

During 1913–1936, the width of the downstream sections of River Gish increased considerably. The reaches G5 (1088.67 m) and G6 (1061.69 m) marked the maximum increase in width. Most interestingly, the downstream section of River Gish from the rail bridge developed a multi-channel system. It is evident that the increased sediment discharge altered the bed configuration and created a favorable condition for the formation of mid-channel bars. This has resulted in the change from a single channel system to a multi-channel system (Biswas et al., 2024). For River Chel, a considerable increase in the width is observed for reach C2 (414.98 m) and C3 (460.21 m). Most interestingly, the reaches C1, C6, and C7 mark a decrease in the river width (−69.08 m, −30.08, and −286.81 m). In the next tenure (1936–1943), a decrease in the width is observed in few reaches of all the three considered rivers. Significant among them are L2 (−66.5 m), L6 (−50.8 m), G7 (−42.76 m), and C4 (−42.96 m). It is to be noted that the trend of decreasing width for L4, G4, and C4 continued in the next tenures as well. During 1943–1965, the width of these three reaches of the three rivers decreased by 108.65 m, 59.89 m, and 61.45 m, respectively. The increase in the width of the piedmont zone for Rivers Gish and Chel became negligible (0.59 m and 5.6 m, respectively) during this period (Biswas et al., 2024) (Figures 2.8 and 2.9).

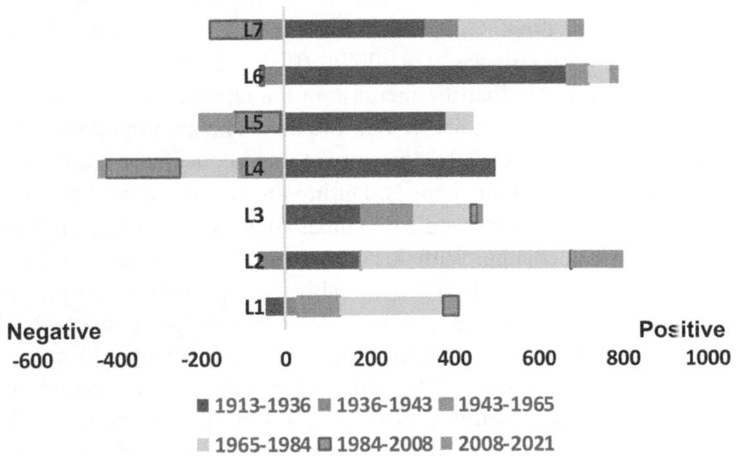

FIGURE 2.8 Negative and positive changes with erosion/deposition of River Lish from 1913 to 2021.

Based on Biswas et al., 2024.

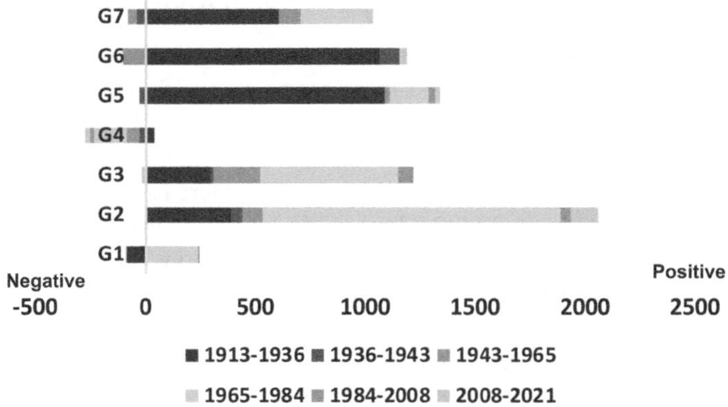

FIGURE 2.9 Negative and positive changes with erosion/deposition of River Gish from 1913 to 2021.

Based on Biswas et al., 2024.

However, a noticeable decrease in the width is marked between the rail and road bridges of the three rivers, i.e., at the reaches G4, G4, and C4 (−108.65 m, −59.89 m, and −61.45 m, respectively). In the next tenure (1965–1984), the decreasing trend continues between the road and rail bridges of the three rivers. It is marked as −140 m, −148 m, and −110 m for Rivers Lish, Gish, and Chel, respectively. But the width at the G2 reach, i.e., the far upstream section of River Gish significantly increased (1354 m). In the next tenure (1984–2008), the width of Chel at C6 section increased significantly (552 m). A decreasing trend was observed for the sections of L4, G4, and C4. In the tenure between 2008 and 2021, the far upstream sections of Lish and Gish (L2 and G2) showed an increase of about 125 m. The reaches between the road and rail bridges for the three rivers, i.e., L4, G4, and C4 decreased (−18 m, −18 m, and −85 m, respectively) (Biswas et al., 2024) (Figure 2.10).

The work thus indicates that the increase in the channel width is a result of the overbank sediment aggradation and retreat of bank. The anthropogenic impact, i.e., the construction of road and rail bridges on the three considered rivers, has a direct influence on the channel width changes. During the construction of the bridges, the width of the reaches was constricted with embankment which is reflected in the temporal analysis of width change. With due course of time, the reaches at and between the rail and road bridges decreased considerably. On the contrary, an increase in the width at upstream and downstream sections from the bridges is marked. The planform change is manifested on the transition from single channel to multi-channel, braiding pattern, dynamic nature of the mid channel bars, dynamic nature of the sediment deposition, and degradation process in connection with the hydraulic conditions of the channel. The piers of the bridges and the constricted channel have restricted the easy flow of water and sediments downstream which has resulted in the overbank sedimentation. This further resulted in bank retreat and an increase of channel width. The increase of channel width is more prominent for River Gish, followed by Rivers Lish and Chel.

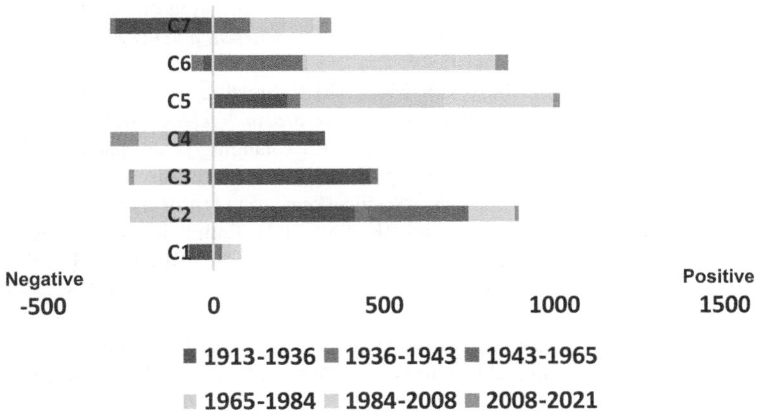

FIGURE 2.10 Negative and positive changes with erosion/deposition of River Chel from 1913 to 2021.

Based on Biswas et al., 2024.

REFERENCES

Biswas, M., and Raha, A. 2022. An overview of open channel monsoon hydraulics of Himalayan foothill rivers, West Bengal, India. *Modeling Earth Systems and Environment*, 9(2): 2969–2979. https://doi.org/10.1007/s40808-022-01481-9

Biswas, M., Raha, A., & Roy, S. 2024. Impacts of bridge constructions on channel dynamics and sediment aggradation: An integrated discussion, West Bengal, India. *Geomatica*, 76(2): 100021.

Brice, J.C. 1975. Air photo interpretation of the form and behaviour of alluvial rivers. Tech. Rep. U.S. Army Research Office.

Bridge, J.S. 1977. Flow, bed topography, grain size, and sedimentary structure in open-channel bends: a three-dimensional model. *Earth Surface Processes*, 2: 401–415.

Burroughs, B.A., D.B. Hayes, K.D. Klomp, J.F. Hansen, and J. Mistak. 2009. Effects of Stronach Dam removal on fluvial geomorphology in the Pine River, Michigan, United States. *Geomorphology*, 110: 96–107.

Chakraborty, A. 2017. Impact of changing river course on livelihood adjustment of the riverine island dwellers in the downstream of Farakka Barrage, India. Ph.D. thesis, University of Gour Banga, Malda, West Bengal, India, pp. 1–189.

Constantine, J.A., T. Dunne, J. Ahmed, C. Legleiter, and E.D. Lazarus. 2014. Sediment supply as a driver of river meandering and floodplain evolution in the Amazon Basin. *Nature Geoscience*, 7(12): 899–903.

Das, R., and G. Samanta. 2022. Impact of floods and river-bank erosion on the riverine people in Manikchak Block of Malda District, West Bengal. *Environment, Development and Sustainability*, 14:1–23.

Engelund, F. 1973. Flow bed topography in channel bend. *Journal of Hydraulic Division, ASCE*, 100(11): 1631–1648.

Gaeuman, D., J.C. Schmidt, and P.R. Wilcock. 2005. Complex channel responses to changes in stream flow and sediment supply on the lower Duchesne River, Utah. *Geomorphology*, 64(3–4): 185–206.

Gogoi, C., and D.C. Goswami. 2014. A study on channel migration of the Subansiri River in Assam using remote sensing and GIS technology. *Current Science*, 106(8): 1113–1120. http://www.jstor.org/stable/24102308

Gu, J., Y. Lu, R. Ji, Z. Wang, L. Jia, and S. Mo. 2020. Channel response to low water levels in the Pearl River Delta: a multi-decadal analysis. *Marine Geology*, 429: 106290.

Gurnell, A.M., W. Bertoldi, and D. Corenblit. 2012. Changing river channels: The role of hydrological processes, plants, and pioneer fluvial landforms in humid temperate, mixed-load, gravel-bed rivers. *Earth Science Review*, 111: 129–141.

Hasanuzzaman, M., B. Bera, A. Islam, and P.K. Shit. 2021. Estimation and prediction of riverbank erosion susceptibility and shifting rate using DSAS, BEHI, and REBVI models: Evidence from the lower Ganga River in India. https://doi.org/10.21203/rs.3.rs-1097959/v1

Hasanuzzaman, M., B. Bera, A. Islam, and P.K. Shit. 2024. Exploring GIS-based modeling for assessing social vulnerability to Ganga Riverbank erosion, India. *Natural Hazards Research.* https://doi.org/10.1016/j.nhres.2024.08.001

Heller, P.L., and C. Paola. 1996. Downstream changes in alluvial architecture: An exploration of controls on channel-stacking patterns. *Journal of Sedimentary Research*, 66(2): 297–306.

Heyvaert, V.M.A., J. Walstra, P. Verkinderen, H.J.T. Weerts, and B. Ooghe. 2012. The role of human interference on the channel shifting of the Karkheh River in the lower Khuzestan plain (Mesopotamia, SW Iran). *Quaternary International*, 251: 52–63. https://doi.org/10.1016/j.quaint.2011.07.018

Hooke, J. 2008. Temporal variations in fluvial processes on an active meandering river over a 20-year period. *Geomorphology*, 100(1–2): 3–13. https://doi.org/10.1016/j.geomorph.2007.04.034

Kong, D., E.M. Latrubesse, C. Miao, and R. Zhou. 2020. Morphological response of the Lower Yellow River to the operation of Xiaolangdi Dam, *China. Geomorphology*, 350: 106931.

Lai, Y.G., and B.P. Greimann. 2008. Modeling of erosion and deposition at meandering channels. *World Environmental and Water Resources Congress*, 2011: 1–11. https://doi.org/10.1061/40976(316)471.

Leddy, J. O., Ashworth, P. J., & Best, J. L. 1993. Mechanisms of anabranch avulsion within gravel-bed braided rivers: observations from a scaled physical model. *Geological Society, London, Special Publications*, 75(1): 119–127.

Luchi, R., G. Zolezzi, and M. Tubino. 2011. Bend theory of river meanders with spatial width variations. *Journal of Fluid Mechanics*, 681: 311–339.

Maetens, W., M. Vanmaercke, J. Poesen, B. Jankauskas, G. Jankauskiene, and I. Ionita. 2012. Effects of land use on annual runoff and soil loss in Europe and the Mediterranean: A meta-analysis of plot data. *Progress in Physical Geography*, 36: 599–653.

Matsuda, I. 2004. River morphology and channel processes. In James C.I. Dooge (ed.), *Fresh Surface Water*, Encyclopedia of Life Support Systems (EOLSS), 12. EOLSS Publishers.

Mohrig, D., P.L. Heller, C. Paola, and W.J. Lyons. 2000. Interpreting avulsion processes from ancient alluvial sequences: Guadalope-Matarranya system (northern Spain) and Wasatch Formation (western Colorado). *Geological Society of America Bulletin*, 112(12): 1787–1803.

Monegaglia, F., M. Tubino, and G. Zolezzi. 2019. Interaction between curvature-driven width oscillations and channel curvature in evolving meander bends. *Journal of Fluid Mechanics*, 876: 985–1017. https://doi.org/10.1017/jfm.2019.574

Montgomery, D.R. 2007. Soil erosion and agricultural sustainability. *Proceedings of the National Academy of Sciences*, 104: 13268–13272.

Neal, C.W.M., and A. Anders. 2015. Suspended sediment supply dominated by bank erosion in a low-gradient agricultural watershed, Wildcat Slough, Fisher, Illinois, United States. *Journal of Soil and Water Conservation*, 70(3): 145–155.

Odgaard, A.J. 1981. Transverse bed slope in alluvial channel beds. *Journal of Hydraulic Division, ASCE*, 107(12): 1677–1694.

Poeppl, R.E., T. Coulthard, S.D. Keesstra, and M. Keiler. 2019. Modeling the impact of dam removal on channel evolution and sediment delivery in a multiple dam setting. *International Journal of Sediment Research*, 34: 537–549.

Saleem, A., A. Dewan, and M.M. Rahman. 2019. Spatial and temporal variations of erosion and accretion: A case of a large tropical river. *Earth System and Environment*, 4: 167–181. https://doi.org/10.1007/s41748-019-00143-8

Seminara, G. 2006. Meanders. *Journal of Fluid Mechanics*, 554(1): 271. https://doi.org/10.1017/s0022112006008925

Sharma, B., U. Amarasinghe, X. Chen, D. de Condappa, T. Shah, A. Mukherji, et al. 2010. The Indus and the Ganges: River basins under extreme pressure. *Water International*, 35(5): 493–521. https://doi.org/10.1080/02508060.2010.512996

Sonowal, G., G. Thakuriah, and S. Hazarika. 2022. Role of channel migration and influencing hydro-geomorphic attributes in Dibru River Basin using remote sensing and GIS. *Nature Environment and Pollution Technology*, 21(5): 2035–2054.

Suryawanshi, S.B., J.B. Sapkale, and V.A. Chougule. 2015. Impact of dam and weirs on cross-sectional characteristics of Urmodi Channel, Maharashtra: An approach to geoinformatics. *International Journal of Engineering and Technical Research (IJETR)*, 2(12): 267–271.

Thorne, C.R., S.R. Abt, and S.T. Maynord. 1995. Prediction of near-bank velocity and scour depth in meander bends for design of riprap revetments. In C.R. Thorne, S.R. Abt, F.B.J. Barends, S.T. Maynord, and K.W. Pilarczyk, (eds.), *River, Coastal, and Shoreline Protection: Erosion Control Using Riprap and Armourstone*. Wiley, 115–136.

Vanacker, V., N. Bellin, A. Molina, and P. Kubik. 2014. Erosion regulation as a function of human disturbances to vegetation cover: a conceptual model. *Landscape Ecology*, 29: 293–309.

Vanacker, V., F. Von Blanckenburg, G. Govers, A. Molina, J. Poesen, J. Deckers, and P. Kubik. 2007. Restoring dense vegetation can slow mountain erosion to near natural benchmark levels. *Geology*, 35: 303–306.

Vanmaercke, M., J. Poesen, G. Govers, and G. Verstraeten. 2015. Quantifying human impacts on catchment sediment yield: A continental approach. *Global and Planetary Change*, 130: 22–36.

Varrani, A., M. Nones, and R. Gupana. 2019. Long-term modelling of fluvial systems at the watershed scale: Examples from three case studies. *Journal of Hydrology*, 574: 1042–1052.

Walling, D. 1999. Linking land use, erosion, and sediment yields in river basins. *Hydrobiologia*, 410: 223–240.

Walling, D. 2006. Human impact on land–ocean sediment transfer by the world's rivers. *Geomorphology*, 79: 192–216.

Zheng, S., B. Wu, K. Wang, G. Tan, S. Han, and C.R. Thorne. 2017. Evolution of the Yellow River delta, China: Impacts of channel avulsion and progradation. *International Journal of Sediment Research*, 32: 34–44.

3 River Planform Changes and Hydrological Consequences

3.1 INTRODUCTION

A continuously moving river is inherently dynamic, which means that anti-erosion measures cannot remain effective for extended periods (Mitra, 2015). The alterations in the river's flow and the shifting of its banks often have significant impacts on the communities living along the river (Bose & Navera, 2017). The geomorphic shifting of river channels along with planform changes can result from many factors, including floods, crustal movement, channelization, and other autogenic causes as well as human constructions such as bridges, dams, and reservoirs. However, the major changes in the river morphology can be seen right after large floods, especially in tropical areas of the world (Saleem et al., 2020). Constructing structures such as dams or bridges can serve as an effective method to address the shifting riverbanks, as these constructions help to minimize the flood-prone areas of the waterway in reaction to hydrological impacts (Lin, 2011; Biswas, 2010; Uddin et al., 2024). In that case, the thalweg lines play a great role in shaping the direction of the riverbed. The thalweg lines are the deepest flow path of the main stream of the river. Immovability of the thalweg lines is a collective effect of hydro-morphological processes in spatio-temporal spaces. The geographical location of the Himalayan foothills is well known as the piedmont strip—narrow W-E extended lands that commonly include alluvial fans and floodplains. The N-S flowing major rivers like Teesta, Jaldhaka, Torsa, and their tributaries carry a huge amount of water volume during the monsoon, resulting in floods. River inundation because of flooding and river bank shifting leads to the lateral disaster and rigorously related to lives, lands, and economy of a region or country. Human beings always imply their knowledge in constructions like bridges and minimize the bridge width for cost benefit. During the monsoon, a huge amount of water passes through the under bridge narrow passage, resulting in floods and bank erosion in furthermost reaches (Figure 3.1).

The free surface flow of a channel that varies seasonally is the focus of open channel hydraulics. The Himalayan foothill rivers' monsoonal hydrological features reveal a harsh situation in terms of stream strength, velocity, and shear stress. Changes in these features have been clearly seen upstream, between, and after the bridges of rivers such as the Lish, Gish, and Chel. On a temporal scale, the bridges have significantly altered the riverine topography expression with channel shifting close to, upstream, and downstream of the bridges (1913–2021). During the monsoon season, the stream strength and velocity rise close to the Chel River's rail and road bridge compared to those of the Gish and Lish Rivers.

DOI: 10.1201/9781003562979-3

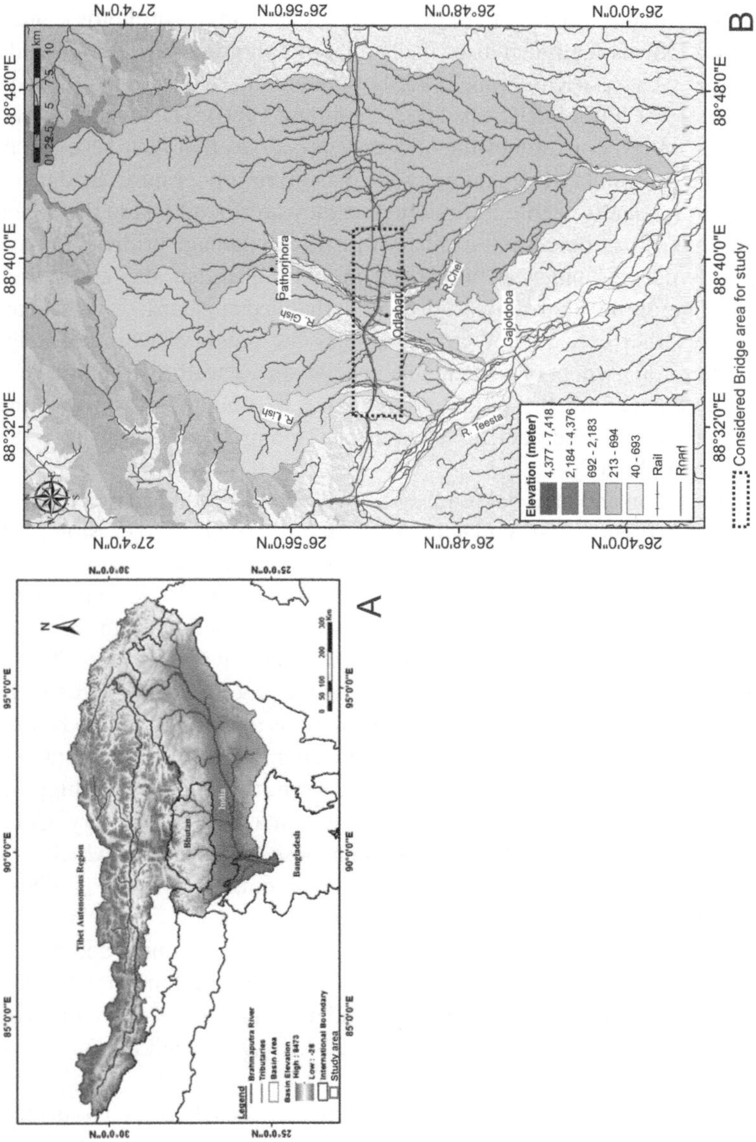

FIGURE 3.1 (A) Location of the study area in Brahmaputra basin. (*Source*: Indian Institute of Technology, Guwahati, India.) (B) Detail map of Lish, Gish, and Chel watershed with specific demarcation of the study area. (C) Elevation map of the entire catchment and lower elevated area. (After Biswas & Raha, 2023.)

3.2 BRIDGE CONSTRUCTION, EFFECTS, SCOURING, CHANNEL DYNAMICS IN UPSTREAM AND DOWNSTREAM OF BRIDGES

Kemp (2010) asserts that a number of elements, including topographic settings, tectonics, flow behavior, sediment character, climate, and the kind of vegetation found in floodplains, contribute to the creation of channel morphology and channel floodplains. Downstream increases in channel morphology have been noted in spite of these many causes (Knighton, 1999; Leopold et al., 1964; Leopold & Maddock, 1953; Park, 1977; Pitlick & Cress, 2002; Schumm, 1960). The construction of dams and bridges on rivers is one of the allogenic processes that have recently been the focus of the morphological examination of riverine topography. Bridges over the rivers are intrinsically feeble to flooding, as they are found inside the locale of the riverbed. In terms of outcomes, ordinary threats can damage bridges fundamentally (in this way causing physical hurts), but in addition it can also cause travel time delays and rerouting that may lead to underhanded mishaps. Any bridge dissatisfaction, whether essential or utilitarian, has the potential to impose significant consequences on owners or dependable specialists, as well as frantic costs. In this manner, understanding the potential influence of flooding to bridges may well be a compelling requirement of communities living in ranges of high flood risk (Pregnolato et al., 2020).

The excavation and material removal from both banks and bed of a river is termed as "scouring" (Prendergast & Gavin, 2014; Hamill, 1999). The erosion of the sediment around a bridge foundation (piers and abutments) due to water is known in engineering as scour. The erosive activity of water that flows, which moves and erodes sediments from the banks and bed of rivers as well as from the area around bridge piers and abutments, is what Breusers and Raudkivi (1991) described as scour. Melville and Coleman (2000) describe scour as the process of water erosion reducing a riverbed to the point where a bridge's foundations are more likely to be exposed.

The erosive action of flowing water causes scour, which is the removal of material from the banks and bed of streams as well as from the area surrounding the bridge piers and abutments. The rates at which various materials scour vary. Cohesive or cemented soils are more resistant to scour than loose granular soils, which are quickly eroded by rushing water. On the other hand, final scour in cemented or cohesive soils can reach depths comparable to those of sand-bed streams.

When flow conditions are steady, scour will reach its maximum depth in hours for sand and gravel bed material, days for cohesive bed material, months for glacial till, sandstones, and shale, years for limestone, and millennia for solid granite. To achieve maximal scour, many floods could be required with flow conditions similar to those found in real bridge crossings.

The fact that some scour processes are cyclical makes it difficult to determine the extent of scour. Near a flood's height, scour can be at its deepest, but it becomes barely perceptible as floods subside and silt fills scour holes.

Road bridges carry importance in national transport system (Carnacina et al., 2019). Due to engineering practices without considering riverscape geomorphology and anthropogenic impacts in the river catchment area along with climate change (Carnacina et al., 2019; Gill et al. 2007; Transportation Research Board and National Research Council, 2008; Wright et al., 2012), new challenges emerge in respect to bridge designs. For instance, Muzzammil and Siddiqui (2009) assessed that the probable consequences of climate change on road infrastructure alone on the African continent was $180 billion. Due to increased precipitation events, storm surges, sea-level rise, and hurricane activity, recent studies on climate change have cautioned about the risks associated with an increase in the magnitude of riverine and coastal flooding (e.g., Goldenberg et al., 2001; Transportation Research Board and National Research Council, 2008; Saunders & Lea, 2005; Solomon et al., 2007; Lyddon et al., 2018a, 2018b).

During the 1987 spring floods, scour destroyed or seriously damaged 17 bridges in New England and New York. In 1985, 73 bridges were destroyed by floods in West Virginia, Virginia, and Pennsylvania. Of the 383 bridge collapses brought on by catastrophic floods, 25% involved pier damage and 75% involved abutment damage, according to a 1973 nationwide survey conducted by the Federal Highway Administration (FHWA) (FHWA 1973). According to a second, more thorough study conducted in 1978, local scour at bridge piers is roughly on par with abutment scour issues (FHWA 1978). Transportation Research Record 950 contains several case studies on the causes and effects of scour at large bridges (TRB 1984).

According to the information available, 23 bridges failed during the 1993 flood in the upper Mississippi basin, resulting in an estimated $15 million in damage. Two lateral bank migrations, one debris load, three pier and abutment scour, two pier scour, fourteen abutment scour, and one unidentified reason were the modes of bridge failure.

Three main factors are taken into account when calculating total scour at a highway crossing:

1. Degradation of the riverbed over time or general scour
2. The bridge's contraction scour
3. Localized scour at the abutments or piers

3.2.1 Aggradation and Degradation

General scour is related to the overall erosion and accretion of the riverbed which is consequent upon the changes of hydraulics and sediment interaction of the channel. The quantity of sediment is important in this regard (Prendergast & Gavin, 2014; Forde et al., 1999). The length of the river where the bridge is situated may be impacted by erosion and degradation, which are long-term changes in streambed elevation brought on by either natural or man-made factors. Although it is not regarded as a part of total scour, aggradation is the deposition of material that has been eroded from the channel or watershed upstream of the bridge. A lack of sediment supply from upstream causes degradation, which lowers or scours the streambed over comparatively long lengths and adds to overall scour.

3.2.2 Contraction Scour

A lowering of the streambed across the stream or waterway bed at the bridge is known as contraction scour. Due to the reduction of the cross-sectional area of a channel, contraction scour can happen. Construction of bridges with piers and abutments facilitates this type of scouring. The channel hydraulic in terms of velocity and shear stress plays a major role here (Prendergast & Gavin, 2014). Due to the construction of piers and abutment, the velocity of the river increases. In order to carry the existing volume of water discharges, through the contracted area within a specific unit of time, the velocity of the channel increases considerably. The shear stress also increases at the piers as a result of the increase in velocity which helps in the detachment of sediments and scouring. It is noted that when the shear stress exceeds the threshold value of the bed's shear stress, the mobilization of sediments takes place (Briaud et al., 1999). The depth of scour may be deeper in some areas of the cross-section, or this lowering may be consistent throughout the bed. When the flow is contracted (or constricted), material is removed from the bed throughout all or most of the channel width, causing contraction scour. Contraction scour differs from long-term deterioration in that it happens close to the bridge or constriction, may be cyclical, and/or is connected to floods.

3.2.3 Local Scour

The third type of scour mainly occurs at the individual bridge piers. Removal of material from the vicinity of piers, abutments, spurs, and embankments is known as local scour. It is brought on by flow acceleration and the vortices that are created when flow obstacles are present. Downward flow is noticed at the upstream of bridge piers which result in localized scouring (Prendergast & Gavin, 2014; Hamill, 1999).

The creation of vortices at the base of piers or abutments is the fundamental process producing local scour at these locations. The bed material surrounding the pier's base is removed by the vortex's movement. A scour hole forms when the rate of the sediment movement out of the base region is higher than the rate of sediment transport into the region. The intensity of the horseshoe vortex decreases with increasing scour depth, which lowers the transport rate from the base region. The phrase "clear-water scour" refers to a condition in which the river does not supply any sediments, whereas "live-bed scour" refers to a situation in which there is interaction between the material's scour process and transportation (Brandimarte et al., 2006). Local scour has an immense influence on the bridge failure as it acts around the piers (Zaid et al., 2018). It leads to the situation of continuous maintenance of bridges in order to avoid structural damages, bridge collapse, and loss of life. Many researchers have attempted to analyze the depth and scenario of scouring through various numerical models and equations. Computer simulations have also been used to map the degree of scour and its areal extent and depth. Many equations are proposed by past researchers based on their experimental findings. Melville (1992) has observed that during 1960–1984, in New Zealand, 29 out of 108 bridge failures occurred due to abutments scour. To address the issue of local scour, it is important to study or note the bed materials, bed configuration, hydraulic and other properties of the flow water, and the geometry of piers (Zaid et al., 2018). All these parameters have a specific impact on the depth of local scour.

The following factors influence the amount of local scour depth at piers and abutments: (1) approach flow velocity, (2) flow depth, (3) pier width, (7) angle of attack of the approach flow to a pier or abutment, (8) pier or abutment shape, and (9) bed configuration.

To understand the nature and degree of scouring at the bridge piers on the Rivers Lish, Gish, and Chel, the following equation of CSU has been adopted:

$$\frac{d}{y_1} = 2.0K_1K_2K_3K_4\left(\frac{b}{y_1}\right)^{0.65} \text{Fr}^{0.43}$$

where

- d = maximum depth of scour, measured below bed elevation (m)
- y_1 = flow depth directly upstream of pier (m)
- b = pier width (m)
- Fr = approach Froude number
- $K_1.K_4$ = correction factors
- K_1 = correction factor for pier nose shape

For angle of attach > 50, K_1 = 1.0

- For angle of attach < 50
- Square nose K_1 = 1.1
- Circular K_1 = 1.0
- Group of cylinders K_1 = 1.0
- Sharp nose K_1 = 0.9
- K_2 = correction factor for angle of attach of flow

$$K_2 = \left(\cos\theta + L / b\sin\theta\right)0.65 \tag{3.1}$$

where

- L = length of pier (along flow line of angle of attach) (m)
- b = pier width (m)
- θ = angle of attach (degrees)
- K_3 = correction factor for bed conditions
- K_4 = correction factor for armoring of bed material
- K_4 varies between 0.7 and 1.0
- K_4 = 1.0 for D50 < 60 mm, or for Vr > 1.0
- K_4 = [1 – 0.89(1 – Vr)2]0.5, for D50 > 60 mm

It is noteworthy to mention that the scouring is maximum at the Chel River and it is increasing gradually. On the other hand, River Lish has the least amount of scouring at the piers. Scouring has been calculated for the third year using the CSU formula.

In the year 2023, River Lish shows decreased scouring with values of 0.66 m and 0.78 m at the rail bridge piers and road bridge piers, respectively. For River Gish, the values of scouring have increased slightly and become 1.11 m and 1.236 m for the rail and road bridge, respectively. River Chel has been marked with progressively increasing level of scouring with the values of 2.299 m and 3.6 m for the rail and road bridge, respectively. It is to note that for River Lish, the scouring has decreased by

0.01 m from 2021 to 2022 and by 0.13 m from 2022 to 2023 for the rail bridge. For the road bridge, the scouring has increased by 0.02 m from 2021 to 2022 and decreased by 0.1 m from 2022 to 2023 for the road bridge. It is evident that bed aggradation is profound at the River Lish which is progressively causing the riverbed to elevate itself. For the rail bridge of River Gish, the scouring has decreased by 0.01 m from 2021 to 2022 and decreased by 0.1 m from 2022 to 2023. For the road bridge of Gish, the scouring has increased by 0.05 m from 2021 to 2022 and decreased by 0.014 m from 2022 to 2023. The scouring of River Chel in case of rail bridge has increased by 0.111 m from 2021 to 2022 and by 0.069 m from 2022 to 2023. For the road bridge of Chel, the scouring has increased by 0.018 m from 2021 to 2022 and 0.09 m from 2022 to 2023. It is thus evident that the increased velocity and stream power at the bridge piers of River Chel are causing profound bed degradation and successive scouring.

3.3 RIVERBED CONFIGURATION, PLANFORM DYNAMICS, AND RIVERINE LANDFORM REFRAMING SYSTEM

Understanding water level in river and riverbed configuration is essential in the transmission of flood flow and channel braiding pattern. Discharge and lateral velocity predictions are compared to observe data with special emphasis on practical applications, calibration technique, and effect of scale. Relative magnitude of the equation in terms of depth is considered. The role of friction, lateral shear, and secondary flows state its relation to the observed flow structure and local site characteristics. Riverbed configuration is a mechanism of combined effect and resulted procedure of geo-hydro-sedimentological characteristics.

For thousands of years, people have been fascinated and inspired by landscapes. Individuals' sensory reactions to landscapes differ significantly. Many people have a sense of belonging that may be tinged with nostalgia when they think of spiritual links. Others may experience amazement coupled with feelings of fear or estrangement.

Artists use poetry, prose, paintings, and other media to try and convey the spirit of landscapes. The environments in which we live and play frequently shape our life experiences. Associations and relationships change throughout time and between locations. When observations are contrasted with experiences from other places, new experiences can lead to new insights. Landscapes are dynamic, interconnected systems. Responses or management actions to disturbances at one location and time may have off-site effects over a range of time periods. This is not always the case, even if these interactions are usually scale-dependent (small impacts have negligible effects that are limited to closer [proximal] areas; e.g., local disruption may cause off-site responses that exceed threshold requirements). These correlations are frequently predictable. River systems' upstream-downstream connections are primarily driven by sediment flux and gravity-induced flow. However, there may occasionally be unexpected results. For instance, the river upstream may be impacted by headcut activity and bed incision that erode valley floor deposits. From watershed to catchment, connections' efficacy and efficiency differ significantly.

The river planform pattern is highly variable, shaped by a combination of independent and dependent factors. Independent factors, such as geology, climate, and human activities, play a significant role in determining the overall structure and behavior of rivers. Dependent factors, on the other hand, adapt to these independent influences. As noted by Surian (2002), Rinaldi (2003), and Surian and Cisotto (2007), the rate of sediment deposition in various river segments, particularly in foredeep regions, is closely linked to factors such as the channel gradient, hydraulic jumps, and changes in altitude.

To fully comprehend the dynamics of sediment transport and deposition, it is essential to analyze the impact of human activities on river morphology and flow behavior. Various forms of engineering constructions on rivers—such as dams, bridges, and sand mining—are examples of anthropogenic interventions that alter the river's planform and dynamics (Biswas & Banerjee, 2018).

Over the past few decades, human-induced changes such as sediment flux, bridge scouring, and structural engineering interventions have been major contributors to river degradation and alterations in channel dynamics (Collins and Dunne, 1989; Rinaldi and Simon, 1998). The construction of bridges, in particular, has had significant consequences on flow patterns, sediment transport, and the overall morphology of riverbeds, causing both aggradation and degradation in certain sections of the river (Brandt, 2000; Phillips, 2001; Petts and Gurnell, 2005; Graf, 1984; Sanyal, 2017). These changes, which modify the natural processes and dynamics of rivers, highlight the importance of understanding the interaction between natural and anthropogenic factors in shaping river systems.

As human activities continue to reshape the landscape, it becomes increasingly important to study the long-term impacts of these changes on river behavior. The integration of hydraulic engineering, environmental science, and river management will be essential in mitigating the adverse effects on river ecosystems and ensuring sustainable management of these vital water systems. These alterations are evident in both the upstream and downstream sections of rivers that are influenced by the presence of bridges. In particular, rail and road bridges across the Lish, Gish, and Chel Rivers play a significant role in shaping their spatial characteristics, especially during the monsoon season. The impact of these anthropogenic factors is evident in how they influence the hydraulic dynamics of the river channels both upstream and downstream (Moussa, 2003; Starkel et al., 2008). Research has shown that the extent of flood hazard zones has expanded downstream as a result of the construction of rail and road bridges.

The Himalayan foothill region, specifically the Darjeeling-Jalpaiguri area, is drained by the Teesta River and its tributaries. This study specifically focuses on the three left-hand tributaries of the Teesta River, located in the Himalayan foothills (Figure 3.1). These rivers exhibit distinct hydro-sediment discharge characteristics, particularly in a monsoon-dominated environment. However, these rivers have experienced varying bed elevation changes (Figure 3.1C) due to the construction of rail and road bridges across them. The channel hydraulics, influenced by these engineering interventions, have significantly impacted the sediment aggradation and degradation processes at the riverbeds.

Previous studies have employed numerical models (Wang et al., 2019) to simulate the velocity profiles of rivers, revealing a decline in velocity from the foothill regions to the confluence or mouth of the rivers. However, this natural flow pattern has been altered by the construction of bridges, leading to shifts in the sediment transport dynamics and channel behavior.

The current study aims to investigate the characteristics of monsoon hydraulics, with a focus on spatial variations before and after the construction of bridges. Additionally, it will examine the behavior of suspended sediment loads in these rivers during the monsoon period, providing insights into how these engineering structures impact sediment dynamics and overall river morphology in a seasonal context. This research contributes to a deeper understanding of how human interventions, particularly bridge construction, can affect natural river processes, especially in regions prone to heavy monsoon rains.

3.4 RIVER HYDROLOGICAL BEHAVIOR IN MONSOON AND POST-MONSOON

Hydrological and sediment data for the three rivers under investigation were collected through field surveys carried out during the monsoon season. The study focused on critical hydraulic parameters, including velocity (m/s), shear stress (τ), and stream power (w). Velocity was a key factor, as it plays a direct role in the processes of sediment aggradation and degradation, thereby influencing the river's flow characteristics. In open channel hydraulics, shear stress quantifies the frictional force exerted by the streamflow on objects within the flow (Knighton, 1998).

Suspended sediment samples were taken using three suspended sediment samplers (Punjab Type) at four designated locations along each river: the foothill, beneath the rail bridge, beneath the road bridge, and 300 m downstream. Sampling occurred every 15 minutes as the team moved between sites. During the monsoon months (June–August), sediment samples were collected three times daily at 06:00, 12:00, and 18:00 hours, with additional collections at 6:15, 12:15, and 18:15 hours, following a random sampling method based on precipitation events. The samples were then evaporated in a hot air oven to accurately determine the suspended sediment concentration (g/m^3).

Velocity was measured using a digital current meter (water velocity logger, model: DVI-K13) where

$$V = m^{-\text{sec}} \text{ computed as} \left(0.3606 * \left(\frac{\text{Rotation}}{60} \right) \right) + 0.003 \tag{3.2}$$

Shear stress is calculated as

$$\tau = \rho g h s \tag{3.3}$$

where ρ is the density of water, g is the acceleration of gravity, s is the slope gradient, and h is the depth of flow.

Stream Power (Hairsine & Rose, 1992)

$$w = \rho g s q \quad (3.4)$$

where ρ is the density of water, g is the acceleration of gravity, s is the slope gradient, and q is the unit discharge.

The selected sampling sites spanned from the foothill region to the confluence of each river. Hydrological measurements were recorded at various points: 300 m, 200 m, 100 m, and 50 m upstream of the rail bridge; at the base of the rail bridge; between the rail and road bridges; below the road bridge; 50 m, 100 m, 200 m, and 300 m downstream of the road bridge; 4 km downstream from the road bridge; and at the confluence. Data were recorded twice a day at 07:00 and 12:00 hours.

This combination of field data and spatial analysis offers a comprehensive view of the hydraulic and sediment dynamics in the area, particularly focusing on the impact of the rail and road bridges on river behavior during the monsoon season. The structured methodology provides an in-depth understanding of how engineering structures affect sediment transport and river morphology over time.

In the context of open channel hydraulics, the variation in flow behavior and sediment transportation capacity is reflected in the spatial distribution of water velocity as it moves from the foothill region to downstream areas. During the monsoon season, the Chel River stands out with a significantly higher average velocity compared to both the Gish and Lish Rivers. In fact, the maximum velocity observed in the foothills of the Chel River reaches 1.67 m/s. After passing through the foothill region, the velocity decreases, but it increases again near the base of both the rail and road bridges. This velocity increase generates turbulent flow conditions, causing swirls and eddies near the bridge piers, which can influence sediment transport and deposition patterns.

When compared to the Chel River, the Gish River exhibits lower velocities along the examined stretches. The highest velocity recorded in the Gish River occurs in the foothill region at 1.529 m/s. However, the velocity decreases sharply as the river flows near the rail (1.325 m/s) and road (1.331 m/s) bridges, reflecting the influence of these structures on the flow dynamics. Interestingly, the velocity picks up again downstream of the road bridge, reaching a maximum of 1.475 m/s. This change suggests that the bridges may create localized changes in flow patterns, influencing both sediment transport and the physical structure of the riverbed.

In contrast to the Chel and Gish Rivers, the velocity at the observed sites along the Lish River is notably lower. The river's flow decreases gradually from the foothill area, reaching as low as 1.247 m/s below the rail bridge and 1.265 m/s below the road bridge. This suggests that the Lish River has a less dynamic flow regime compared to the Chel and Gish Rivers, with slower velocities potentially influencing sediment deposition and the overall riverbed morphology (Figure 3.2).

The differences in velocity across the three rivers Chel, Gish, and Lish demonstrate how environmental features, such as the topography, and human infrastructure,

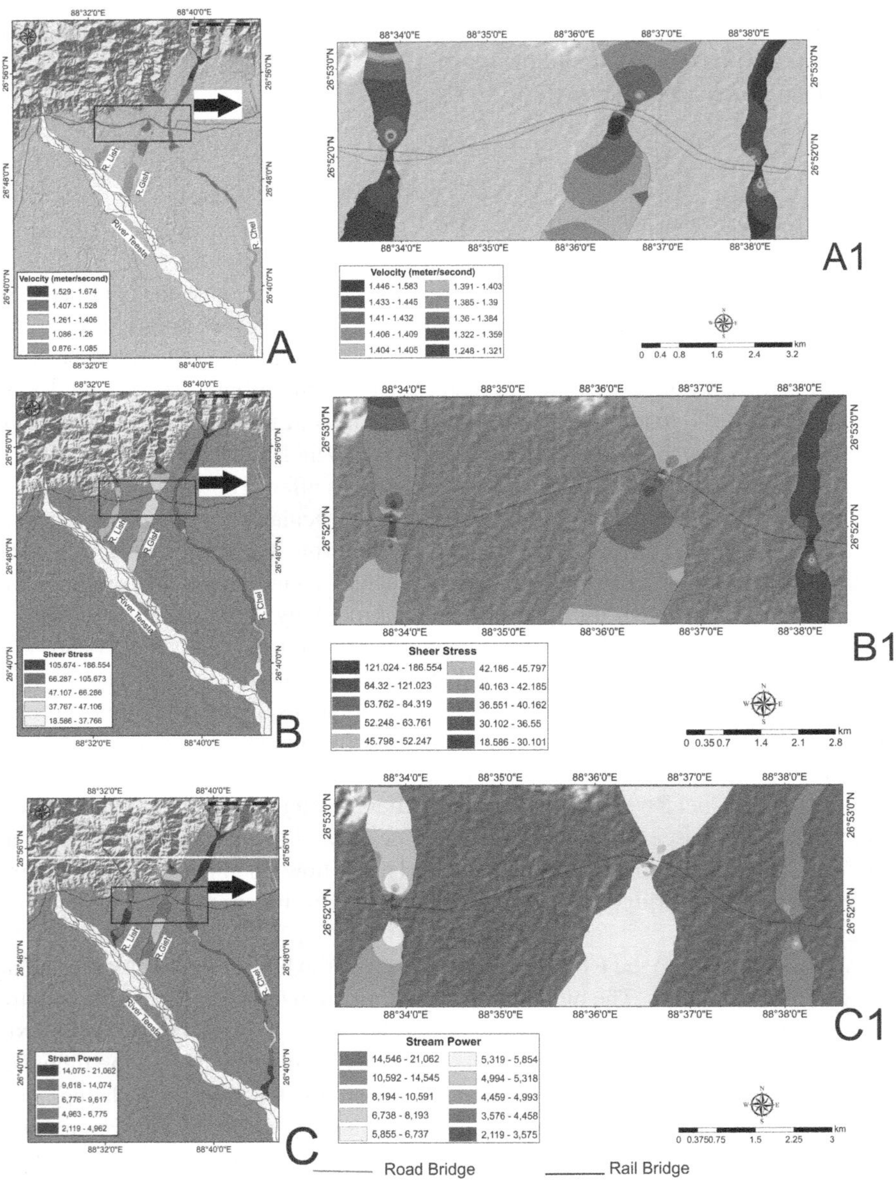

FIGURE 3.2 (A) Spatial distribution of monsoon water velocity (v/s) of three respective rivers. A1. Changes in velocity were observed upstream of the road bridge, between the rail bridge and the road bridge and downstream of the road bridge in Rivers Lish, Gish, and Chel. The velocity ranges from 1.446 to 1.248. (B) Distribution of shear stress along the channels ranges from 37.766 to 186.55. B1. The inset details the shear stress variation in the studied reaches of the area; the variation in values observed at different scales for Rivers Lish, Gish, and Chel ranges from 37.766 to 18.586. (C) Spatial variation of stream power during the monsoon that covers the entire channel. C1. Details of stream power distribution in the bridge area in three considered rivers (Biswas & Raha, 2023).

like bridges, can significantly influence river flow behavior. These variations in velocity can, in turn, affect sediment transport dynamics, which play a crucial role in shaping the riverbed morphology and determining flood hazard potential. The study highlights the complexity of river hydraulics, where even small changes in flow conditions can have broader implications for sediment movement, deposition, and erosion, particularly during the monsoon season when flow rates and sediment transport are at their highest. From the foothill zone (109.76) to 100 meters upstream of the rail bridge, the shear stress in the Chel River decreases significantly, reaching a value of 73.5. However, it increases to 156.8 below the rail bridge and continues to rise, peaking at 190.512 below the road bridge. Further downstream, the shear stress drops to 37.485, suggesting a sediment aggradation process in this area. Notably, the Chel River exhibits the highest shear stress values compared to the other two rivers, indicating its strong erosional forces in certain sections.

The Gish River, on the other hand, shows a more varied distribution of shear stress along its course. Below both the rail and road bridges, the shear stress drops sharply to 28.224 and 26.303, respectively. However, downstream of these bridges, the shear stress increases sharply to 48.70, suggesting significant changes in flow conditions that influence sediment movement and deposition patterns. In contrast, the Lish River consistently displays the lowest shear stress values at both the rail and road bridge locations, compared to the Chel and Gish Rivers. The reduced shear stress in this river indicates less erosional force, which could result in less sediment transport and slower changes in river morphology.

The elevated shear stress values observed in the foothill region and 300 meters upstream of the rail bridge signify significant sediment degradation, with particles being broken down and transported downstream. As the flow moves further downstream and the shear stress decreases, these sediments are deposited, highlighting the interplay between shear stress and sediment transport/deposition. This variation in shear stress along the rivers suggests that while high shear stress leads to sediment erosion and transport, lower values result in the deposition of eroded material, influencing the overall sediment dynamics and riverbed structure. The differences in shear stress patterns across the three rivers Chel, Gish, and Lish underscore the complex interactions between flow dynamics, sediment transport, and deposition that shape river morphology over time. Understanding these variations is crucial for assessing flood risks, sediment management, and the potential impact of infrastructure on river systems. The Chel River exhibits significantly higher stream power compared to the other two rivers, due to variations in the riverbed slope and velocity profile, which result in fluctuating values at different stretches of the river. Notably, below the rail bridge (20,489.06) and the road bridge (21,308.21), the stream power increases substantially. This increase in stream power contributes to intense scouring at the bridge piers, which can lead to significant riverbed erosion. Conversely, when the stream power is lower, the degraded sediments from these more erosive reaches are carried downstream and deposited in areas with reduced energy.

In the Gish River, the stream power reaches its highest value in the foothill region (14,751.45), after which it steadily decreases as the river approaches the rail bridge (4,320.73) and continues to drop further under the road bridge (3,479.09). The reduction in stream power near the bridge piers hinders the smooth flow of river discharge and the transportation of sediments. As a result, these conditions favor sediment aggradation, where eroded material is deposited rather than transported. The overall pattern of stream power in the Gish River promotes less erosive activity, encouraging sediment deposition and leading to an increase in the channel width over time.

The Lish River, in comparison, demonstrates even lower stream power values than both the Chel and Gish Rivers. The stream power reaches its lowest points below the rail bridge (2,454.23) and road bridge (2,094.92). However, further downstream, the stream power starts to increase again, reaching a value of 7,264.20. This pattern suggests that while the Lish River has relatively low erosive energy in certain stretches, the increasing stream power downstream could promote higher sediment transport and potential channel reshaping (Figure 3.2).

These variations in stream power across the three rivers highlight the complex interactions between river flow dynamics, sediment transport, and riverbed morphology. The Chel River, with its high stream power, is more erosive and prone to significant scouring, especially around infrastructure such as bridges. In contrast, the Gish and Lish Rivers exhibit lower stream power values, which encourage sediment deposition and changes in channel morphology, such as widening. These differences are critical for understanding the effects of varying stream power on sediment dynamics, channel formation, and the long-term evolution of river systems. Furthermore, the presence of bridges and other anthropogenic structures can significantly influence stream power and, by extension, the sediment transport processes, which in turn impacts flood hazards, river management, and infrastructure stability.

Suspended sediment refers to the particles transported by streamflow through the turbulence created by eddies in water (Crawford, 1991; Parsons et al., 2015). These particles are typically small in size, usually less than 25 mm, particularly clay and silt, which allows them to remain suspended in the water column. The concentration of the suspended sediment varies across different rivers and regions. For instance, the Chel River has a relatively low average suspended sediment concentration of 0.95 gm/l. In the foothill region, the suspended sediment concentrations for Chel, Gish, and Lish are 1.28 gm/l, 1.3 gm/l, and 1.49 gm/l, respectively. The suspended sediment levels in these rivers exhibit notable fluctuations. For the Gish River, a significant spike in sediment concentration is observed near the rail bridge (located at 26.875°N and 88.609°E), which can be attributed to the influx of hydro-sediment discharge from the right-side tributaries during the monsoon season. These tributaries contribute to additional sediment load, increasing the overall concentration of the suspended material in the main river.

For both the Lish and Gish Rivers, sediment concentrations decrease significantly near the road bridge, dropping to 1.475 gm/l and 0.764 gm/l, respectively. This reduction is due to substantial deposition of suspended sediments in these regions, where the flow velocity is likely lower, leading to sediment settling. In contrast, the Chel River exhibits lower suspended sediment concentrations at both the rail and road bridges, with values of 1.047 gm/l and 0.902 gm/l, respectively. This suggests that the Chel River may have more stable flow conditions that result in less sediment transport and deposition at these bridge locations.

These variations in suspended sediment concentrations across the three rivers can be linked to several factors, including tributary inflows, changes in flow velocity, and the influence of engineering structures like bridges. In areas where the flow is slower or where there is a convergence of tributaries, sediment tends to accumulate, leading to higher concentrations. Conversely, in regions where the flow velocity increases, suspended particles are more likely to remain in suspension and be transported downstream. Understanding these dynamics is crucial for managing sediment transport, river morphology, and the ecological health of the river systems particularly during the monsoon season when sediment loads are at their peak.

The hydraulics of a river are directly related to its scouring. The degree and extent of scouring at the bridge piers are significantly influenced by the stream power, velocity, and shear force. In addition, the river's bed structure, slope, flow, and sediment load influence its hydraulics. Because the river's energy is not depleted carrying the sediment load, it is noteworthy that rivers with a steep bed slope and little sediment load typically have faster velocities. Additionally, it serves as a mechanism for the bed's deterioration and deposition.

The bed tends to be eroded by higher velocity and stream force, and vice versa. It should be noted that the bed's current slope is altered by deterioration, which contributes to the bed's further degradation. However, the reduced stream power and velocity aid in the aggradation process, changing the bed's slope and therefore aiding in the bed sedimentation process.

It is crucial to remember that not all rivers have the same level of scouring at the piers. Pier scouring varies in intensity and extent along the Lish, Gish, and Chel Rivers. The highest scouring rate is seen in the Chel River, and it appears to be rising with time. According to the data, scouring has eroded the bed to a significant depth around the piers of the road and rail bridge.

According to the river's hydraulics, the scouring close to the bridge piers is caused by the Chel's increased velocity during the monsoon and post-monsoon seasons. The river bottom is more severely degraded because of the swirling pattern created by the rising velocity close to the bridge piers. This scouring process is also made easier by the increased stream power. The Lish River's hydraulics are not conducive to scoring; instead, sedimentation and aggradation cause the bed elevation to rise close to the piers. Both during the monsoon and post-monsoon seasons, the river's velocity is seen to be lower. Additionally, compared to the other two rivers, the stream strength is noticeably lower. The Lish River has a considerable volume of suspended silt. One may argue that this increased volume of suspended silt is carried by the stream power, which helps to aggrade the material at the bed, particularly close to the bridge piers. According to the hydraulic data, the velocity, stream power, and shear stress all appear to decrease close to the Lish River's rail and road bridge.

The scouring condition and river hydraulics fall halfway between the Lish and Chel ranges. Near the train bridge, the velocity and stream power drop, then rise slightly before falling once more. Nevertheless, the levels of velocity, stream power, and shear stress are significant enough to promote scouring at the bridge piers, despite the decreasing trend of these parameters close to the spans. As a result, scouring is seen at the piers of both road and rail bridges.

Overall stability of the bridges is affected by the scouring (Gaudio et al., 2012; Melville & Coleman, 2000; Muzzammil & Siddiqui, 2009; Zarrati et al., 2004). In order to combat scouring, structural and hydraulic countermeasures can be viable options (Prendergast & Gavin, 2014; NCHPR, 2009). By preventing abrupt changes in flow direction brought on by blunt pier faces blocking the flow, hydraulic countermeasures help to avoid fast flow expansion or contraction. These abrupt shifts in flow have the potential to produce the vortices that cause scour. They can be avoided by simplifying pier designs and preserving wider bridge openings throughout the design phase. Debris like fallen trees and other items that frequently become stuck in bridge openings and impede the flow must be removed in order to maintain clear openings.

There are various strategies that can be used to lessen the bridge local scour effect. One strategy is to ensure that bridge abutments are adequately protected and that countermeasures are applied correctly, which will reduce the bridge local scour effect. Specifically, channel size, soil condition control, channel morphology curbing, and vegetation preservation are used to reduce the bridge's local scour effect. It appears that protecting bridge piers against scouring is necessary to guarantee that the local scour effect on bridges is minimized. In order to get the intended outcomes, different researchers utilize a variety of countermeasures and techniques (Ettema et al., 2006). Furthermore, the suggested techniques are distinctly categorized under several headings, such as armoring and flow-altering countermeasures. Evidently, a collar is a form of modifying countermeasures which ensures that the water's downflow is deflected, therefore entirely controlling any scouring around the pier (Figure 3.3).

The entire study discloses some distinct imprints of engineering construction on channel hydraulics which is related to the sediment aggradation and bed scouring processes. It is essential to remember that each of the three rivers under consideration is affected differently by the construction of bridges. The river width (m) near bridges was drastically reduced after the bridge construction on the Rivers Lish and Gish, which led to a bottleneck situation in both rivers. Especially the River Gish flows in a very narrow passage in recent times resulting in huge sedimentation in the upstream of the rail bridge. According to the analysis, the channel hydrological factors induce the process of sediment aggradation near the rail, road bridges, and upstream section of both rivers, which is considerably responsible to modify the channel planform as well as increasing the riverbed elevation due to the aggradation of fine sediments

Year	River Lish		River Ghish		River Chel	
	Rail Bridge	Road Bridge	Rail Bridge	Road Bridge	Rail Bridge	Road Bridge
2021	0.8	0.86	1.1	1.2	2.119	3.492
2022	0.79	0.88	1.09	1.25	2.23	3.51
2023	0.66	0.78	1.11	1.236	2.299	3.6

FIGURE 3.3 Temporal-scale scouring near rail and road bridges from 2021 to 2023 with photographic evidences.

near rail bridge piers of River Lish. Huge amounts of sediment discharge that become unable to pass through the narrow passage between the bridges have deposited upstream and are substantially widening the channel in the upstream reach higher than downstream. The condition is mostly prevalent in River Gish. The channel hydraulic parameters have displayed varied condition in River Chel, as the parameters have created a condition of degradation process especially scouring at bridge piers and subsequent lowering of the bed. Thus, the entire study highlights the major problems related to bridge construction on the rivers, which alters the spatial distribution of hydrological parameters at different reaches of the rivers especially before, between, and after the bridges.

The analysis encompassed both the expansion and contraction of channel width between the years 1913 and 2021. It also involved determining the impacted area, considering scenarios with or without bridge structures or bridge relocations. Additionally, field verification was conducted, which included a comparative assessment with adjacent rivers, ensuring the accuracy of the findings. When considering the seven section lines, it becomes evident that the River Gish predominantly experiences a bottleneck condition among them. Notable patterns in the shifting of river channels emerge over time: between 1913 and 1936, a positive shift occurred in both the Rivers Chel and Lish. However, from 2008 to 2021, minimal channel shifting was observed in the River Chel, and a similar trend was noted in the period from 1984 to 2008 for the River Lish. Notably, the Gish River exhibits its narrowest width in the vicinity of the bridge section (G4). The most significant deposition took place in the G2 section (upstream) between 1965 and 1984, particularly following the construction of the road bridge. These patterns are largely mirrored in the case of the River Lish as well. Comparing other surrounding river cross-section areas of bridges from the DEM data and survey, it is notified that this section is more dynamic in the river planform evolution mentioned in Chapter 2.

The prepared cross-sections by the Total Station survey disclose the aggradation and scouring processes in river platform dynamics in 2022 and 2023 with the photographic sequences. The cross-section prepared by Total Station explain the river topographic changes where sediment aggradation is mostly common below the rail bridge pier of River Lish. Near the rail bridge of River Chel, huge scouring occurred during the monsoon and it is increasing (Figure 3.4).

During the British period, the rail bridge was constructed to facilitate the transportation of tea leaves from the factories in North Bengal foreland plain. In the later period during the 1960s, the road bridge has been constructed in a parallel alignment with the rail bridge to increase the accessibility. With due course of time, the river morphology along with its bed topography, sediment aggradation process, bank and bed erosion, and channel form have been altered and modified significantly in the upstream and downstream sections of the considered rivers. The changes of channel width on a temporal scale (from 1913 to 2021) are consequent upon the construction of engineering structures, i.e., bridges on the rivers. The widths of the bridges were considerably minimized for the structural efficiency which is affecting the river hydraulics to a greater extent as a huge volume of river discharge has to pass through this narrow passage during the monsoon. Below the rail and road bridge, the velocity of the River Chel abruptly increases (1.51 and 1.58 m/s, respectively), along with an

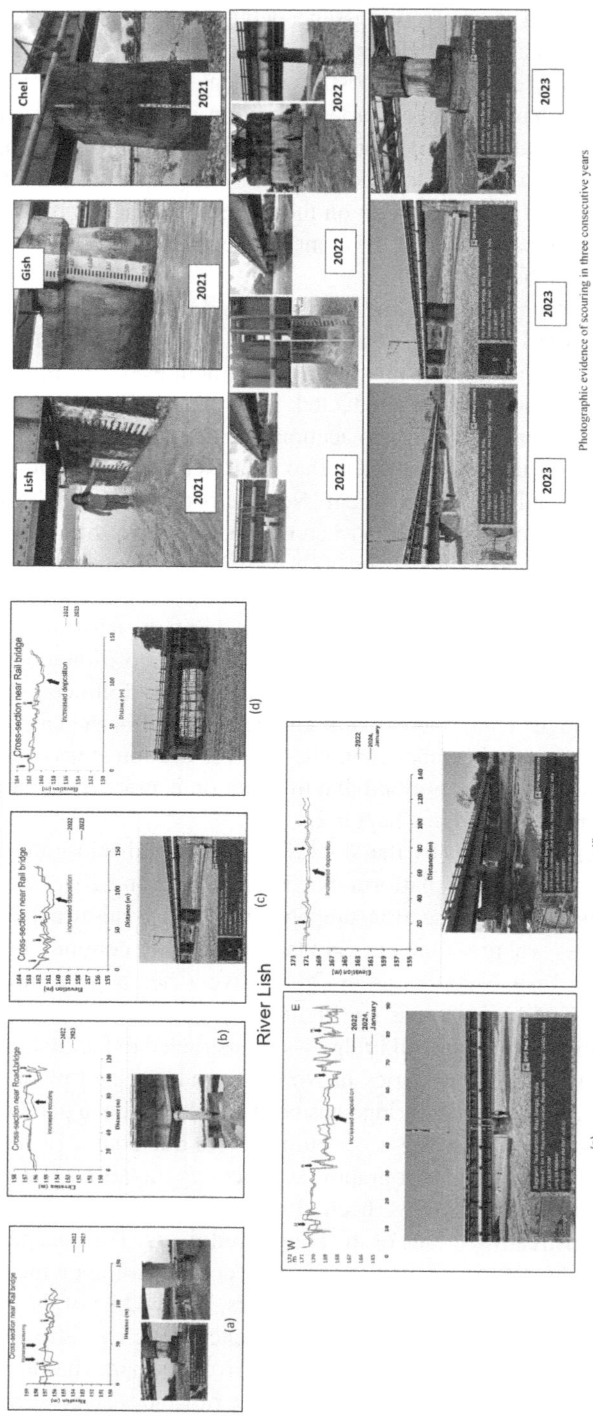

FIGURE 3.4 Cross-sections by the Total Station Survey of 2022–2023 in Rivers Lish, Gish, and Chel with photographic evidences and changes from 2021 to 2023.

increase in shear stress (156.8 and 190.51). The shear stress affects the bed load and sediment transport movement. It should be noted that sediments start to migrate from their position when the drag force outweighs the gravitational force that keeps the sediment particles in place. It is interesting to note that increased shear stress acts as an agent to transport the sediments from bed and bank and it moves from its place leading to the process of degradation. Also, lower shear stress will lead to the process of aggradation of sediments.

The bottom of the two bridges has also been marked with the maximum stream power (20489.06 and 21308.20), which inhibits the process of sediment aggradation. Instead, it has been observed that high velocity and stream power accelerate the process of sediment degradation, which results in scouring, close to the piers of the bridges (2.119 m at rail bridge and 3.492 at road bridge). Thus, the depth of the riverbed greatly increases close to the two bridges in River Chel (more than 1 meter). Due to the substantial volume of water that has to flow through the narrow channel width (70.36 m) between the two bridges, the velocity of River Chel is considerably higher. Due to decreased stream power (6849.055), the velocity decreases (1.37 m/s) further downstream of Chel from the road bridge where the depth of the water decreases and the channel width increases. This has contributed to considerable sediment aggradation and the formation of mid-channel bars in downstream sections. It is also noted that the amount of suspended sediment in Chel gradually decreases from the foothills to downstream, which implies that the aggradation takes place in the upstream reach from the rail bridge. But the rate of deposition in River Chel is not the same as Gish and Lish. It may be assumed that the distance from the foothills to the bridge is maximum (10.5 km) in River Chel and the carried sediments from the catchment area get sufficient space along the upstream reach of the river to deposit. Below the road and rail bridge, the amount is much lesser, which further helps to decrease the carrying load of the streamflow and thus helps to increase the potentiality of the streamflow to move with a higher velocity leading to the scouring and sediment degradation process. The temporal shifting in channel of River Gish suggests a huge aggradation process upstream of the bridges where the river width has increased from 249.47 m (1913) to 2309 m (2021). The sediments coming from the mountainous region and other upstream catchment tributaries have deposited in the upstream reaches of Gish River from the rail bridge, resulting into lateral expansion of the riverbed and subsequently an increase in the channel width. The aggradation process has resulted into the increase in bed elevation especially near the bridges (1.0 m from 1965 to 2021). The discharge and velocity of River Gish is less (mean velocity is 1.37 m/s) than River Chel that signify the suspended aggradation process in the riverbed and thus encouraging sediment deposition and lesser degree of scouring. At the bottom of the rail and road bridge of Rivers Lish and Gish, the velocity decreases significantly that encourages the sediments to be deposited at the bottom of the piers. The bed elevation has increased in Rivers Lish and Gish, which is accompanied by low depth of streamflow. Approximately 1.2 m and 1.0 m increase in bed elevation has been marked in Rivers Lish and Gish, respectively, from 1965 to 2021. The stream power (4320.73 and 3479.09) decreases considerably beneath the bridges along with lower shear stress (28.22 and 26.30) which further accelerates the sediment aggradation process. The median grain size of the bed sediment of Rivers Lish and Gish is

0.55 mm and 1.00 mm, respectively. The velocity is least in River Lish compared to the other two rivers. The velocity decreases from the foothill zone and becomes minimum below the rail and road bridge. The streamflow becomes sluggish near the bridges, which consequently leads to the sedimentation process under the rail and the road bridge. It has been noted that the depth of water becomes least here signifying the gradual increase in the bed elevation due to the aggradation process. The shear stress and stream power are also marked to be minimum near the bridge, which discourages the degradation process and further accelerates the deposition of suspended and bed load sediments. The problem of scouring is practically absent in the case of River Lish. Rather the gradual rise of bed elevation due to the sediment aggradation process has led to the decrease in gap between the bridges from the riverbed. It has been noted that the amount of suspended sediment increases from the foothill to the rail bridge, which further decreases near the road bridge, implying the deposition of it in between.

REFERENCES

Biswas, M., and A. Raha. 2023. An overview of open channel monsoon hydraulics of Himalayan foothill rivers, West Bengal, India. *Modeling Earth Systems and Environment*, 9(2): 2969–2979.

Biswas, S.K. 2010, August. Effect of bridge pier on waterways constriction: a case study using 2-D mathematical modelling. In *IABSE-JSCE Joint Conference on Advances in Bridge Engineering-II*, 369–376.

Bose, I., and U.K. Navera. 2017. Flood maps and bank shifting of Dharla River in Bangladesh. *Journal of Geoscience and Environment Protection*, 5(9): 109–122.

Brandimarte, L., A. Montanari, J.L. Briaud, and P. D'Odorico. 2006. Stochastic flow analysis for predicting river scour of cohesive soils. *Journal of Hydraulic Engineering*, 132(5): 493–500.

Brandt, S.A. 2000. Classification of geomorphological effects downstream of dams. *Catena*, 40(4): 375–401.

Breusers, H.N.C., and A.J. Raudkivi. 1991. *Scouring Hydraulic Structures Design Manual.* A.A. Balkema.

Briaud, J.L., F. Ting, H.C. Chen, R. Gudavalli, S. Perugu, and G. Wei. 1999. SRICOS: prediction of scour rate in cohesive soils at bridge piers. *Journal of Geotechnical and Geoenviromental Engineering*, 125(4): 237–246.

Carnacina, I., S. Pagliara, and N. Leonardi. 2019. Bridge pier scour under pressure flow conditions. *River Research and Applications*, 35(7): 844–854. https://doi.org/10.1002/rra.3451

Collins, B.D., and T. Dunne. 1989. Gravel transport, gravel harvesting, and channel-bed degradation in rivers draining the southern Olympic Mountains, Washington, USA. *Environmental Geology and Water Sciences*, 13(3): 213–224.

Ettema, R., G. Kirkil, and M. Muste. 2006. Similitude of large-scale turbulence in experiments on local scour at cylinders. *Journal of Hydraulic Engineering*, 132(1): 33–40.

Forde, M.C., D.M. McCann, M.R. Clark, K.J. Broughton, P.J. Fenning, and A. Brown. 1999. Radar measurement of bridge scour. *NDT and E International*, 32(8): 481–492.

Gaudio, R., A. Tafarojnoruz, and F. Calomino. 2012. Combined flow-altering countermeasures against bridge pier scour. *Journal of Hydraulic Research*, 50(1): 35–43. https://doi.org/10.1080/00221686.2011.649548

Gill, S.E., J.F. Handley, A.R. Ennos, and S. Pauleit. 2007. Adapting cities for climate change: the role of the green infrastructure. *Built Environment* 33(1): 115–133. https://doi.org/10.2148/benv.33.1.115

Goldenberg, S.B., C.W. Landsea, A.M. Mestas-Nuñez, and W.M. Gray. 2001. The recent increase in Atlantic hurricane activity: causes and implications. *Science*, 293(5529): 474–479. https://doi.org/10.1126/science.1060040

Graf, W.L. 1984. A probabilistic approach to the spatial assessment of river channel instability. *Water Resources Research*, 20(7): 953–962.

Hamill, L. 1999. *Bridge Hydraulics*. E&FN Spon.

Kemp, J. 2010. Downstream channel changes on a contracting, anabranching river: the Lachlan, southeastern Australia. *Geomorphology*, 121: 231–244. https://doi.org/10.1016/J.GEOMORPH.2010.04.018

Knighton, D. 1999. Downstream variation in stream power. *Geomorphology*, 29: 293–306.

Leopold, L., and J.T. Maddock. 1953. *The Hydraulic Geometry of Stream Channels and Some Physiographic Implications*. US Government Printing Office. https://doi.org/10.3133/PP252

Leopold, L., M.G. Wolman, and J.P. Miller. 1964. *Fluvial Processes in Geomorphology*. Dover Publications.

Lin, Q. 2011. Influence of dams on river ecosystem and its countermeasures. *Journal of Water Resource and Protection*, 03(01): 60–66.

Lyddon, C., J.M. Brown, N. Leonardi, and A.J. Plater. 2018a. Flood hazard assessment for a hyper-tidal estuary as a function of tide-surge-morphology interaction. *Estuaries and Coasts*, 41(6): 1565–1586. https://doi.org/10.1007/s12237-018-0384-9

Lyddon, C., J.M. Brown, N. Leonardi, and A.J. Plater. 2018b. Uncertainty in estuarine extreme water level predictions due to surge–tide interaction. *PLoS One*, 13(10): e0206200. https://doi.org/10.1371/journal.pone.0206200

Melville, B.W. 1992. Local scour at bridge abutments. *Journal of Hydraulic Engineering*, 118: 615–631.

Melville, B.W., and S.E. Coleman. 2000. *Bridge Scour*. Water Resource Publications.

Mitra, S. 2015. Shifting courses of Ganga River, it causes and resultant hazards of Manikchak block, Malda district, West Bengal. *International Journal of Humanities and Social Science Studies*, 2: 348–350.

Moussa, R. 2003. On morphometric properties of basins, scale efects and hydrological response. *Hydrol Process* 17(1): 33–58. https://doi.org/10.1002/hyp.1114

Muzzammil, M., and N.A. Siddiqui. 2009. A reliability-based assessment of bridge pier scour in non-uniform sediments. *Journal of Hydraulic Research*, 47(3): 372–380. https://doi.org/10.1080/00221686.2009.9522008

Park, C.C. 1977. World-wide variations in hydraulic geometry exponents of stream channels: an analysis and some observations. *Journal of Hydrology*, 33: 133–146.

Petts, G.E., and A.M. Gurnell. 2005. Dams and geomorphology: Research progress and future directions. *Geomorphology*, 71(1–2): 27–47.

Pitlick, J., and R. Cress. 2002. Downstream changes in the channel geometry of a large gravel bed river. *Water Resources Research*, 38: 1216. https://doi.org/10.1029/2001WR000898

Phillips, J. 2001. Sedimentation in bottomland hardwoods downstream of an east Texas dam. *Environmental Geology*, 40(7): 860–868.

Pregnolato, M., A.O. Winter, D. Mascarenas, A.D. Sen, P. Bates, and M.R. Motley. 2020. Assessing flooding impact to riverine bridges: an integrated analysis. *Natural Hazards and Earth System Sciences Discussions*, 2020: 1–18.

Prendergast, L., and K. Gavin. 2014. A review of bridge scour monitoring techniques. *Journal of Rock Mechanics and Geotechnical Engineering*, 6(2): 138–149. https://doi.org/10.1016/j.jrmge.2014.01.007

Rinaldi, M. 2003. Recent channel adjustments in alluvial rivers of Tuscany, Central Italy. *Earth Surface Processes and Landforms: The Journal of the British Geomorphological Research Group*, 28(6): 587–608.

Sanyal, J. 2017. Predicting possible effects of dams on downstream river bed changes of a Himalayan river with morphodynamic modelling. *Quaternary International*, 453: 48–62.

Saleem, A., A. Dewan, M.M. Rahman, S.M. Nawfee, R. Karim, and X.X. Lu. 2020. Spatial and temporal variations of erosion and accretion: a case of a large tropical river. *Earth Systems and Environment*, 4: 167–181.

Saunders, M.A., and A.S. Lea. 2005. Seasonal prediction of hurricane activity reaching the coast of the United States. *Nature*, 434: 1005–1008. https://doi.org/10.1038/nature03454

Schumm, S.A. 1960. *The Shape of Alluvial Channels in Relation to Sediment Type*. US Government Printing Office, 17–30.

Solomon, S., D. Qin, M. Manning, Z. Chen, M. Marquis, K.B. Averyt, … H.L. Miller. 2007. *Climate Change 2007: The Physical Science Basis*. Cambridge University Press.

Starkel, L., S. Sarkar, R. Soja, and P. Prokop. 2008. *Present-day evolution of the Sikkimese-Bhutanese Himalayan piedmont* (Vol. 219). IGiPZ PAN.

Surian, N. 2002. Downstream variation in grain size along an Alpine river: analysis of controls and processes. *Geomorphology*, 43(1–2): 137–149.

Transportation Research Board, National Research Council. 2008. *Potential Impacts of Climate Change on U.S. Transportation*. National Research Council Transport Research Board.

Uddin, M.J., A. Shakik, and F.B. Mim. 2024. Understanding the impact of bridge structure on river morphology through geospatial techniques: case on Teesta River, Bangladesh. *Discover Environment*, 2: 91. https://doi.org/10.1007/s44274-024-00136-z

Wang, W., W. Huai, S. Li, P. Wang, Y. Wang, and J. Zhang. 2019. Analytical solutions of velocity profilein flow through submerged vegetation with variable frontal width. *Journal of Hydrology* 578: 124088. https://doi.org/10.1016/j.jhydrol.2019.124088

Wright, L., P. Chinowsky, K. Strzepek, R. Jones, R. Streeter, J.B. Smith, and W. Perkins. 2012. Estimated effects of climate change on flood vulnerability of U.S. bridges. *Mitigation and Adaptation Strategies for Global Change*, 17(8): 939–955. https://doi.org/10.1007/s11027-011-9354-2

Zaid, M., Z. Yazdanfar, H. Chowdhury, and F. Alam. 2018. A review on the methods used to reduce the scouring effect of bridge piers. *Proceedings 2nd International Conference Energy Power (ICEP2018)*, 13–15 December 2018, Sydney, Australia.

Zarrati, A.R., H. Gholami, and M.B. Mashahir. 2004. Application of collar to control scouring around rectangular bridge piers. *Journal of Hydraulic Research*, 42(1): 97–103. https://doi.org/10.1080/00221686.2004.9641188

4 Sediment Genesis, Transportation, and Deposition

4.1 INTRODUCTION

The term "sediment" typically denotes fine materials that settle at the bottom of a river. "Sediment transport" is a broad term utilized for the movement of materials (e.g., silt, clay, sand, gravel, and boulders) in rivers and streams. The material that is carried is referred to as the sediment load. When the material in the bed is rough and non-cohesive, it is primarily the submerged weight of the sediment particles that opposes the movement. The coarse particles move as individual entities. However, when the material is fine and includes silts and clays, cohesive forces dominate and hinder motion. In such a material, clumps of particles move together as a single unit. The transport regime of sediments consists of phases such as rest, entrainment, motion, and deposition. These stages are established by the relationship between tractive and resistance forces that influence sediment particles. The major sediment sources are soil erosion, shore erosion, slope sliding, catchment degradation, and catchment deterioration. The motion of sediment particles is dependent on competency, lift concept, and critical tractive forces. The nature of the bed surface and the water surface varies in response to changes in sediment characteristics, flow characteristics, and fluid characteristics in alluvial channels. These varieties of the bed and water surfaces are categorized based on their characteristics and are referred to as regimes of flow.

The organized study of changes in river morphology over time provides insight into the dynamic features of rivers (Gurnell et al., 1994; Graf, 1984; Lewin, 1987). The characteristics of channel morphology typically include width, depth, velocity, and discharge (Leopold & Maddock, 1953). The current state of channel morphology results from various geomorphic processes that have influenced the riverscape for many years (Ghosh & Mukhopadhyay, 2021; Khan et al., 2018). Rivers, particularly in tropical regions, are marked by considerable modifications in hydro-sediment discharge (Latrubesse et al., 2005; Dingle et al., 2019), flow dynamics (Biswas & Raha, 2022), channel planform alterations (Bandyopadhyay et al., 2014), and river morphological modifications. Changes in river morphology may be initiated by natural elements such as recurring flood events, significant sediment deposition from river catchment areas, as well as human-induced factors like the construction of engineering structures (bridges, dams, reservoirs) on rivers.

So the mode of transport depends on the contact load that includes the nature of the bed surface, and the water surface varies as sediment characteristics, flow characteristics, and fluid characteristics are altered in alluvial channels. These types of bed

DOI: 10.1201/9781003562979-4

and water surfaces are categorized based on their features and are referred to as regimes of flow and, secondly, on saltation load specifically defines the material bouncing along the surface or shifting directly or indirectly due to the influence of bouncing particles creating a small path. Thirdly, suspended load that is related to the turbulent fluctuations in a fluid keep the stuff moving in it suspended. Wash load consists mostly of silt and the finest clay particles; it does not aid in the silting process and will go downstream along the channel. "Bed load" is the collective term for contact load and saltation load. Therefore, everything carried on or next to the bed is considered bed load.

High denudation and weathering rates, as well as catastrophic rainfall, have a major impact on the catchment sediment generation process in tropical regions. These mountain ranges are the source of many rivers. Because of the low temperature, soil formation proceeds slowly. As a result, alpine soil is extremely thin and poorly grounded.

The Himalayas are regarded as a youthful mountain range that was created during the mountain-building process 10 million years ago and continues to grow, expanding by roughly 10 mm every year. Because sedimentary rocks predominate in these mountain ranges, there is a great deal of sediment transport and deposition worldwide. High levels of deforestation in catchment areas may serve as the source of sediments, which over time may change the roughness and morphology of the channel (Lin et al., 2020; Dahl et al., 2018; Kuriqi et al., 2019; Ardıçlıoğlu & Kuriqi, 2019; Bertrand & Liébault, 2018). Landslides in the mountainous catchment areas of the rivers, such as Lish, Gish, and Chel, which are the only foothill tributaries that joined Teesta, have a significant impact on the denudation process. The rivers' silt concentrations and extremely enormous size of sediments are caused by their tremendous velocity that carries sediment loads of all sizes, ranging from big (boulders, cobbles, and pebbles) to tiny (sand, silt, and clay). In India, 90% of the Himalayan silt is quartz. The mineral composition of the river loads is dominated by illegal clays and detrital sands. The point source delivery of the sediments from the landslide-prone area is favored by the rivers' close proximity to the hillslope and valley floor (Korup et al. 2004). Riverbed slope replacement and complex channel morphodynamics are also observed in the Meghalaya Plateau and Western Ghats, as well as the eastern Himalayan Mountain region, which includes the Teesta system and its tributaries, which carry piles of sediments deposited over time during the monsoon (Mandal & Mondal, 2019). Slope, rainfall, lithology, and high strain are brought on by the Indo-Eurasian Plate collision (Singh et al., 2022), as well as seismic occurrences. Frequent earthquakes in the Darjeeling Himalaya cause ongoing instability and disturbance, which leads to frequent landslides that are associated with numerous factors that contribute to the sediment genesis of Himalayan regions.

4.2 RIVER SEDIMENT TRANSPORT

Irregular non-stationary two-phase flow is defined by the redistribution of solid matter in the riverbed or water area of a water body. This involves processes such as sedimentation, bottom erosion, and the movement of sediment. Sediment can be transported within river currents through various mechanisms, including the dragging or rolling of particles along the riverbed (known as bed load sediment), the bouncing of particles through a process called saltation, the suspension of particles in the water column (referred to as suspended sediment), and the movement of particles across the water's surface due to surface tension (flotation). The maximum quantity of solid material that a given water discharge can transport is termed the "flow transporting capacity." This transporting capacity plays a crucial role in the redistribution of sediment within a river channel, serving as a primary factor influencing channel dynamics. Algorithms for calculating sediment transport, which have been developed independently for both suspended and bed load sediment in river flows, present certain challenges regarding calculation accuracy. Sediment transport is influenced by the energy of the flow and the size of the particles that can be transported. The transport potential of a river flow determines the quantity (mass) of particles that can be moved. Depending on the distribution of particle sizes within the transportable material, some sediment rises within the water column and moves with the flow, while other particles are dragged along or rolled over the riverbed. Furthermore, the proportion of the suspended sediment to bed load transport is highly variable, influenced by the hydrodynamic flow patterns and the influx of solid materials from external sources, such as the river basin or soil dumping into the river channel. The hydrodynamic flow patterns, in turn, are shaped by factors such as channel irregularities, variations in slope, the size of bottom sediments, and the water and sediment discharges from upstream sources. In the context of river flow, one can examine the variation in flow velocity within a specific water discharge (referred to as phase hydraulic space) or its mean value (the average flow velocity within that phase hydraulic space) for the current water discharge. In the former scenario, the term "phase hydraulic space" is applicable. The range of potential changes in flow velocity at a given water discharge allows for the velocity to align with various critical values. Critical flow velocities in rivers are defined as those velocities at which the dynamics of the water flow and the sediment it transports undergo significant alterations (Shmakova, 2022). These critical velocities determine the movement patterns of both the liquid phase encompassing laminar, turbulent, subcritical, and supercritical flows and the solid phase of the flow. Consequently, different categories of critical velocities are established, which influence sediment transport.

- Non-eroding velocity refers to an extreme speed at which the primary layer of bottom sediment remains stationary.
- Breakaway velocity signifies the initiation of mass movement among particles.
- Non-silting velocity denotes an extreme speed at which particles remain suspended in the water (Figure 4.1).
- "v1—velocity lesser than minimal non-eroding velocity (no sediment movement, G = 0)" (Shmakova, 2022);
- "v2—non-eroding velocity (start of suspended sediment movement, Gsuspend)" (Shmakova, 2022);
- "v3—breakaway velocity (start of bed load and suspended sediment movement (Shmakova, 2022), Gsuspend + bed load)" (Shmakova, 2022);
- "v4—non-silting velocity (transporting capacity of a flow, Gmax)" (Shmakova, 2022).

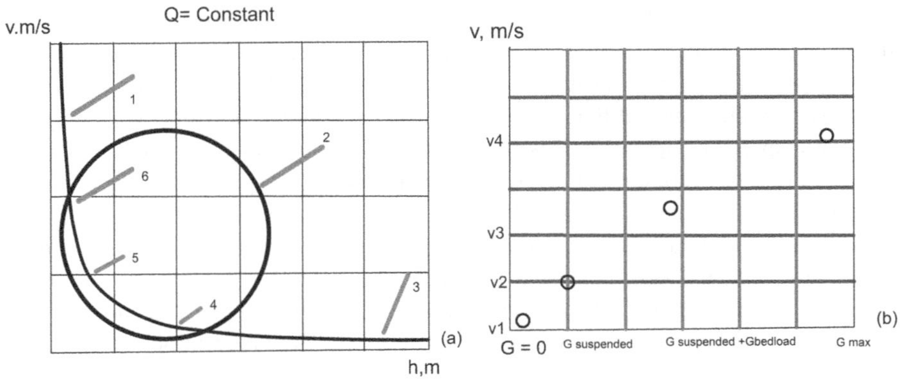

FIGURE 4.1 (a) The hydraulic phase space of a flow within a channel cross-section for a constant discharge (Q) includes the following elements: (1) region representing the maximum suspended load within the flow; (2) region of physically feasible values; (3) defined flow area; (4) non-eroding velocity (the threshold for the movement of suspended sediment); (5) break-away velocity (the initiation of movement for both suspended and bed load sediment); (6) non-silting velocity (based on Shmakova, 2022). (b) Correlation of critical velocities and type of sediment transport (based on Shmakova, 2022).

The hydraulic phase of a space is defined by the channel's morphometry and the characteristics of the underlying surface. Additionally, the flow's transporting capacity is influenced by the volume of the solid material that enters the flow. It is evident that water discharge determines the potential to attain specific critical velocities. Consequently, for given water discharge, various ratios of hydraulic variables within the flow state can be established. This ratio is significantly influenced by the influx of solid materials from the catchment area. The flow velocities that can be achieved within the phase hydraulic space for a constant water discharge have been previously discussed. Now, let us focus on the average velocity associated with a particular water discharge. When sediment in the river flow is generated exclusively through channel deformation, we refer to these as channel-forming water flows. In this scenario, the average rates at which the channel deformation process begins and concludes correspond to specific critical values.

- Total sediment discharge: The primary equation of the mathematical model governing the movement of water and solid matter in river flow is founded on the balance of forces that influence the motion of sediment particles within the "water flow—bottom sediment—sediment" system (Shmakova, 2022).

$$\vec{F}_{flow} + \vec{F}_{s} = 0 \tag{4.1}$$

Bottom sediment may be formulated as "water flow—bottom sediment—sediment"

$$\vec{F}_{grav} + \vec{F}_{inert} + \vec{F}_{A} + \vec{F}_{grav\ s} + \vec{F}_{resist} + \vec{F}_{intert\ s} = 0 \tag{4.2}$$

where

$\vec{F}_{grav}$ = shear projection of the flow gravity

$\vec{F}_{inert}$ = The force of inertia of the moving volume of water enclosed between the sections.

$\vec{F}_{A}$ = Archimedes force

$\vec{F}_{grav\ s}$ = A retaining projection of gravity acting on a sediment particle moving in a flow.

$\vec{F}_{resist}$ = Bottom resistance force

$\vec{F}_{intert\ s}$ = The inertia force of a particle traveling within a flow is described. The bottom resistance force, Fresist, is formulated similarly to the established equation that illustrates the linear correlation between soil shear resistance and normal load.

$$F_{resist} = Rf + cS \tag{4.3}$$

where

F = Internal friction coefficient, dimensionless;
C = Adhesion of soil particles during shear kg/(m.s^2) (for disconnected soil $c = 0$);
S = The area of force application, m^2; in this case, the load R represents the system of loads (F_{flow}) acting on soil particles from flow side (Shmakova, 2022).

$$R = \vec{F}_{flow} + \vec{F}_{grav} + \vec{F}_{inert} + \vec{F}_{A} \tag{4.4}$$

So, the equation of water and solid matter movement may be derived as:

$$(1-f)\left(mg\left[I-\frac{dh}{dx}\right]-m\frac{dv}{dt}\right)-N_{act\,m_s}\frac{dv_s}{dt}+N_{act\,m_{sg}}-cS=0 \tag{4.5}$$

$$\frac{dh}{dt}+h\frac{dv}{dx_{v\frac{dh}{dx}}}=0 \tag{4.6}$$

$$V_s=\sqrt{v^2}+A=w^2 \tag{4.7}$$

$$\frac{dE_{flow}}{dt}-\frac{dE_s}{dt}=0 \tag{4.8}$$

where

m = The mass of the volume of water enclosed between the two cross-sections, kg;
g = acceleration of gravity, m/s^2;
I = Bottom slope, dimensionless;
V = Flow velocity, m/s;
h—flow depth, m; x—longitudinal coordinate, m; S—force application area, m^2; m_s—particle mass, kg; v_s—particle velocity, m/s; N_{act}—the number of moving particles in the flow; ω—hydraulic particle size, m/s; E_{flow}—kinetic energy of the flow, kgm^2/s^2; E_s—kinetic energy of moving particles, kgm^2/s^2; f—coefficient of internal friction, dimensionless; c—the parameter of adhesion of soil particles during shear, kg/(ms^2).

4.3 CASE STUDY OF CATCHMENT SEDIMENT GENERATION AND TRANSPORTATION

The Teesta River's eastern foredeep section is drained by numerous left-hand tributaries, such as Lish, Gish, and Chel, which have shown notable changes in channel geometry over the past century (Biswas & Raha, 2022). The catchment of these three rivers is prone to multiple landslides, according to earlier studies (Mandal & Mandal, 2017; Mandal & Mondal, 2019). The transportation of debris and degraded sediments from landslide-prone areas downstream of the main river is caused by landslides, which are linked to first-, second-, and third-order streams, according to topographic maps and satellite images of the affected areas. Because of the favorable conditions under which the transported sediments are deposited, the channel's morphology changes. However, an integrated discussion of the mechanisms underlying channel morphological changes, transportation, and catchment sediment generation has not yet been developed. Significant sedimentation at the riverbed is revealed by the temporal variation in the channel width and the outcomes of channel dynamics. To gain insight into the catchment sediment production, a landslide vulnerability map has been created. Figure 4.2a shows the five landslide susceptibility zone classifications: very high, high, moderate, low, and very low (Biswas et al., 2024). The catchment areas for the three rivers—Lish, Gish, and Chel—fall within

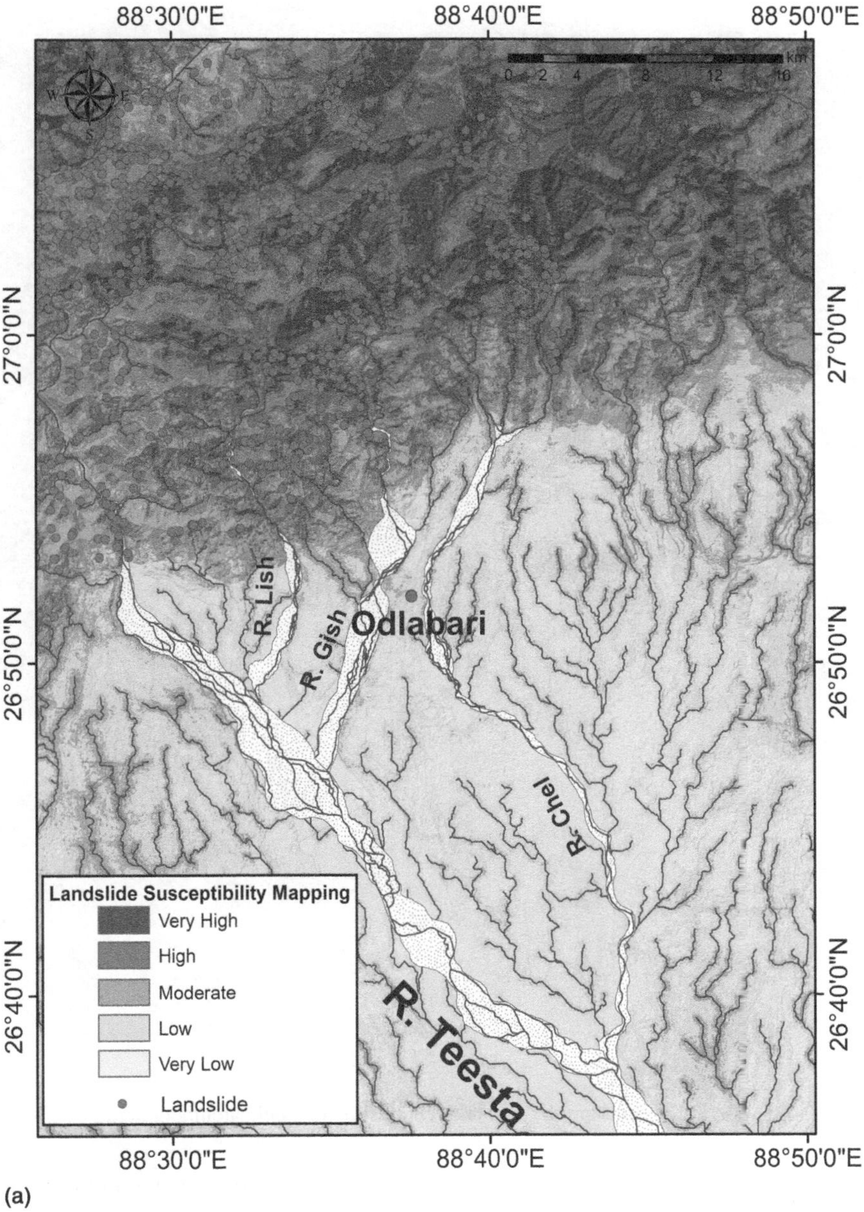

FIGURE 4.2 (a) A map that displays the susceptibility of landslides in the catchment area, with five classes: very high, high, moderate, low, and very low.

(Continued)

(b)

FIGURE 4.2 (b) Field photograph of an active landslide in Chel catchment area.

the high and very high vulnerable zones and it should be noted that some of them are active (Figure 4.2b). Because of the steep gradient and higher elevation, landslides are more likely to occur in accordance with the intense rainfall during the monsoon. Furthermore, it has been determined that the catchment area contains specific lineaments that have made the area vulnerable to erosion. The construction of roads along hillslopes, especially in the Gish and Chel Rivers' catchment area, is one human-caused factor that has increased the region's susceptibility to landslides. The landslide's cause is also influenced by the watershed areas' more than 3399 mm of annual precipitation. The Geological Survey of India has plotted historical landslide events to show how these landslides relate to the first- and second-order streams of the three rivers under consideration. Numerous landslides (https://bhukosh.gsi.gov.in/Bhukosh/MapViewer.aspx) in the catchment area are directly related to the first- and second-order streams, as shown in Figure 4.3. The aforementioned condition is also confirmed by field visits. It is evident that landslides' sediments primarily loose debris, such as sand, cobbles, pebbles, and silts. These sediments are carried down to the first- and second-order streams that are connected (Figures 4.3b and 4.4).

After debouching onto the plains of the foothill section, these sediments are then carried downstream and deposited. A Sediment Transport Index (STI) map has been created to confirm that the catchment sediments have been transferred to the rivers under consideration. The sediment transport index (STI) derived from stream power theory and combined of upslope contributing area (As) and slope (B) (Moore & Burch, 1986).

$$\text{STI} = (m+1) * (As/22.13)^m * \sin\left(\frac{B}{0.0896}\right)^n \tag{4.9}$$

where *As* is the specific catchment area, *B* is the local slope gradient, *m* is usually set to 0.4, and the slope extent "*n*" is usually set to 1.4.

Figure 4.2b shows five different STI classes. It is significant to remember that the upstream and downstream portions of Rivers Gish and Lish have the highest STI values (> 991.37) (Figure 4.2b) (Biswas et al., 2024).

A Sediment Transport Index (STI) map has been created to confirm that the watershed sediments have been transferred to the rivers under consideration. Figure 4.5 shows five different STI classifications. It is significant to observe that the upstream and downstream portions of Rivers Gish and Lish have the highest values (> 991.37) of STI. It can be observed that the sediments, primarily consisting of loose debris, sand, cobble, pebbles, and silt from the landslides, are conveyed down to the associated first- and second-order streams. These sediments are then moved further downstream and are deposited after flowing onto the plains of the foothill section. Mainly during pre- and monsoon seasons, huge rainfall also triggers the landslide events that increase sediments flux near the bridges. The collected data from the nearest tea garden, Pathorjhora, reveal that from 2018 to 2022, maximum rainfall occurred in June–August (Figure 4.6).

The recent study of the entire Sikkim-Darjeeling Himalaya discloses that heavy downpours during the initial phase of the monsoon (i.e., from the second week of June to the first week of July) have elevated the sub-surface pore water pressure and diminished the shear strength of slope materials, resulting in slope failure. Rainfall data gathered from 10 June to 9 July from various rain gauge stations situated in Gangtok, Pakyong, Mangan, Gyalsing, Soreng, Tadong, Mazitar, Namthang, Ranipool, Darjeeling, Kalimpong, and Kurseong indicate that the highest rainfall

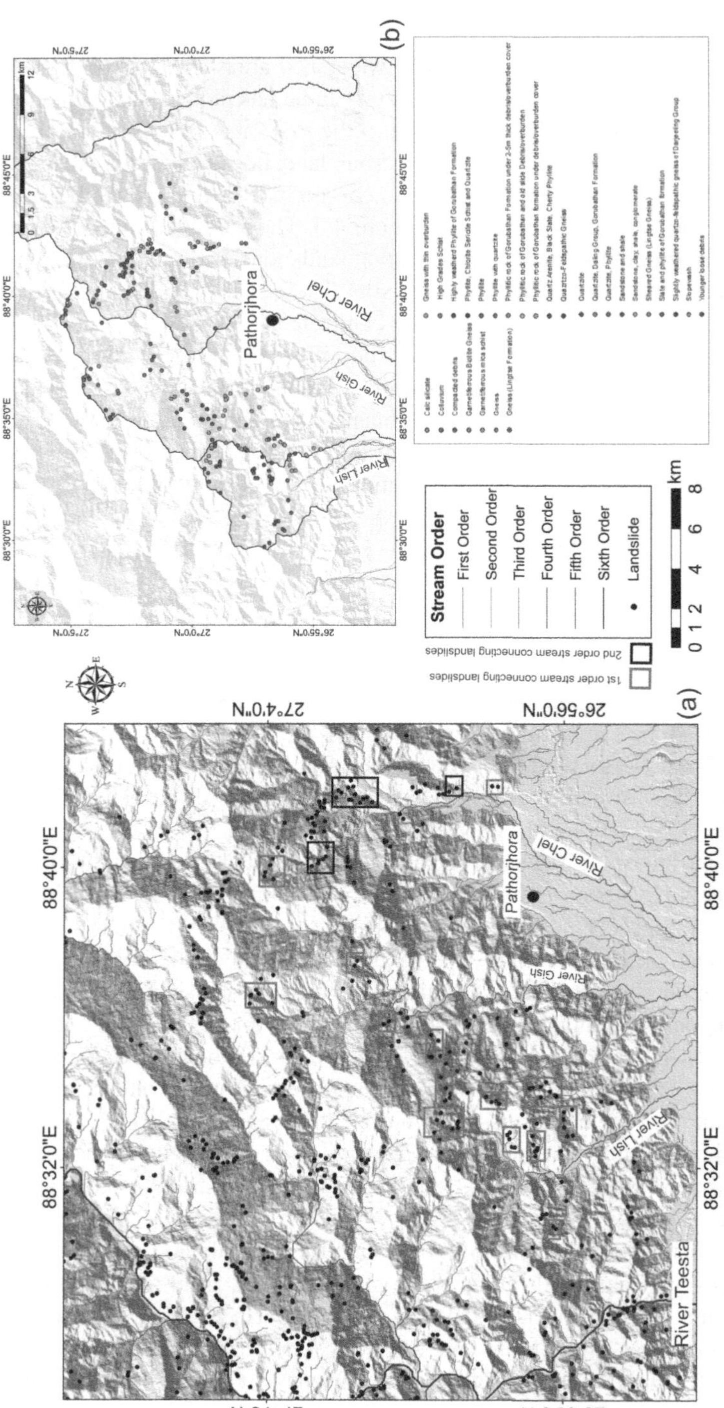

FIGURE 4.3 (a) First- and second-order streams connecting active landslides in the catchment area (https://bhukosh.gsi.gov.in/Bhukosh/MapViewer.aspx). (b) Rock and mineral composition of the landslides (https://bhukosh.gsi.gov.in/Bhukosh/MapViewer.aspx).

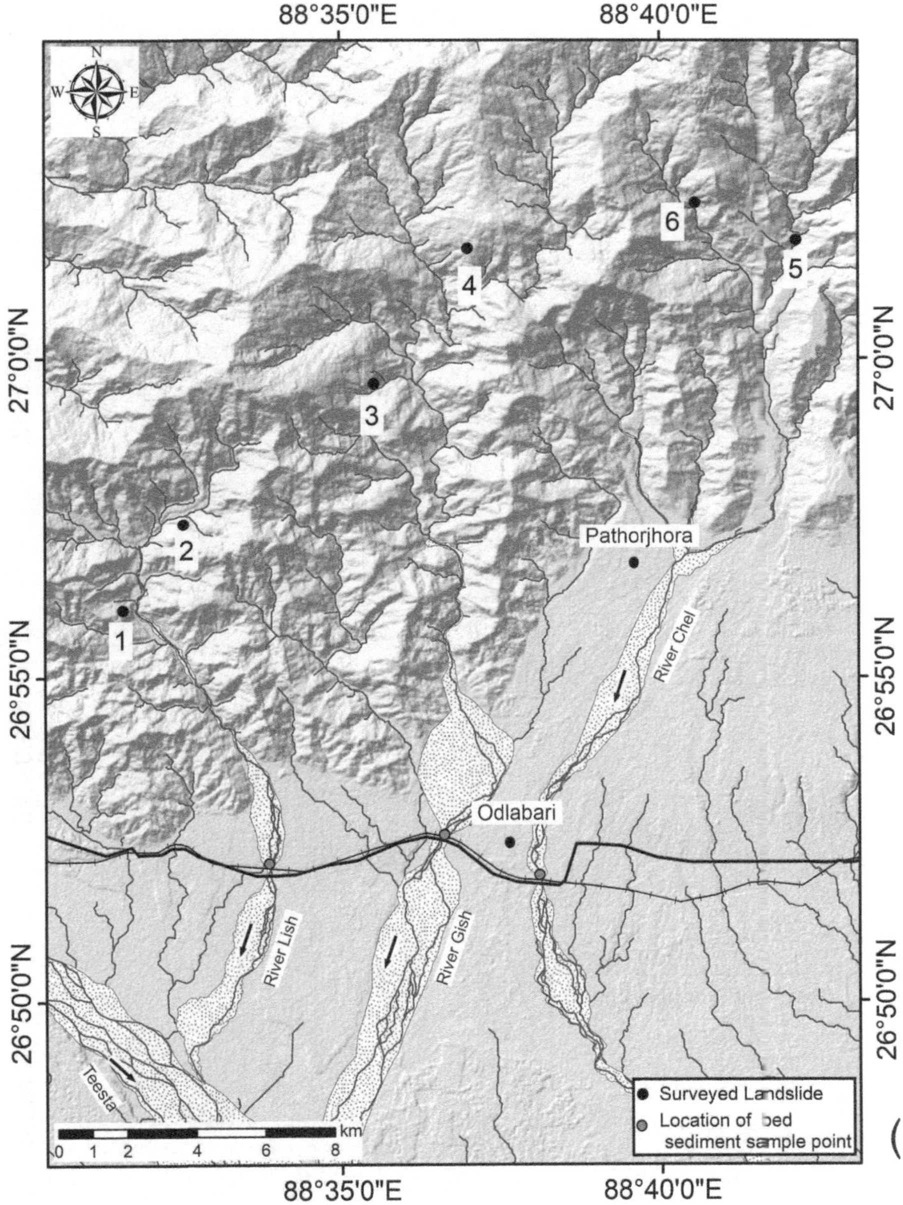

FIGURE 4.4 Locations of six considered sample collecting points (marked in red circle) in the catchment (1, 2, River Lish, 3, 4, River Gish and 5, 6, River Chel) and three sample points of collecting riverbed sediments (marked in yellow color) for microstructure analysis.

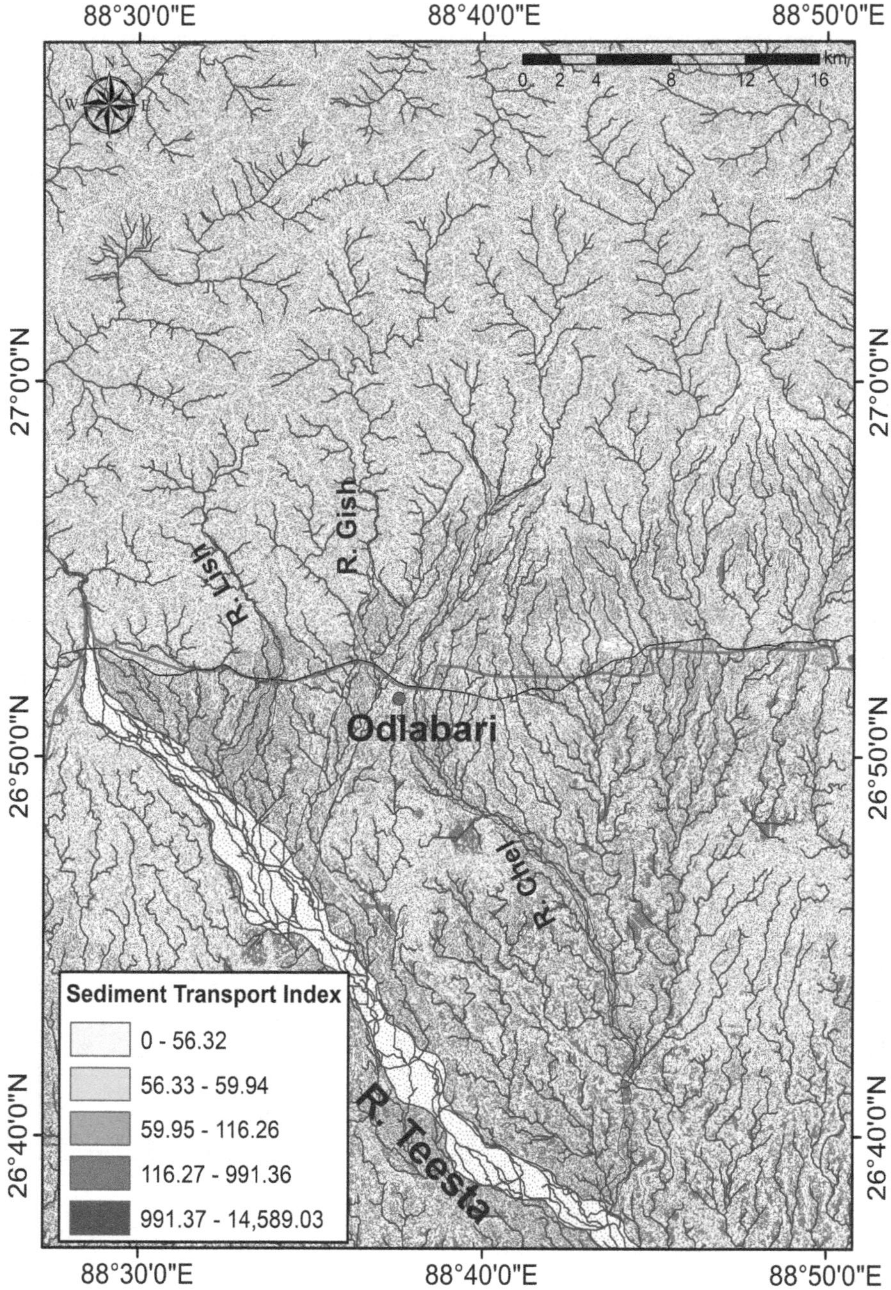

FIGURE 4.5 Sediment transport index map of the catchment and downstream area.

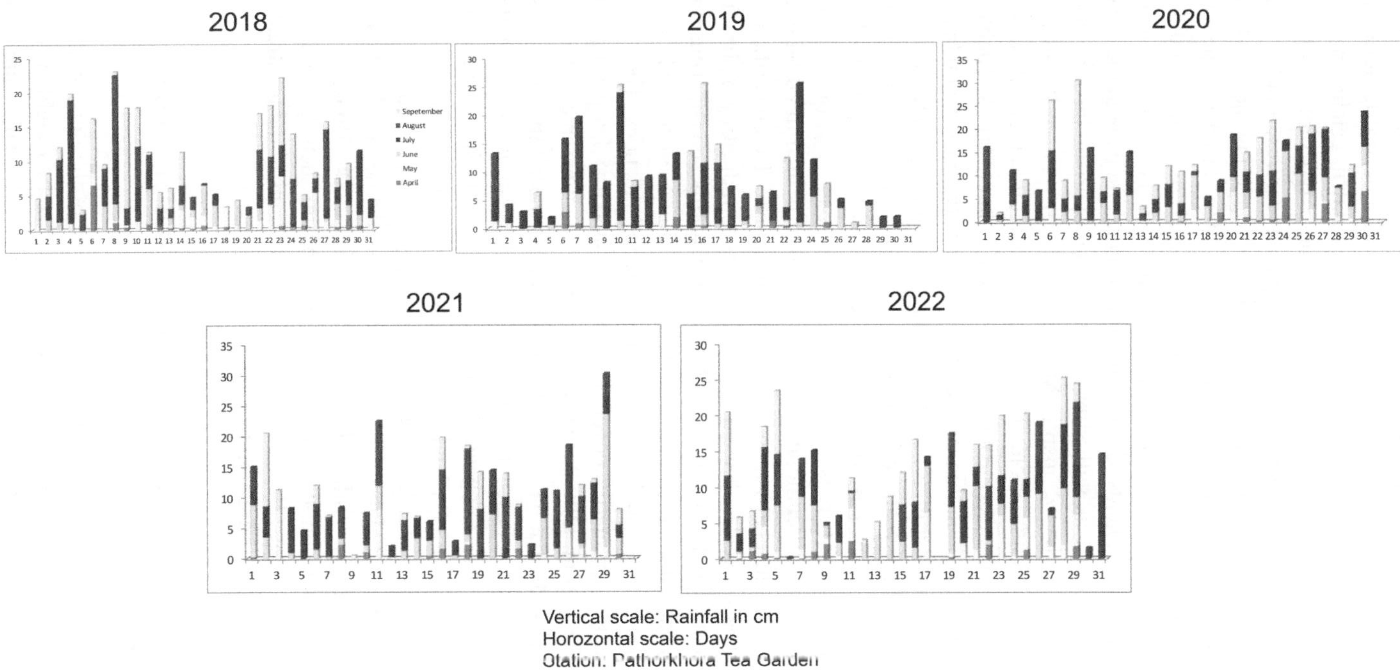

FIGURE 4.6 Distribution of rainfall from April to August at Pathorjhora tea garden in 2018–2022.

(220 mm) in a single day (13 June 2024) occurred at Mangan in Sikkim State. During this perilous 30-day period, the total rainfall reached 755 mm. Nevertheless, the average yearly rainfall in Sikkim is 2600 mm, with 70% of it occurring during the monsoon season (IMD, 2023). This total foothill area along with its catchment experiences landslides in recent times. The vulnerability to these types of geo-hydrological occurrences in this area is influenced by topographical features, lithology, and structural breaks, with rainfall and earthquakes serving as triggering factors (Dutta et al., 2021; Singha & Sarkar, 2024, 2025). This research intends to detail the early monsoon hazard situation of the Sikkim-Darjeeling Himalayas, which could assist in reducing risk in the future. It might be due to active tectonics and earthquake events in surrounding areas. So, landslide susceptibility is concerned about the area's active tectonics; earthquake events with stress and strain comparison have been practiced using the "b" value in the "z" map in Matlab 18 (2020) version. It will explain the magnitude/frequency correlation. The frequency-magnitude distribution (FMD), also known as the Gutenberg-Richter relationship (Gutenberg & Richter, 1944), gives the following explanation of the relationship between the number of earthquakes (N) and their magnitude (M) that occur in a specific geographic area (Gupta & Biswas, 2023). The low "b" value of 0.77 ± 0.04, which is revealed by a detailed earthquake analysis of the past 100 years, indicates a higher frequency of low-magnitude landslide events. There is a significant earthquake in terms of time with cumulative moments (nm) and cumulative events. This plot is then fitted with a straight line and Mmin is the level at which data fall below the line. Detail earthquake analysis of last 100 years reveals a low "b" value 0.77 ± 0.04 indicating higher frequency of landslide events with low magnitude. With respect to time, considering cumulative moments (nm) and cumulative events, the major earthquake occurred in 2011. The mountainous top part of the Teesta Basin is prone to landslides. The mountainous part of the basin is primarily covered by Sikkim and the Darjeeling Himalaya. According to Basu et al. (2009), the area is extremely vulnerable to landslide incidents. Every landslide that has been observed happens during the rainy season, particularly during periods of exceptionally high rainfall. When it rains a lot, the earth's substance becomes less compact. As a result, the safety factor value decreases, increasing the likelihood of a landslide (Agarwal & Narain, 1991). The years with the most rainfall and landslides are shown in Figure 4.6 (Source: http://www.indiawaterportal.org/met_data). The frequency of landslides is significantly influenced by the severity of rainfall (Dahal & Hasegawa, 2008; Bhusan et al., 2014). However, only annual rainfall data is accessible, and no such data is available for individual landslide events. Because of this, this study aims to demonstrate that high rainfall has a significant impact in landslide occurrences.

4.4 PHYSICAL AND CHEMICAL CHARACTERISTICS OF SEDIMENTS

The particle-size distribution of sediment grains is one of the most significant characteristics of sediment deposits. When estimating the specific weight of sediment and forecasting its behavior, the distribution is crucial. When choosing samples for laboratory analysis and researching deposits in the field, several safety measures need to be followed. Many field problems cannot be solved by laboratory studies. In the field, issues like choosing which beds or deposits to sample, as well as figuring out the origin of deposits and the rate of deposition, must be resolved. It is necessary to interpret field and lab data regarding the sediment's texture and nature, as well as its relationships to other formations, soils, and to land use. A technique for fine-grain separation involves monitoring the rate at which sediment particles settle in a water column. After thoroughly mixing a suspension of the sediment sample with a deflocculate, like sodium carbonate, it is placed in a graduated cylinder with an 800-mm-high water column. A siphon is used to remove the top portion of the suspension after ten minutes. Grain sizes of 1/16 mm and larger make up the coarse sediment that stays at the bottom. In order to achieve a clean separation, this procedure is typically repeated roughly four times. It is then possible to treat and examine the coarse and fine separates independently. The degree to which the constituent particles in a sediment deposit are similar or dissimilar determines the degree of sorting. Size, shape, specific gravity, and mineral or chemical composition are just a few of the attributes that can be compared. The classification refers to size distribution in the majority of sediment studies. The phrase "well graded" in engineering refers to the deposit being poorly sorted and having a variety of size grades. The environments where sediment deposits form are listed in the following classification, roughly ranked from the worst-sorted deposits to the best-sorted ones. Like many others, there are a number of exceptions to this classification, some of which are listed in the summary descriptions that follow. It is impossible to draw clear distinctions between the different groups of sediment deposits because they all grade into one another.

Piedmont or alluvial-fan deposits of the study area are characterized by a broad range of sizes; as a result, they constitute one of the groups of sediments that are poorly sorted. The majority of the load of vigorous and rapid streams is rapidly deposited at the base of steep slopes due to a lower gradient. With little to no stratification, large rock blocks and boulders are frequently combined with pebbles, sand, silt, and clay. Comparing the six considered sites, D50 values of River Chel increased near foothill 2.36–2.38 mm in 2021–2022. Toward downstream, it decrease as 2.16–2.26 mm and again increase between bridge area as 2.73–2.76 mm and decreased toward further mouth. In River Gish, D50 values in monsoon/post-monsoon in 2021–2022 increase between the bridges 1–1.4 mm. River Lish strikingly has a different view where D50 values continuously decrease and near bridges, D50 varies 0.5–0.355 μ. Grain size above 256 mm in Zingg Shape Classification shows that most of the shapes are under disc and blade, followed by rod and sphere. The calculated D50 values of six considered sites specify alteration in D50 values in Rivers Chel and Gish between upstream of the bridge and downstream of the bridge. In River Lish, there are no such alterations, but mostly fine < 75 μ are deposited near the bridge as the D50 value is 0.5 μ in both monsoon and post-monsoon seasons. Grain size above 256 mm in Zingg Shape Classification shows that most of the shapes are under dise and blade, followed by rod and sphere. Bed sediments were collected in the monsoon season from the six study sites, i.e., foothill, upstream, between the bridges, downstream, far downstream, and confluence of the three rivers—Lish, Gish, and Chel. It is to be noted that the D50 is very less (0.355 mm) for River Lish between the bridges. The finer materials are getting deposited under lower velocity and decreased stream power near the bridges of Lish. On the other hand, the D50 of the sediments of Chel between the bridges is 2.4 mm. Gish also has greater D50 than River Lish. The grain size collected from the considered six study sites of Rivers Lish, Gish, and Chel demonstrates that the grain size between the bridges has decreased from 0.5 to 0.355 mm during 2021–2023. It remained the same in the downstream and far downstream reaches but has increased (0.25 mm) in 2023 from 2021 to 2022 (0.212 mm). It is interesting that the grain size near the foothill of Gish has increased in 2023 (2.38 mm) from 2021 (0.85 mm) to 2022 (0.9 mm). Between the bridges, the grain size has become 1.8 mm in 2023 which was 2.36 mm in 2022 and 1.00 mm in 2021. At the downstream reach, the grain size is the same (0.85 mm) during 2021–2023. In the case of Chel, the grain size near the foothill has increased in 2023 (3.35 mm) from 2021 to 2022 (2.83 mm). In the upstream of the bridges and between the bridges also, the grain size has increased slightly in 2023 (2.58 and 2.4 mm, respectively) which was 2.36 mm in 2021–2022. Figure 4.7 shows the comparative grain size distribution of three considered rivers in six field sites. The Zingg classification defines that most of the grains are blade and dise in shape from foothill to far downstream (Figure 4.8).

Therefore, the sediment character that is coming down primarily during the monsoon and deposited in the low sloped downstream near bridges is justified by the microscopic thin section analysis of landslide and riverbed samples. Very fine quartz, feldspar, chlorite, and other micaceous minerals (26°57.409′ N, 88°32.552°′ E) make up Lish Landslide Point (LLP) 1 (Figure 4.9a). The micaceous minerals and chlorite define the foliation in the sample. Quartz, feldspar, and biotite make up the majority of the sample in LLP 2 close to 26°56.060′N/88°32.52′E (Figure 4.9b). Medium-sized grains have recrystallized, and biotite is found near their edges. The biotite defines the foliation, which is generally coarsely foliated. Compared to the bed sample, it is made up of opaque minerals, rock fragments, garnet, biotite, chlorite, quartz, and altered feldspar (Figure 4.9c). In essence, feldspar and quartz are present in the rock fragments. This sample's grain size ranges from extremely fine to coarse. Grain feldspar undergoes a substantial transformation into clay minerals. Every grain is oriented at random. Between the bridges, the bulck of bed materials are least in amount. Near 26°59.60′N/88°36.991′E, Gish Landslide Point 1(GLP 1) rock is characterized by a fine-grained texture, primarily composed of quartz, feldspar, chlorite, and opaque minerals such as ilmenite, with minor amounts of fine garnet and muscovite present (Figure 4.9d). It displays a crude foliation and has undergone a process of recrystallization. In Gish Landslide Point 2 (GLP 2), the rock primarily consists of chlorite and muscovite, along with quartz, feldspar, and minor amounts of opaque minerals. It displays an extremely foliated and mylonitic texture, with quartz ribbons oriented parallel to the foliation (Figure 4.9e) and polygonal boundaries indicating static recrystallization. These distinct textures collectively suggest that the rock has undergone significant deformation and is associated with a shear zone most probable for Gish transverse fault. The Gish riverbed sample is composed of quartz, feldspar, rock fragments, garnet, biotite, chlorite, and opaque minerals (Figure 4.9f). The feldspar grains are fine to course in size and are altered. Garnet grains are angular and relatively smaller yet fractured. Some rare occurrence of rounded metamorphic zircon has been documented in this sample. Rock fragments are coarser in size and essentially composed of quartz, feldspar, biotite, and muscovite. Abundance of rock

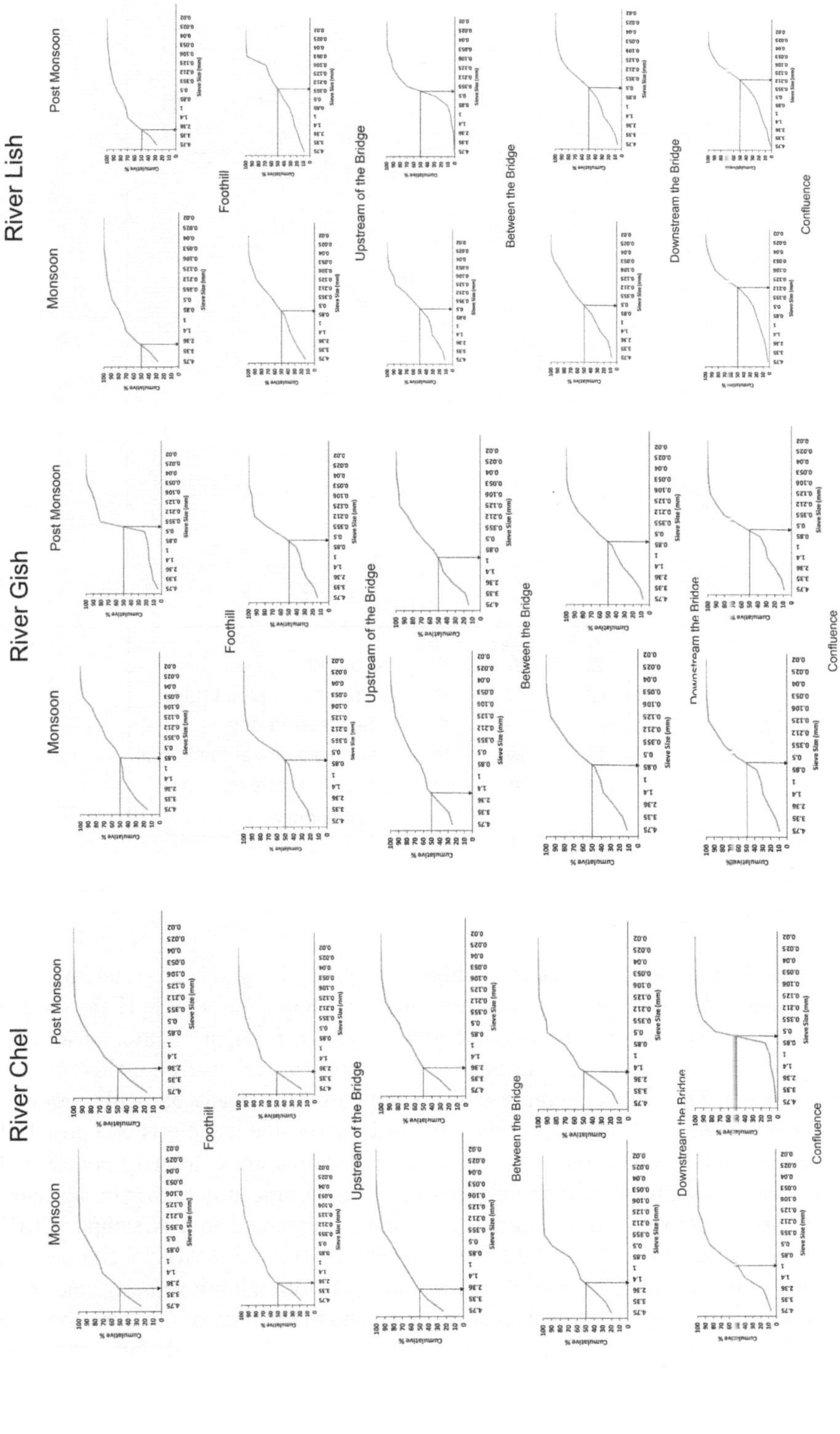

FIGURE 4.7 Graphical representation of grain size distribution of three rivers at six study sites in monsoon and post-monsoon seasons, 2022.

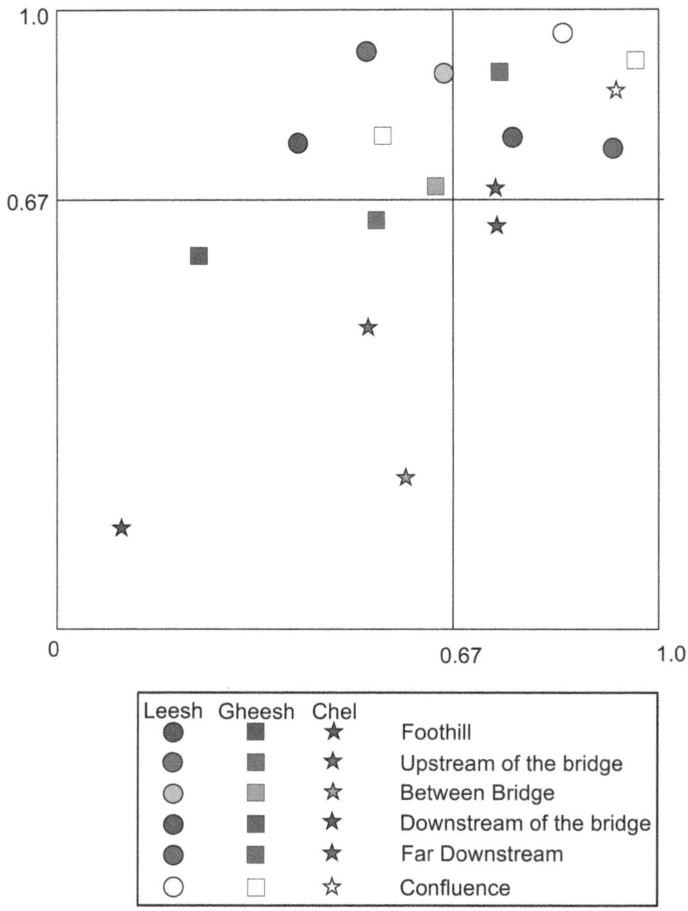

FIGURE 4.8 Graphical representation of grain size distribution, according to Zingg classification.

fragments in this sample is relatively higher. Overall, the grains are randomly oriented. Landslides of River Chel catchment near 27°1.836′N/88°42.134′E, the sample CLP1 is composed of extremely coarse-grained quartz, feldspar, biotite, muscovite, and garnet. However, muscovite and garnet grains are relatively smaller in size (Figure 4.9g). The grains are recrystallized and do not show any particular orientation. The second landslide connecting first-order river, the sample is composed of extremely coarse quartz, feldspar. The textural features are extremely course and grains are recrystallized with abundance of mica is significantly higher in this sample. Coarse cryptocrystalline quartz grains are documented in this sample CLP2 (Chel Landslide Point 2, 27°2.421′N/88°40.555′E) (Figure 4.9h). In comparison with the Chel riverbed sediment also composed of chlorite, biotite, quartz, and feldspar (Figure 4.9i). Opaque minerals are present, and in majority of the cases they are magnetite. Catchment landslides in monsoon that are connected with first- and

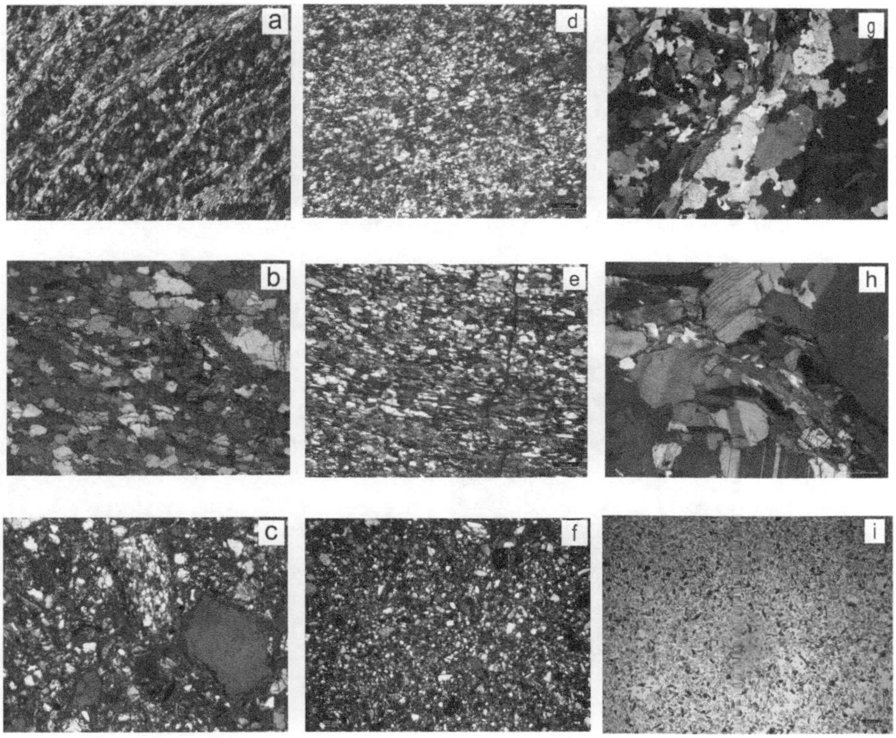

FIGURE 4.9 "Petrography and digenetic features of (a) LLP1: base rock consists of fine quartz, feldspar, chlorite. (b) LLP2: is the recrystallized, medium-sized grains have biotite along their grain borders. (c) Lish riverbed sample: orientation of every grain is random. Rock shards are composed of quartz and feldspar. (d) Mylonitic texture that is strongly foliated, with quartz ribbons extending parallel to the foliation and irregular borders that indicate kinetic recrystallization. (e) Quartz ribbons are oriented parallel to the foliation and have a mylonitic structure. (f) The feldspar grains are fine to course in size and are altered Huge fractures are available. (g) During recrystallization, granules do not exhibit any certain orientation. (h) Coarse cryptocrystalline quartz grains. (i) Opaque minerals with angular shape" (Biswas et al., 2024).

second-order streams carried the sediments (Figures 4.10a–d). Grains are of fine to medium size and angular as it covers maximum length from the origin point to the bridge crossing. The three rivers under consideration appear to be dynamic, according to the temporal analysis of the channel shift. Since the shifting between the bridges became limited due to the narrowing of the reach, the bridge construction is credited as one of the controlling factors of this dynamic nature. From 1913 to 2021, every river exhibited a similar pattern of decreasing width between the rail and road bridge, which can be attributed to human influence on the channel planform (Biswas & Raha, 2022). Yet, the enormous increase in channel width in the rivers' upstream and downstream sections points to a significant sedimentation mechanism originating in the catchment. For the Gish River, the far upstream section (G4) from the rail

FIGURE 4.10 (a and b) Landslides connected with streams in River Chel. (c) Landslide in Gish catchment. (d) Landslide in Lish catchment.

bridge is where the majority of the sediments from the catchment area are deposited. According to Biswas and Raha (2022), the hydraulic conditions of the river's velocity, stream power, and sheer stress allow the depositional condition, which has aided in the river's lateral expansion and increased width. The lateral expansion of the channel, which results from the widening of the channel upstream and downstream of the rail and road bridges, respectively, is another indication that River Lish exhibits the same sediment aggradation mechanism at its bed. It is clear that the engineering construction obstructs the rivers' natural sediment flow. However, the trend of the rise in the River Chel is less noticeable.

4.5 NATURE OF SUSPENDED SEDIMENTS IN MONSOON AND POST-MONSOON

Suspended sediment refers to the particles carried by the turbulent eddies in the stream flow, as described by Crawford in 1991 and Parsons et al. in 2015. It's worth noting that these particles are of small size, typically less than 25 mm, especially consisting of clay and silt, which allows them to remain suspended in water. The quantity of the suspended sediment is relatively lower, averaging 0.95 grams per liter for the Chel River. In the foothill region, the suspended sediment quantities for

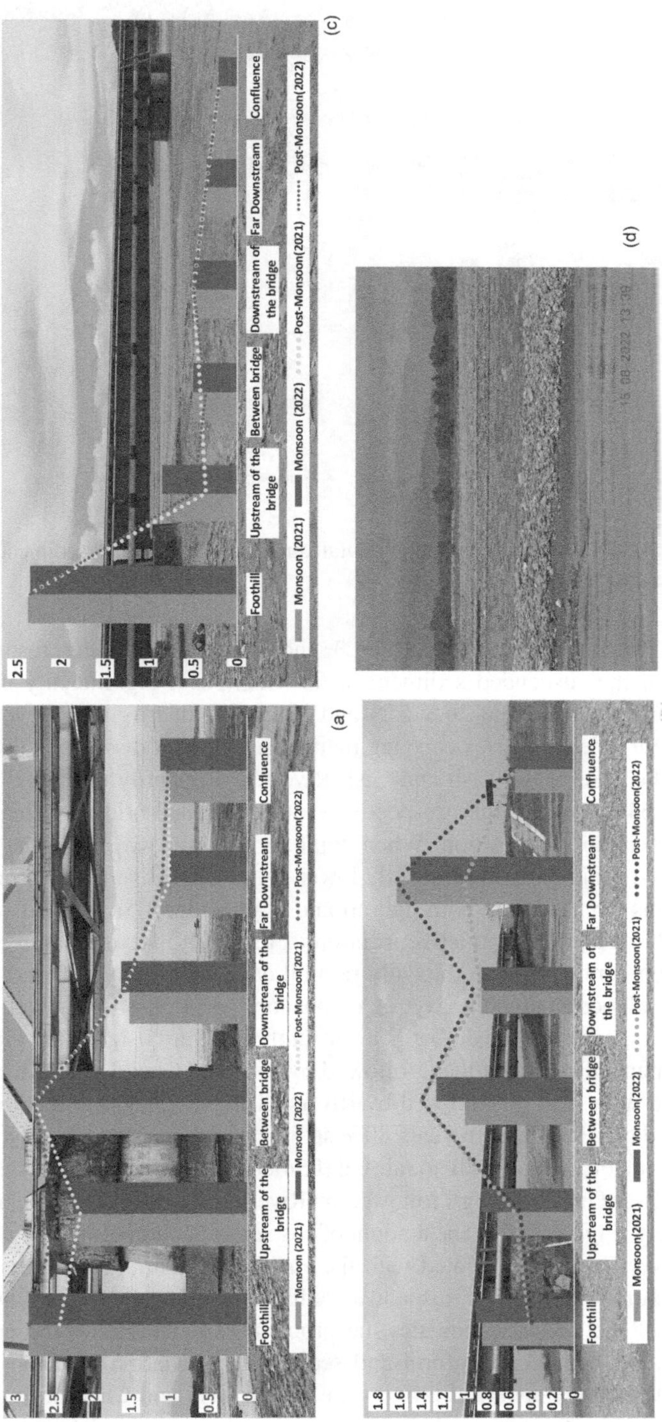

FIGURE 4.11 Field photograph of landslides in the catchment area of Rivers Chel, Gish, and Lish.

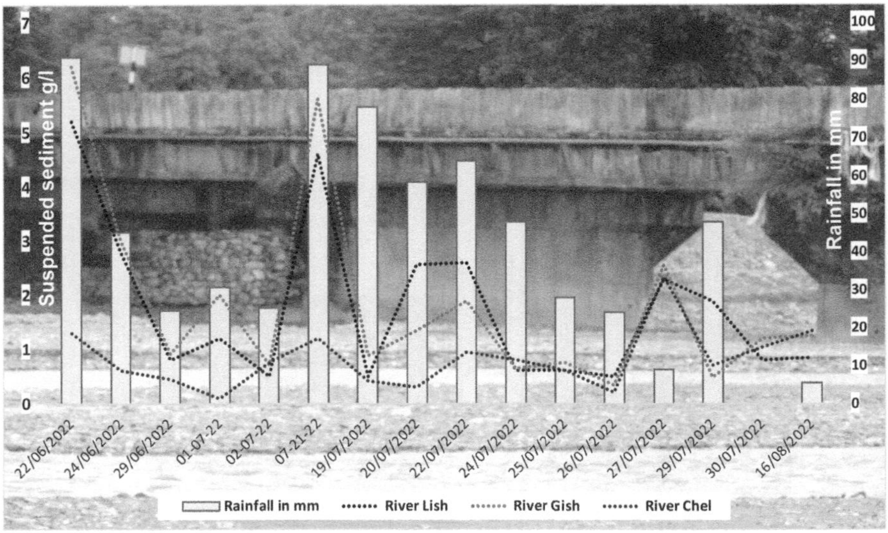

FIGURE 4.12 Graphical representation of rainfall amount and suspended sediment in 2022, with maximum rainfall days in July (approx.).

the Chel, Gish, and Lish Rivers are 1.28, 1.3, and 1.49 grams per liter, respectively. The quantity of the suspended sediment in the Gish River significantly increases near the rail bridge (located at 26.875°N and 88.609°E) due to the inflow of sediments from right-side tributaries carrying their own sediment discharges during the monsoon. In the case of the Lish and Gish Rivers, the quantity decreases notably (1.475 and 0.764 grams per liter, respectively) near the road bridge, primarily due to substantial sediment deposition. When comparing these two rivers, the amount of the suspended sediment is significantly lower in the Chel River, measuring 1.047 and 0.902 grams per liter near both the rail and road bridges. Suspended sediment samples near bridge (most distinctive location of the study) are measured in gram/liter. It has been compared with the daily rainfall data. This study is based on 80% maximum rainfall occurred in the days of a monsoon season. It is to be noted that the sediment load increases on the day of heavier rainfall. River Gish has the maximum amount of suspended sediment load, followed by Lish and Chel. On 11th July, River Gish is marked with 6.56 g/l followed by Rivers Lish and Chel with 5.88 and 3.47 g/lum rainfall days. The study reveals 80% sediments come during heavy rainfall, respectively. Even on the day with no rainfall (05/08/2023), the amount of suspended sediments of River Gish is 1.15 g/l followed by Rivers Lish (0.89 g/l) and Chel (0.45 g/l), respectively (Figure 4.11). The amount of suspended sediments near the bridges of the three rivers is directly related with the amount of rainfall (mm). River Gish brings the maximum amount of suspended sediments followed by Lish and Chel. The outcome of the research indicates the problem associated with the construction of bridges on the river planform and sediment aggradation mechanism. It is noticeable that the three considered rivers, i.e., Lish, Gish and Chel, have different impacts of bridge construction on them. River Lish experiences significant sediment

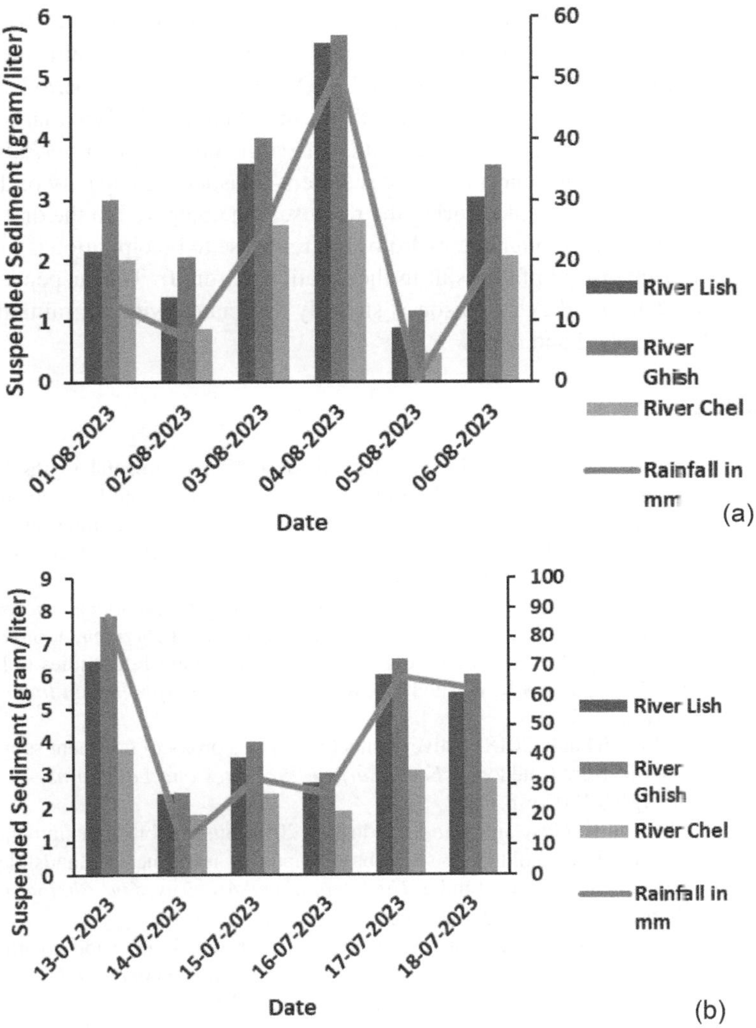

FIGURE 4.13 Graphical representation of rainfall amount and suspended sediment in 2023, with maximum rainfall days in July and August (approx.).

aggradation near the rail and road bridges which has resulted in the rise of bed elevation. It is noticed that a huge amount of fine sediments (mostly less than 40 micron) get deposited near the bridge pier under the hydraulic conditions of lesser velocity, decreased stream energy. River Gish in spite of the huge suspended sediment load deposits a lesser amount of sediment near the bridge which can be contributed to the higher velocity and stream power than that of River Lish. Scouring is also observed in Gish. However, the problem of scouring is massive in River Chel.

Rainfall serves as a fundamental catalyst for erosion and the movement of sediment. When precipitation hits bare soil, it has the potential to dislodge particles, and the ensuing runoff transports these particles downstream, thus elevating the volume of the suspended sediment in rivers and streams. Suspended sediment load is the primary type of particles transported by a river and constitutes a significant part of the overall sediment transport (Li et al., 2024). During the monsoon period, precipitation occurrences were more common and severe, which resulted in a large runoff flow and the movement of a greater quantity of sediments into the river. Likewise, because of the significant runoff flow, the sediment-transporting capacity of the river flow was comparatively greater during the rainy season compared to the dry season. This suggests that even though the hydrological response to precipitation is minimal, occurrences of intense rainfall result in the creation of runoff. The suspended sediment produced during the dry season is strongly associated with the rainfall as we observed (Figures 4.12 and 4.13).

REFERENCES

Agarwal, A., and S. Narain. 1991. Floods, flood plain and environmental myths (State of India's Environment: A Citizens' Report, 3). Centre for Science and Environment.

Ardıçlıoğlu, M., and A. Kuriqi. 2019. Calibration of channel roughness in intermittent rivers using HEC-RAS model: Case of Sarimsakli Creek, Turkey. *SN Applied Sciences*, 1. https://doi.org/10.1007/s42452-019-1141-9

Bandyopadhyay, S., N.S. Kar, S. Das, and J. Sen. 2014. River systems and water resources of West Bengal: a review. *Geological Society of India Special Publication*, 3, 63–84.

Basu, S.R., S.K. De, and B. Bera. 2009. Progress of research in landslide studies in India. In S. Singh, H.S. Sharma, S.K. De, and V.S. Kale (eds.), *Geomorphology in India.* Prayag Pustak Bhawan, 171–198.

Bertrand, M., and F. Liébault. 2018. Active channel width as a proxy of sediment supply from mining sites in New Caledonia. *Earth Surface Processes and Landforms*, 44: 67–76. https://doi.org/10.1002/esp.4478

Bhusan, K., S. Kundu, K. Goswami, and S. Sudhakar. 2014. Susceptibility mapping and estimation of rainfall threshold using space-based input for assessment of landslide hazard in Guwahati city in North East India. *The International Archives of the Photogrammetry, Remote Sensing and Spatial Information Sciences*, 40: 15–19.

Biswas, M., and A. Raha. 2022. An overview of open channel monsoon hydraulics of Himalayan foothill rivers, West Bengal, India. *Modeling Earth Systems and Environment*, 9(2):2969–2979. https://doi.org/10.1007/s40808-022-01481-9

Biswas, M., A. Raha, and S. Roy. 2024. Impacts of bridge constructions on channel dynamics and sediment aggradation: An integrated discussion, West Bengal, *India. GEOMATICA*, 76(2): 100021. https://doi.org/10.1016/j.geomat.2024.100021

Dahal, R.K., and S. Hasegawa. 2008. Representative rainfall thresholds for landslides in the Nepal Himalaya. *Geomorphology*, 100(3–4): 429–443. https://doi.org/10.1016/j.geomorph.2008.01.014

Dahl, T.A., A.D. Kendall, and D.W. Hyndman. 2018. Impacts of projected climate change on sediment yield and dredging costs. *Hydrological Processes*, 32: 1223–1234. https://doi.org/10.1002/hyp.11486

Dingle, E.H., E.C. Paringit, P.L.M. Tolentino, R.D. Williams, T.B. Hoey, B. Barrett, H. Long, C. Smiley, and E. Stott. 2019. Decadal-scale morphological adjustment of a lowland tropical river. *Geomorphology*, 333: 30–42. https://doi.org/10.1016/j.geomorph.2019.01.022

Dutta, K., N. Wanjari, and A.K. Misra. 2021. Study of qualitative stability analysis and rainfall thresholds for possible landslide occurrence: A case study of Sikkim Himalaya. *Journal of Taibah University for Science*, 15(1):407–422. https://doi.org/10.1080/16583655.2021.1984701

Geological Survey of India. n.d. Bhukosh map viewer. https://bhukosh.gsi.gov.in/Bhukosh/MapViewer.aspx

Ghosh, B., and S. Mukhopadhyay. 2021. Channel planform dynamics, avulsion and bankline migration: A study in the monsoon-dominated Dwarkeswar River, eastern India. *Arabian Journal of Geosciences*, 14. https://doi.org/10.1007/s12517-021-07270-5

Graf, W.L. 1984. A probabilistic approach to the spatial assessment of river channel instability. *Water Resources Research*, 20(7):953–962. https://doi.org/10.1029/wr020i007p00953

Gupta, S., and M. Biswas. 2023. Seismo-tectonic and morphological study of the north-east Himalaya. *Geosciences Journal*, 27(1): 1–21. https://doi.org/10.1007/s12303-022-0016-z

Gurnell, A.M., S.R. Downward, and R. Jones. 1994. Channel planform change on the River Dee meanders, 1876–1992. *Regulated Rivers: Research & Management*, 9(4):187–204.

Gutenberg, B., and C.F. Richter. 1944. Frequency of earthquakes in California. *Bulletin of the Seismological Society of America*, 34: 185–188.

IMD 2023. https://metnet.imd.gov.in/docs/imdnews/ANNUAL_REPORT2023English.pdf

Khan, A., L.A. Rao, A.P. Yunus, and H. Govil. 2018. Characterisation of channel planform features and sinuosity indices in parts of Yamuna River flood plain using remote sensing and GIS techniques. *Arabian Journal of Geosciences*, 11. https://doi.org/10.1007/s12517-018-3876-9

Korup, O., M.J. McSaveney, and T.R.H. Davies. 2004. Sediment generation and delivery from large historic landslides in the Southern Alps, New Zealand. *Geomorphology*, 61: 189–207. https://doi.org/10.1016/j.geomorph.2004.01.001

Kuriqi, A., G. Koçileri, and M. Ardıçlıoğlu. 2019. Potential of Meyer-Peter and Müller approach for estimation of bed-load sediment transport under different hydraulic regimes. *Modeling Earth Systems and Environment*, 6: 129–137. https://doi.org/10.1007/s40808-019-00665-0

Latrubesse, E.M., J.C. Stevaux, and R. Sinha. 2005. Tropical rivers. *Geomorphology*, 70: 187–206. https://doi.org/10.1016/j.geomorph.2005.02.005

Leopold, L.B., and T. Maddock. 1953. The hydraulic geometry of stream channels and some physiographic implications. *Professional Paper*.

Lewin, J. 1987. Historical river channel changes. In *Palaeohydrology in Practice: A River Basin Analysis*, John Wiley and Sons, 161–175.

Li, H., J. Lv, X. He, Y. Bao, and G. Nsabimana. 2024. Precipitation-dependent sensitivity of suspended sediment concentration to turbidity in a mountainous river in southwestern China. *Ecological Indicators*, 159: 111644.

Lin, B., X. Chen, and H. Yao. 2020. Threshold of sub-watersheds for SWAT to simulate hillslope sediment generation and its spatial variations. *Ecological Indicators*, 111: 106040. https://doi.org/10.1016/j.ecolind.2019.106040

Mandal, B., and S. Mandal. 2017. Landslide susceptibility mapping using modified information value model in the Lish river basin of Darjiling Himalaya. *Spatial Information Research*, 25(2): 205–218. https://doi.org/10.1007/s41324-017-0096-4

Mandal, S., and S. Mondal. 2019. Statistical approaches to landslide susceptibility assessment and prediction. https://doi.org/10.1007/978-3-319-93897-4

Moore, I.D., and G.J. Burch. 1986. Physical basis of the length slope factor in the Universal Soil Loss Equation. *Soil Science Society of America Journal*, 50: 1294–1298. https://doi.org/10.2136/sssaj1986.03615995005000050042x

Shmakova, M. 2022. Sediment transport in river flows: New approaches and formulas. In *Modeling of Sediment Transport*. IntechOpen.

Singh, S., A. Joshi, A. Sahu, R.A. Prasath, S. Sharma, and C.S. Dwivedi. 2022. Himalayan landslides–causes and evolution. In S. Kanga et al. (eds.), *Disaster Management in the Complex Himalayan Terrains*, Geography of the Physical Environment, 33–42. https://doi.org/10.1007/978-3-030-89308-8_3

Singha, N., and S. Sarkar. 2024. Modelling landslide susceptibility along major transportation corridor in Darjeeling Himalayas using GIS-based MCDA approaches. *Modeling Earth Systems and Environment*. https://doi.org/10.1007/s40808-023-01942-9

Singha, N., and S. Sarkar. 2025. Early-monsoon extreme rainfall events and associated landslides in Sikkim-Darjeeling Himalayas during June–July 2024.

5 Geo-Meteorological Hazards and Risk

5.1 CONCEPT, SCIENCE ASPECTS OF CAUSAL AND TRIGGERING FACTORS OF HAZARDS, AND RELEVANCE OF GEO-METEOROLOGICAL IN SPATIO-TEMPORAL SCALE

Burton and Kates (1964) offer one of the first definitions of risks, arguing that natural hazards are those aspects of the physical environment that are dangerous to people and brought on by external factors. Although they are not human-caused, some occurrences have negative effects on people and have their origins in the physical world. Although Burton and Kates' definition is accurate in that risks are detrimental to individuals, it ignores how humans contribute to or exacerbate the effects of risks. Floods, for instance, might result from human activities such as deforestation, heavy land use, or the collapse of dams built to regulate flooding, or they can be caused by natural variations in climatic conditions (Haque, 1997).

Similar to this, landslides are frequently caused by earthquakes, volcanic eruptions, and heavy rainfall (sometimes on hills and mountainside areas that have been cleared by wildfires). However, they can also be caused by human activities such as logging, road construction, and house construction, all of which can reveal bare soil (Wisner et al., 2004). Vulnerability, which is a complicated result of several elements including wealth, education, gender, demography, technology, and, most importantly, readiness, is another context in which hazards are frequently characterized. Another potential susceptibility aspect is location.

There are several classifications for hazards. Geo-meteorological dangers have been the main focus of this work. It signals processes, occurrences, and circumstances related to geomorphology, geology, weather, or the environment that might be hazardous or that could endanger human life, property, or health. Floods, erosion, debris flows, rock falls, and other landslides are examples of geohazards. One might also classify human-induced processes as geohazards.

Mass movements on slopes, including some that entail little to no actual sliding, are often referred to as landslides (Varnes, 1984). Lithology, bedding inclination, structural context, and material properties are some crucial elements in determining the likelihood of landslides. It is crucial to note, nevertheless, that some elements, such as intense rain or earthquakes, might serve as catalysts for landslip incidents. Additionally, it has been noted that the landslip danger has increased due to deforestation in hilly areas, particularly on steeper gradients. When slopes experience a drop in the shear strength of the hillside material, either as a result of an increase in shear stress or as a result of a mix of anthropogenic activities and natural ecosystem processes, landslides occur. There are three types of hazards: anthropogenic, natural, and socio-natural. The majority of natural hazards are linked to natural occurrences

DOI: 10.1201/9781003562979-5

and processes. Human actions and decisions are the primary causes of anthropogenic hazards, often known as human-induced hazards. The likelihood of armed conflicts and other socially unstable or tense situations that are governed by national and international humanitarian law are not included in this phrase. A number of risks are socio-natural, meaning they are linked to both natural and man-made elements, such as climate change and environmental deterioration. Geo-metrological hazards include geological as well as hydrological hazards that affect the physical landscape and human life in terms of economic loss. Geological or geophysical hazards originate from internal earth processes. Examples are earthquakes, volcanic activity and emissions, and related geophysical processes such as mass movements, landslides, rockslides, surface collapses, and debris or mud flows. Hydro-meteorological factors are important contributors to some of these processes. Tsunamis are difficult to categorize: although they are triggered by undersea earthquakes and other geological events, they essentially become an oceanic process that is manifested as a coastal water-related hazard. Oceanographic, hydrological, or atmospheric factors can cause hydro-meteorological risks. Examples include drought, heat waves and cold spells, coastal storm surges, floods, especially flash floods, and tropical cyclones, sometimes referred to as hurricanes and typhoons. Other risks such as landslides, forest fires, locust plagues, epidemics, and the movement and spread of toxic materials and volcanic eruption material can also be influenced by hydro-meteorological conditions. In recent times, both geological and hydrological hazards are making crucial environmental imbalance. In 2020–2021, climate-related risks gained even more attention. During those two years, weather and climate-related hazards were responsible for 90.3% of all natural disasters that were reported. This is equivalent to 7.3% (Figure 5.1) due to geological dangers like earthquakes and volcanic activity.

In addition to this, the next-highest percentage of disaster victims is caused by geological hazards and disasters. Over the last 12 years, the proportion who were affected by a climate- or weather-related disaster has held steady between 97% and 98%. On average, about 100 million people are affected every year by climate- and weather-related disasters. Alongside this, disasters linked to geological hazards make up the next largest proportion of the people affected by disasters. In the past six decades, there has been a consistent rise in the number of countries facing simultaneous disasters. Overlapping disasters are described as situations where two disasters occur in a single country, with the onset of the second disaster occurring before the first has concluded. These concurrent disasters can manifest in various ways: for instance, storm-related wind damage alongside a distinct flooding event, or a heatwave occurring in conjunction with a wildfire.

Over the past three decades, an average of 44 countries annually have experienced simultaneous disasters at least once within a year. Additionally, this period has recorded an average of 199 simultaneous disasters each year, a figure significantly greater than that of earlier decades. This elevated frequency of simultaneous disasters is evident, notwithstanding substantial fluctuations from year to year. When two disasters strike simultaneously in the same location, it is reasonable to anticipate that their combined effects will be greater than if they were to occur separately. For instance, response teams that are equipped to manage the consequences of a single disaster may become overwhelmed when faced with two events at once. Similarly,

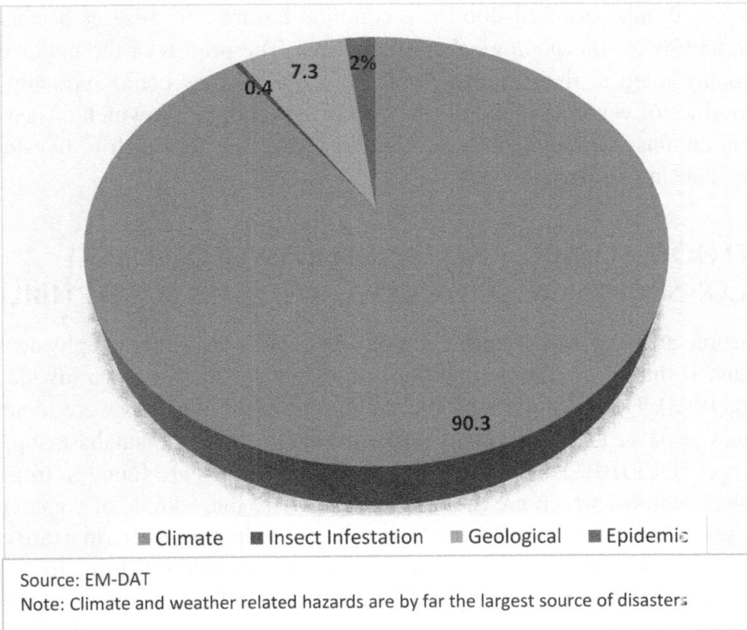

FIGURE 5.1 Detail distribution of natural disasters in 2020–2022 (World Disaster Report, 2022).

communities that possess the resilience to withstand one disaster with limited repercussions may experience significant damage when confronted with the dual challenges of concurrent disasters. So, geological and metrological hazards impact more and increasing over time.

The effects of disasters can be intensified when multiple events happen simultaneously within the same country. However, in larger countries, these disasters may not be geographically proximate. To address this variable, our analysis concentrates on disasters that take place within the same first-level administrative division, such as a province. This approach increases the likelihood that the disasters are spatially adjacent. When examining climatic, hydrological, and meteorological disasters individually, it becomes evident that the simultaneous occurrence of such events within a first-level administrative division of a country, or shortly thereafter, results in a greater number of individuals being impacted. This phenomenon holds true irrespective of the nature of the accompanying disaster. The trend intensifies as the time gap between the two disasters decreases.

India has a long history of being susceptible to flooding. Approximately 40 million hectares of the nation's total geographical area are at risk of flooding, with 7.5 million hectares (25%) experiencing floods annually. River floods account for over 60% of the total flood damage in the country, while the remaining 40% is attributed to other causes. Each year, floods negatively impact around 1.75 million individuals and 30,000 livestock. The proportion of flood-affected land compared to the overall geographical area is significant.

In North Bengal, foothill flood is a common hazard and disaster because of its location and hydro-morphological consequences. The origins of the major rivers in the Himalayas led to their unpredictable behavior, as they could suddenly release large volumes of water and mud during floods triggered by snowmelt or heavy rains in the mountains. This inevitably resulted in significant loss of life, livestock, and property, causing widespread suffering.

5.2 FLOOD, SEISMIC, AND LANDSLIDES: GEOLOGICAL CONSEQUENCES, AREA, COVERAGE, AND SUSCEPTIBILITY

More people are negatively impacted by floods than by any other geophysical occurrence, and it makes up almost one-third of all natural disasters worldwide (Smith & Ward, 1998). Heavy rain, monsoon rain, and tropical cyclones were identified as the causes of 64%, 11%, and 6% of floods, respectively, in the database supplied by Adhikari et al. (2010). The majority of developing nations are found in tropical and subtropical regions, which are the only places where these kinds of weather events may be seen. Additionally, most people reside in floodplains where they are exposed to varying degrees of river erosion and floods (Tingsanchali & Karim, 2005).

Many rivers drain the Himalayan foreland in West Bengal, with a general flow direction of south and southeast (Raha & Biswas, 2022a, 2022b). Along with its underlying geology, neo-tectonics activities, and climatic dominance in terms of increased rainfall, the North Bengal rivers have undergone remarkable changes in the last few centuries with regard to riverbed area, channel patterns, discharge, and channel shifting. These rivers' massive silt deposition, which is linked to monsoonal rainfall and flow changes, has significantly changed the foreland area. The length, intensity, and geographical distribution of rainfall all affect the rise in river water levels (Roy, 2011; Paul & Biswas, 2019; Biswas & Banerjee, 2018). Occasional landslip incidents in the rivers' higher catchment areas cause abrupt water bursting, which ultimately causes the river systems to discharge enormous amounts of water, as was the case with the Teesta River in 1968 (Roy, 2011).

The main rivers that pass through Jalpaiguri and Alipurduar are the Teesta, Lish, Gish, Chel, Neora, Muri, Jaldhaka, Diana, Rohtikhola, and Sukti. The rivers are unpredictable due to factors including slope, flow patterns, and high rainfall (Ghosh, 2018). The three Teesta tributaries on the left bank—Lish, Gish, and Chel—are vulnerable to flooding. However, these three rivers have varying levels of flood coverage. In 1991, the Chikmati embankment of the Chel River was most badly impacted. According to the annual flood statistics for West Bengal, there was a single breach in the Lish embankment in 1992.

In 1996, an 80-meter breach occurred in the armored Manabari embankment on the Chel River near the Odlabari Gramme Panchayat. In the year 2000, rivers such as the Chel and Gish experienced very high velocities because of the flow from higher catchment regions and the continuous rain in these basin areas from the second week of June to the second week of August. Consequently, there were significant landslides in the upper watershed, which increased the river's destructive power with the aid of big trees, logs, and churning water.

In 2002, between 30 June and 1 July 2002, the Gish River in Malbazar shifted its path, wreaking havoc across a large area. The Gish River's course abruptly changed between 30 June and 1 July 2002, impacting about 30,000 acres of property, including two blocks and nine gram panchayats in the Mal and Nagrakata subdivisions. There were two major breaches on the Manabari embankment for around 200 meters on the left side of River Gish at Manabari under block Mal in 2012 as a result of heavy rainfall in Malbazar (219.40 mm) on 11 September. Therefore, it is crucial to research the areas and its zonation's flood vulnerability.

The Brahmaputra floods in 1987, 1988, 1998, 2007, 2010, and 2022 resulted in a significant number of fatalities. A number of researchers have also assessed the likelihood of various Brahmaputra tributaries to propagate flood. Teesta, Jaldhaka, and Torsa are three of the significant tributaries of the Brahmaputra that flow through the Himalayan foreland area of North Bengal (Roy et al., 2021; Roy, 2011; Chakraborty & Datta, 2013). These rivers debouch into the floodplains after originating from the Himalayan region, flow toward south-west direction and possess their own hydro-sediment discharge. Most rivers are recognized as being significantly notorious for their unpredictable nature. Due to extensive and continuous bank erosion, floods, and channel shifting, many people become homeless throughout the rainy season (Roy, 2011). One of the main tributaries of the River Brahmaputra, the River Teesta, causes significant damage to large tracts of land in North Bengal, particularly in the Mal, Jalpaiguri Sadar, and western half of the Maynaguri block (Das et al., 2017). Records indicate that the Darjeeling and Sikkim regions of the Himalayas experience severe, persistent rain, which had caused the floods in 1906, 1968, and 1993. Large-scale flooding and inundation had occurred in Jalpaiguri town in 1952, 1972, 1998, 2000, and 2002 as a result of continuous and intense local rainfall (Roy, 2011). The propagation of flood is also a resultant effect of engineering constructions like rail bridge and road bridge that have squeezed up the river width (Biswas & Banerjee, 2018). During the monsoon season, a huge amount of water pass through the narrow passage and affects both the up/downstream section increasing flood risk and susceptibility (Raha et al., 2023). Lish, Gish, and Chel are three tributaries that flow into the Teesta River on its left bank (Raha et al., 2023; Biswas & Banerjee, 2018). Teesta is joined by the Rivers Lish, Gish, and Chel upstream and downstream of the Gajoldoba barrage, respectively. The increasing volume of water in the monsoon season of these rivers contributes to the rising of the water level of River Teesta and also in the flooding of the adjacent plains of North Bengal. Along with the natural floods, bridges imply a role in upstream and downstream flooding in monsoon. The local economy, infrastructure, and people's livelihoods are all seriously threatened by the frequent floods in the North Bengal plains (Roy et al., 2021). Although it is nearly impossible to completely eliminate the risk of flooding, adopting hazard management techniques can lessen the severity of the negative impacts of floods. In this context, flood hazard rating and susceptibility need to be evaluated for assessing the probable flood coverage area. Biswas and Raha (2023) analyzed the details as follows.

The probable risk zone of flood hazard, as shown in the Flood Hazard Rating map (Foudi et al., 2015), is most applicable.

$$HR = d^{*}(v + 0.5) + DF \tag{5.1}$$

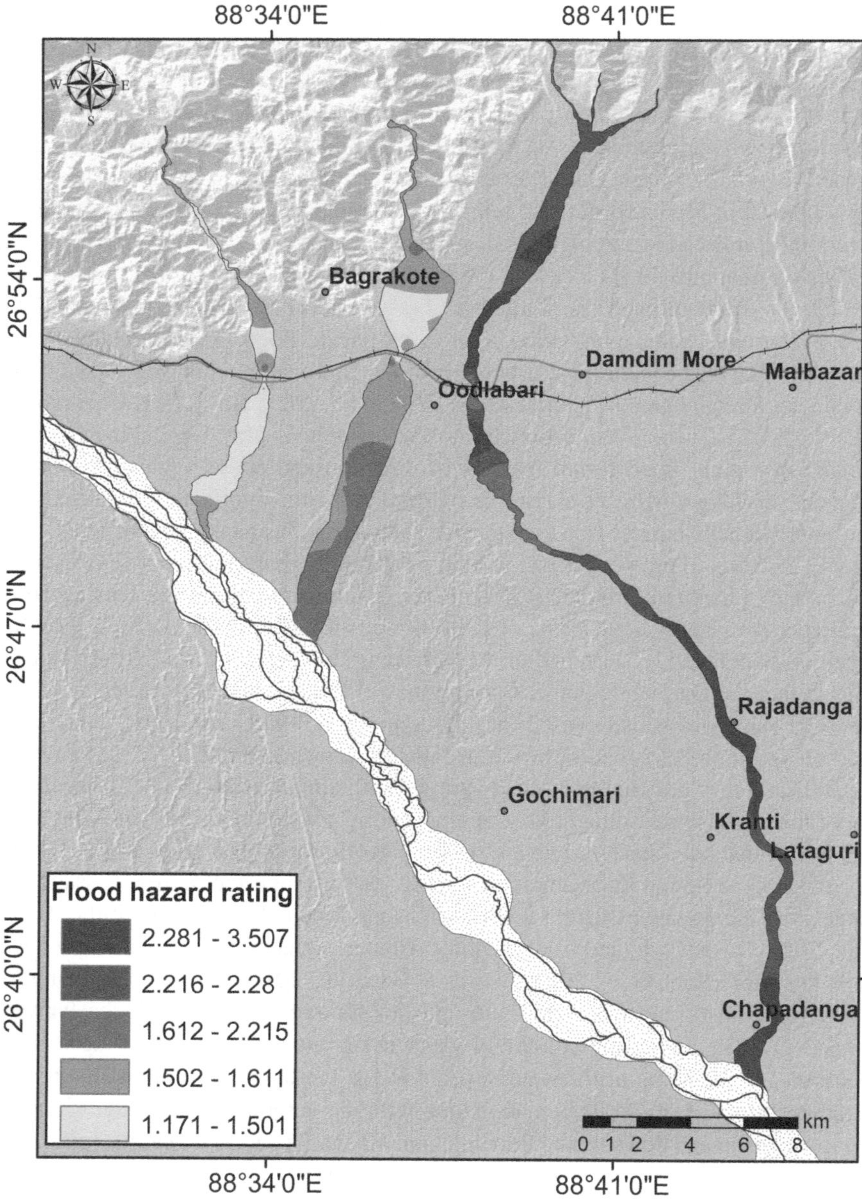

FIGURE 5.2 Flood hazard rating map of three considered tributaries of Teesta (Biswas & Raha, 2023).

where d = depth of river water in m, v = velocity in m/s, DF = Debris Factor (0, 0.5, 1 depending on probability of debris to come). Primary data has been collected from field to execute the analysis.

The flood risk rating map has been created using five categories (Figure 5.2). It is noted that River Chel has been acknowledged for having the highest HR values.

The section of the river immediately upstream and downstream of the bridge for rail and road shows triggering levels (2.281–3.507), indicating a potential for flooding. Greater values are also noted in the lower sections of the Chel near Ranjadanga, Kranti, and Chapadanga. The Gish River has been assessed to have a moderate flood risk, especially in the stretch downstream from the road bridge to where it joins the Teesta River. In contrast, River Lish exhibits slightly reduced vulnerability (1.171–1.501) to the risk of flooding.

Following an examination of earlier literature and onsite research, specific factors have been selected to create the flood susceptibility mapping (Table 5.1). The parameters were calculated on the ArcGIS 10.4.1 platform and subsequently categorized into five classes. The parameters have been allocated weightage through the FUZZY-AHP method, as detailed below.

The fuzzy AHP approach might be viewed as a developed explanatory tactic from the conventional AHP. Researchers that have utilized fuzzy AHP, including Chang (1996), Van Laarhoven and Pedrycz (1983), and Boender et al. (1989), have demonstrated that fuzzy AHP typically delivers a more accurate picture of fundamental leadership. There are several ways to decide the weights in the fuzzy AHP technique. Using TFNs, the degree approach first proposed by Chang (1992) is used to determine fuzzy weights for selecting criteria (Chang, 1996). Chang's fuzzy AHP degree analysis method has been explained by Boutkhoum et al. (2015) and Biswas & Raha (2023).

Phase 1 Fuzzy synthetic extent calculation:

Assuming Y and A are the objects, the goal sets for Y and A are $Y = \{y1, y2, y3\ldots., yn\}$ and $A = \{a1, a2, a3\ldots., am\}$, respectively (Chang, 1992, 1996). For each item that may be obtained, the extent analysis estimates for m are (Meshram et al., 2019; Biswas & Raha, 2023):

$$M^1{}_{ai,} M^2{}_{ai,} M^3{}_{ai,}\ i = 1,2,3,\ldots..,n \tag{5.2}$$

where $M_{ai}^{m}\left(j = 1,2,2,\ldots..m\right)$ are TFNs.

TABLE 5.1
Details of the Inter-Item Correlation Matrix of Four Considered Indicators

	Inter-Item Correlation Matrix			
	Depth	**Manning_n**	**Velocity**	**Debris_p**
Depth	1.000	−.107	.786	.553
Manning_n	−.107	1.000	−.094	−.058
Velocity	.786	−.094	1.000	.577
Debris_p	553	−.058	.577	1.000

According to the ith object, the estimation of the fuzzy synthetic extent is described as:

$$F_i = \sum_{j=1}^{m} M_{ai}^{j} * \left[\sum_{j=1}^{n} \sum_{j=1}^{m} M_{ai}^{j} \right]^{-1} \tag{5.3}$$

$$\sum_{j=1}^{m} M_{ai}^{j} = \left(\sum_{j=1}^{m} \alpha_j \beta_j \gamma_j \right) \tag{5.4}$$

$$\sum_{j=1}^{n} \sum_{j=1}^{m} M_{ai}^{j} = \left(\sum_{j=1}^{n} \alpha_j \beta_j \gamma_j \right) \tag{5.5}$$

Phase 2 Assessment of fuzzy estimates

The degree of likelihood of $M_2 = (\alpha_2 \beta_2 \gamma_2) \geq M_1 = (\alpha_1 \beta_1 \gamma_1)$ is defined as:

$$V(M_2 \geq M_1) = \sup\left[\min\left(\mu M_1(y), \mu M_2(y) \right) \right] \tag{5.6}$$

It can also be stated as:

$$V(M_2 \geq M_1) = tgh(M_1 \cap M_2) \tag{5.7}$$

$$= \begin{cases} 1, if\ \beta_1 \geq \beta_2 \\ 0, if\ \alpha_1 \geq \gamma_2 \\ \dfrac{(\alpha_1 - \gamma_2)}{(\beta_2 - \gamma_2) - (\beta_1 - \gamma_1)}, \text{ if otherwise} \end{cases} \tag{5.8}$$

Phase 3 Priority weight estimation

$M_i(i = 1, 2, 2, \ldots, k)$ can be categorized as:

$$V(M > M_1, M_2, \ldots, M_k) = V(M \geq M_1) \text{and} (M \geq M_2) \text{and} \ldots (M \geq M_k) \tag{5.9}$$

$$V(M > M_1, M_2, \ldots, M_k) = \min V(M \geq M_1), if\ m(P_i) = \min V(S_i \geq S_k) \\ \text{for } k = 1, 2, \ldots, n; k \neq i \tag{5.10}$$

The weight vector can be specified as:

$$W_p = (m(P_1), m(P_2), \ldots, m(P_n))^T \tag{5.11}$$

where $P_i(i = 1, 2, \ldots n)$ are n elements.

Phase 4 Normalizing weight vector

$$W = (W(P_1), W(P_2), \ldots, W(P_n))^T \tag{5.12}$$

where W is a non-fuzzy number.

To comprehend the impacts of flood hazards on various land uses, a LULC map has been developed using secondary data (Karra, Kontgis et al. "Global land utilization/land coverage utilizing Sentinel-2 and advanced learning techniques." IGARSS 2021–2021 IEEE Worldwide Geoscience and Remote Sensing Symposium. IEEE, 2021).

- Geomorphology:
  A diverse range of topographical units can be seen in the study area (Figure 5.3). Mountains and valleys exhibiting moderate to high degrees of dissection dominate the upper catchment region of the rivers. A multitude of alluvial fans has formed from the mountain front extending southward as a result of substantial sedimentation. Numerous rivers in the fore deep region have eroded the alluvial fans, making them highly fragmented. Three of the rivers being studied have formed several mid-channel bars (Channel Islands) due to sediment accumulation in the riverbed. The areas around rivers have been classified as dynamic floodplains. Nonetheless, areas outside the active floodplain have been classified as older floodplains (Figure 5.3).
- Elevation:
  The slope differentiation of the study area is illustrated in the accompanying figure, which classifies the terrain into five distinct categories: Class 1 (0.0–1.44), Class 2 (1.45–1.68), Class 3 (1.69–3.13), Class 4 (3.14–11.9), and Class 5 (12–64.6). It is noteworthy that the slope profile does not exhibit a consistent decline from the hilly regions to the southern floodplains. The dissected alluvial fan reveals slope gradients that range from high to moderate (1.69–11.9), whereas the upper river catchments are characterized by steep slope gradients (12–64.6 degrees). In contrast, the lower floodplains and areas adjacent to the rivers feature very gentle slopes, which hinder the efficient flow of rainwater and contribute to flooding (Biswas & Raha, 2023) (Figure 5.4a).

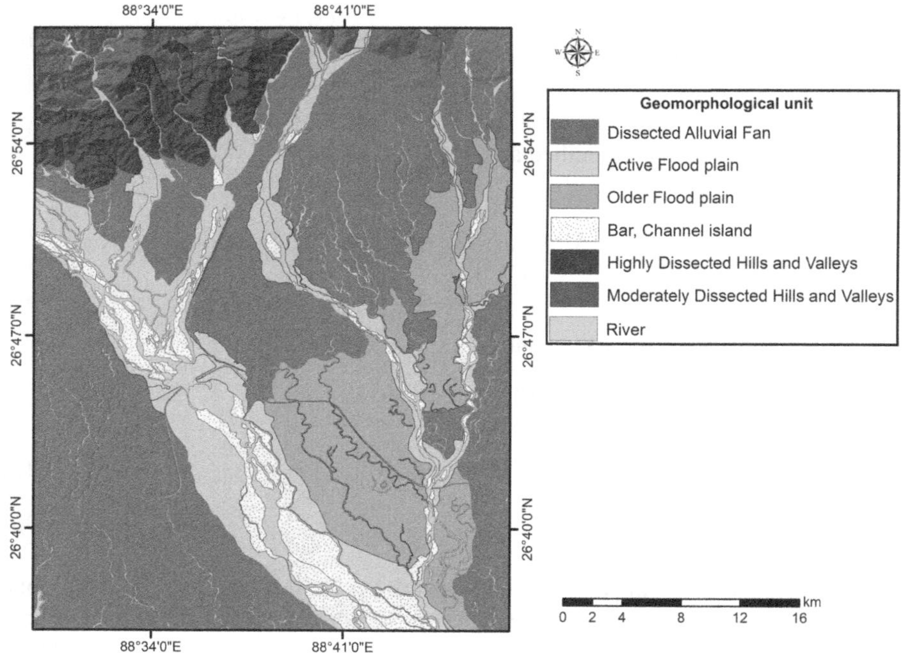

FIGURE 5.3 Detailed geomorphological map of the study area, showing classifications such as dissected alluvial fan, active floodplain, older floodplain, bar-channel island, highly dissected hills and valleys, and moderately dissected hills and valleys (Biswas & Raha, 2023).

- Slope:
  The slope differentiation of the study area is represented in the figure, categorizing it into five distinct classes: Class 1 (0.0–1.44), Class 2 (1.45–1.68), Class 3 (1.69–3.13), Class 4 (3.14–11.9), and Class 5 (12–64.6). It is important to note that the slope profile does not continuously decrease from the hilly region to the southern floodplains. The dissected alluvial fan displays gradients ranging from high to moderate (1.69–11.9), while the upper river catchments show a steep slope gradient (12–64.6 degrees) (Biswas & Raha, 2023). The lower floodplains and areas adjacent to the rivers are characterized by very gentle slopes, which impede the smooth flow of rainwater and contribute to flooding (Figure 5.4b).
- Rainfall:
  The slope differentiation of the study area is illustrated in the accompanying figure, which classifies the terrain into five distinct categories: Class 1 (0.0–1.44), Class 2 (1.45–1.68), Class 3 (1.69–3.13), Class 4 (3.14–11.9), and Class 5 (12–64.6). It is noteworthy that the slope profile does not exhibit a consistent decline from the hilly regions to the southern floodplains. The dissected alluvial fan reveals slope gradients that range from high to moderate (1.69–11.9), whereas the upper river catchments are characterized by steep slope gradients (12–64.6 degrees). In contrast, the lower floodplains

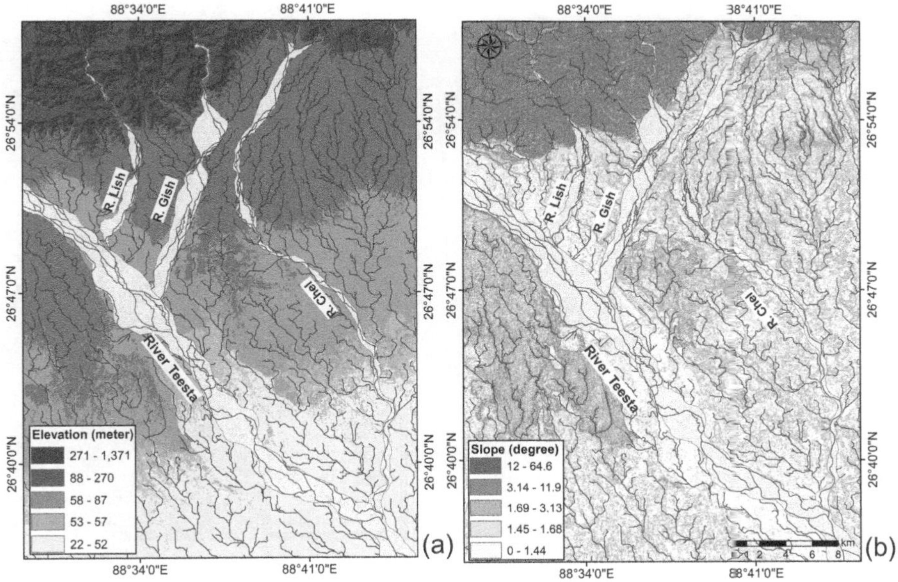

FIGURE 5.4 (a) The major rivers and their tributaries are superimposed on an elevation map of the study area that shows five different classes. (b) The region's general gradient, from the mountainous terrain in the north to the lower floodplains in the south, is depicted on this slope map with five distinct classes (Biswas & Raha, 2023).

and areas adjacent to the rivers feature very gentle slopes, which hinder the efficient flow of rainwater and contribute to flooding (Figure 5.5a).

- Distance from rivers:
  The Himalayan Frontal Province is characterized by a multitude of rivers and their tributaries, making the distance from these water bodies a crucial element in assessing the potential intensity of flood risks. The classification system includes Class 1 (0–1.38 km), Class 2 (1.39–2.11 km), Class 3 (2.12–3.49 km), Class 4 (3.5–6.08 km), and Class 5, which pertains to distances ranging from 6.09 to 10.94 km (Figure 5.5b).
- Topographic Wetness Index (TWI):
  The figure elucidates the variations in topographic wetness across the examined area. This variation is categorized into five distinct classes: Class 1 (13.97–22.41), Class 2 (10.4–13.96), Class 3 (8.16–10.39), Class 4 (6.3–8.15), and Class 5 (2.64–6.29). It is noted that regions in proximity to rivers exhibit elevated TWI values (13.97–22.41), while values decrease progressively as one moves away from these water bodies (Biswas & Raha, 2023). The mountainous regions display particularly low TWI values, attributed to their steep slopes and significant elevation (2.64–6.29) (Figure 5.6a).
- Normalized Difference Vegetation Index (NDVI):
  The region under investigation is distinguished by its extensive deciduous forest coverage. The classification includes Class 1 (−0.15 to 0.1), Class 2 (0.11 to 0.17), Class 3 (0.18 to 0.19), Class 4 (0.2 to 0.26), and Class 5 (0.27

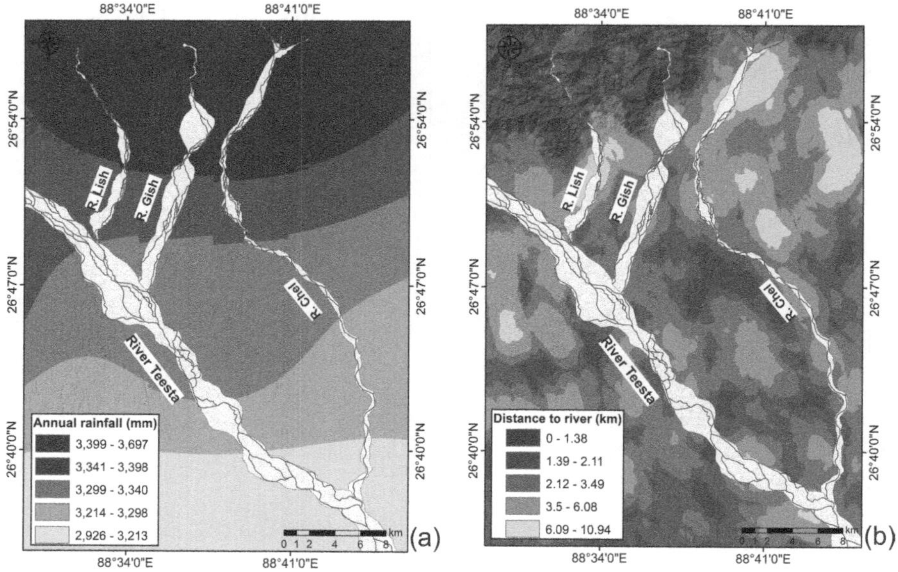

FIGURE 5.5 (a) Regional variations in annual rainfall (mm) across the study area are depicted on the isohyet map (Source: https://sites.uea.ac.uk/cru/data). (b) A five-class map of the distance from rivers (in kilometers) was created using the Arc toolbox on the ArcGIS 10.4 platform (Biswas & Raha, 2023).

to 0.51), as illustrated in the accompanying figure. Within this area, several forest patches demonstrate elevated NDVI values, notably Lish Forest, Churanti Forest, Lethi Forest, Mal Forest, Apalchand Forest, and Baikunthapur Forest, which fall within the range of 0.27–0.51. Conversely, lower NDVI values are observed in the river valleys and their adjacent regions. In the southeastern part of the area, near the confluence of the Chel River and its downstream section, the density of the forest cover diminishes significantly (Biswas & Raha, 2023) (Figure 5.6b).

- Bare Soil Index (BSI):
  The figure illustrates the degree of barrenness of the landscape. The Bare Soil Index is divided into five categories: Class 1 (0.02–0.29), Class 2 (−0.03–0.01), Class 3 (−0.08–0.04), Class 4 (−0.12–0.09), and Class 5 (−0.28–0.13). Notably, forested areas exhibit the lowest BSI values, ranging from −0.28 to −0.13. Additionally, it is important to note that the region surrounding the confluence of the three rivers—Lish, Gish, and Chel—displays higher BSI values, specifically between 0.02 and 0.29 (Biswas & Raha, 2023) (Figure 5.6c).

5.2.1 Flood Susceptibility Mapping

A Flood Susceptibility Map has been developed based on eight key parameters. The accompanying figure illustrates five distinct classes of susceptibility, indicating

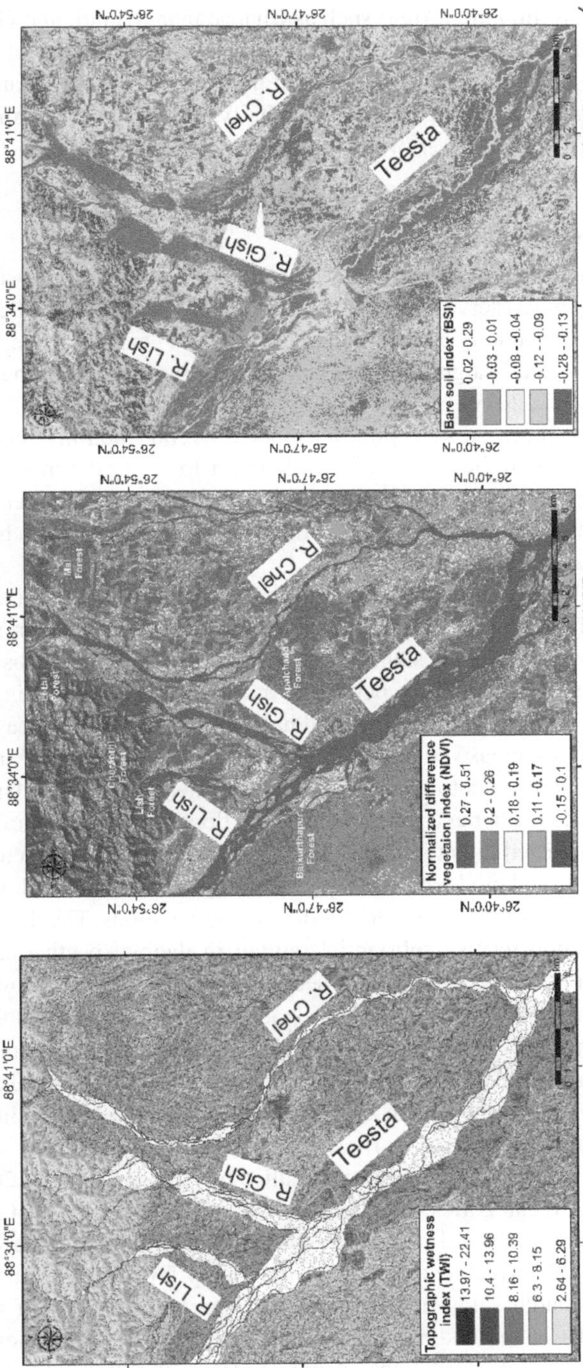

FIGURE 5.6 (a) Map prepared Topographic Wetness Index (TWI) for distance from river. (b) Normalized difference vegetation index (NDVI). (c) Bear soil index (BSI) map.

varying levels of flood risk. It is important to note that the southeastern region of the study area, particularly near the downstream section of the Chel River, is identified as being at a higher risk of flooding. Areas such as Chapadanga, Kranti, and Gochimari are among those vulnerable to flood events. The downstream portion of the Gish River shows an extremely high sensitivity to flooding, while the upstream and middle sections are categorized as having high vulnerability. In contrast, the middle and downstream areas of the Lish River present a moderate to low flood risk (Biswas & Raha, 2023).

5.2.1.1 Validation

It is essential to validate all models and corroborate the findings of the study. The area under the curve (AUC) serves as the most effective instrument for validating the Multi-Criteria Decision-Making Model in the field of spatial earth science, as it provides a comprehensive and straightforward assessment of the model's outputs (Malik et al., 2020; Darabi et al., 2019).

To validate the current flood susceptibility model, several flood inventory points have been incorporated into the AUC analysis. In order to verify the model through AUC, previous research studies and official documents from the West Bengal government's Irrigation and Waterways Department (www.wbiwb.gov.in) have been reviewed.

This study highlights the varying susceptibility of the three rivers examined in relation to flooding. The flood hazard rating map indicates that the Chel River, across its different sections, demonstrates a significant likelihood of flooding. Observations from field visits have revealed that the depth and accumulation of debris in the Chel River are notably high, contributing to elevated hazard ratings. The southeastern part of the study area shows increased vulnerability due to its lower elevation and gentle slope, which allows floodwaters to remain stagnant for extended periods. The lack of vegetation and bare land further exacerbates the risk of flooding. Geomorphologically, the transition from dissected hills to the distal areas of alluvial fans, characterized by a high altitudinal profile and moderate to steep slopes, facilitates the rapid flow of river water, resulting in a lower susceptibility to flooding in these regions. The three foothill rivers under investigation receive substantial rainfall in their respective catchment areas. The combination of increased hydro-sediment discharge in the downstream sections and a reduced capacity to transport this load leads to overflow and subsequent inundation of adjacent areas. Additionally, the construction of bridges has resulted in significant sediment accumulation in Lish and Gish Rivers (Biswas & Raha, 2023), which has diminished the riverbed depth and, consequently, the rivers' carrying capacity. This phenomenon is identified as a primary contributor to flooding.

The accuracy of the AUC curve has been classified by Arabameri et al. (2020) and Rimba et al. (2017) into categories: excellent (0.9–1.0), very good (0.80–0.90), fair (0.70–0.80), poor (0.60–0.70), and weak (0.50–0.60). The AUC calculation yielded a result of 0.832, or 83.2%, which can be considered highly favorable for the model's performance. This research study highlights the varying susceptibility of the three rivers examined in relation to flooding. The flood hazard rating map indicates that the Chel River, across its different sections, demonstrates a significant likelihood of flooding (Biswas & Raha, 2023). Observations from field visits have revealed that the depth and accumulation of debris in the Chel River are notably high, contributing to elevated hazard ratings. The southeastern part of the study area shows increased vulnerability due to its lower elevation and gentle slope, which allows floodwaters to remain stagnant for extended periods. The lack of vegetation and bare land further exacerbates the risk of flooding. Geomorphologically, the transition from dissected hills to the distal areas of alluvial fans, characterized by a high altitudinal profile and moderate to steep slopes, facilitates the rapid flow of river water, resulting in a lower susceptibility to flooding in these regions. The three foothill rivers under investigation receive substantial rainfall in their respective catchment areas (Biswas & Raha, 2023).

The combination of increased hydro-sediment discharge in the downstream sections and a reduced capacity to transport this load leads to overflow and subsequent inundation of adjacent areas. Additionally, the construction of bridges has resulted in significant sediment accumulation in the Lish and Gish Rivers (Biswas & Raha, 2023), which has diminished the riverbed depth and, consequently, the rivers' carrying capacity. This phenomenon is identified as a primary contributor to flooding (Figure 5.7).

- Based on the color of the topsoil, the oxidation index, and the color of the sediments in the weathering zone, the Himalayan foreland portion has been divided into 11 geological units (Figure 5.8) (Gupta, 2008; Biswas & Banerjee, 2018). The Chalsa Formation, Baikunthapur Formation, and Shuangaon Formation cover a sizable portion of the studied area. Small areas of the Himalayan Mountain Frontal zone contain the remaining geologic units. The Chalsa Formation, which is primarily composed of brown soil, forms the lobes of the alluvial fans as it extends southward from the piedmont scarp in the late to middle Holocene (Gupta, 2008). The black soil cover in the floodplains is the result of the Chalsa Formation overlapping the Baikunthapur Formation (Late to Middle Holocene), south of the alluvial fans. The Teesta River's eastern edge includes the Neotectonic activity along the mountain front, where MBT and MFT are linked to significant topogtaphic anomalies in terms of breaks. Lateral, oblique, or tear faults are the characteristics of the Himalayan belt's salient-recess transition, which is a sinuous arc (Marshak, 2004). However, to the east of the Dharan salient-Gorubathan recess lies the Jaldhaka recess, where Quaternary sediments have been deposited as sand, cobbles, and pebbles and the Siwalik formation has not yet been exposed (Nakata, 1989; Kar et al., 2014). Many E-W and N-S trend fault lines, such as MBT and MFT, split these deposited post-Pleistocene quaternary fan surfaces. The northward dip of the approximately 90-meter-thick Matiali scarp is dominated by MBT (Nakata, 1989; Biswas & Raha, 2023).

The region is seismically unstable because of the Himalayan orogeny, which is caused by the Indian plate moving closer to the Eurasian plate. It results from the region's regular occurrence of earthquakes of various magnitudes. It is observed that low-magnitude earthquakes with magnitudes between 3.0 and 5.0 mb often upset the area in question. The risk of landslides in this region has increased due to frequent

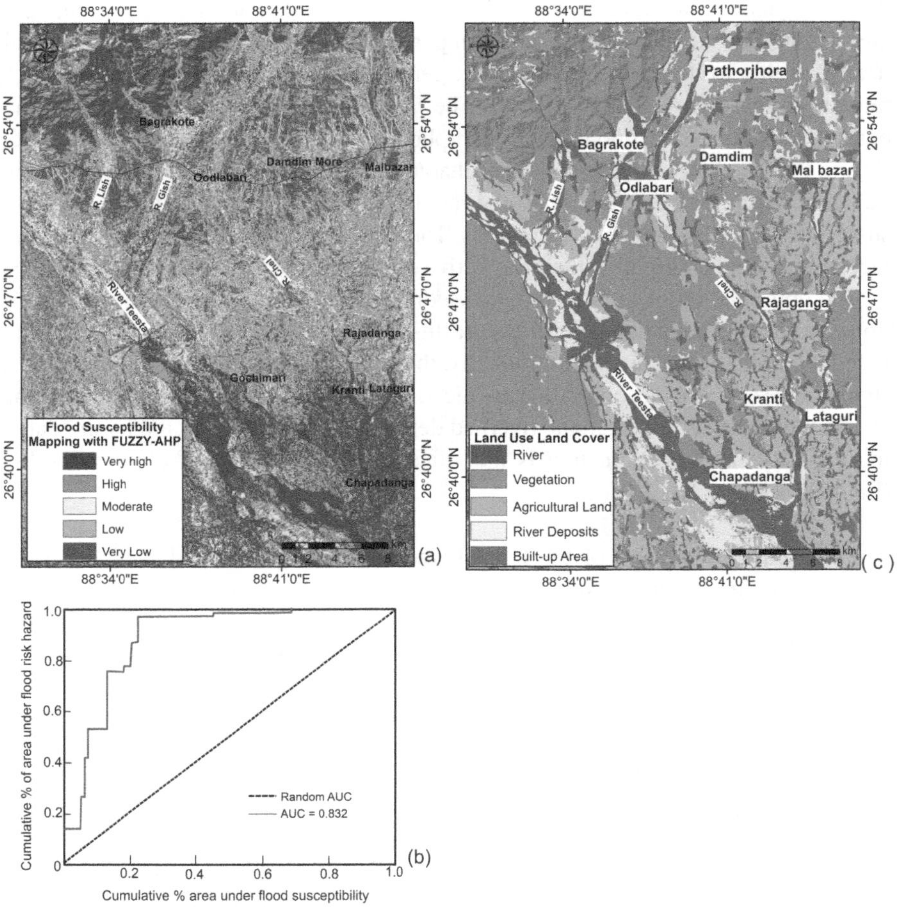

FIGURE 5.7 (a) Flood susceptibility mapping with FUZZY-AHP. (b) Land use/land cover of the concerned area. (c) Computation of the AUC curve for validation (Biswas & Raha, 2023).

earthquakes. To determine the scope of landslip dangers in the catchment areas of the Rivers Lish, Gish, and Chel, the landslip susceptibility zonation is carried out (Biswas & Raha, 2023).

5.2.1.2 Earthquake ("*b*" Value Analysis)

Landslide susceptibility is linked to the active tectonics of a region; the analysis of earthquake events, incorporating stress and strain comparisons, has been conducted using the "b" value in "z" map within the MATLAB 18 (2020) version. This analysis elucidates the correlation between magnitude and frequency. The frequency-magnitude distribution (FMD), commonly referred to as the Gutenberg-Richter relationship (Gutenberg & Richter, 1944), provides an explanation of the relationship between the number of earthquakes (N) and their magnitude (M) within a defined

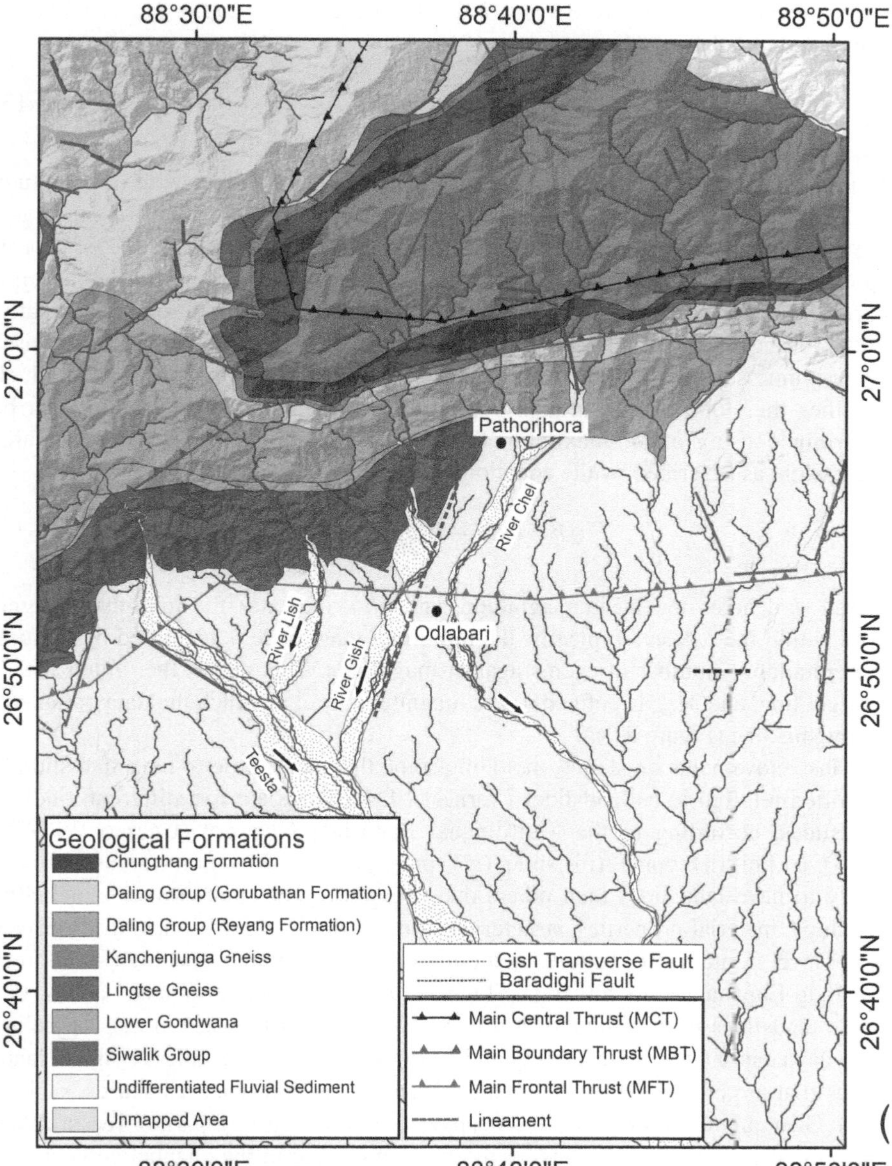

FIGURE 5.8 Detailed geological formations with major thrusts and faults (Biswas & Raha, 2023).

geographical area (Gupta & Biswas, 2023; Biswas & Raha, 2023). The relationship is expressed as follows:

$$\log N = a - bM \tag{5.13}$$

In this equation, N represents the total number of earthquakes with a magnitude of M or greater. The constants "a" and "b" are parameters where "a" indicates the overall seismic activity in a specific area; thus, a higher value of "a" corresponds to increased seismic activity, and vice versa. The parameter "b" is influenced by the stress regime and the tectonic and structural conditions of the study area. Consequently, the number of seismic events (N) is represented on the abscissa (logarithmic scale), while their magnitude is represented by the "b"-value, which signifies the slope of the frequency-magnitude distribution. The b-value can be determined using either maximum likelihood estimation or linear least squares regression, as described by the equation:

$$b = 1 / \bar{M} - M_{min} \log e \tag{5.14}$$

Here, $\bar{M}$ denotes the mean magnitude, and M_{min} indicates the minimum magnitude within the dataset. Typically, the minimum magnitude is identified by plotting the cumulative number of events against magnitude. This plot is then fitted with a straight line, and M_{min} is defined as the magnitude level at which the data points fall below this line (Figure 5.9).

Mass movements on slopes, including some that entail little to no actual sliding, are often referred to as landslides (Varnes, 1984). There are five different kinds of landslides, according to the Multilingual Landslide Glossary (Unesco WP/WLI 1993): (i) fall, (ii) topple, (iii) slide, (iv) spread, and (v) flow. Some areas are more likely to have landslides than others due to a variety of circumstances, including lithology, material properties, structural setting, bedding inclination, and subsurface water level. A number of factors, including lithology, rainfall, slope, high strain from the Indo-Eurasian Plate collision (Singh et al., 2022), and earthquake events, can cause landslides in the Himalayan regions. Frequent earthquakes in the Darjeeling Himalaya cause ongoing instability and disruption, which frequently results in landslides that are connected to the streams. Many of the left-hand tributaries, such as Lish, Gish, and Chel, drain the foredeep portion of the eastern part of Teesta River. Over the past century, there has been a notable change in the channel geometry of these tributaries (Biswas & Raha, 2023). Previous studies have demonstrated that there are numerous landslides within the catchment of these three rivers (Mandal & Mandal, 2017, 2019). Earlier research (Lee & Pradhan, 2006) has been examined to determine the parameters needed to create a landslide-susceptibility map that illustrates the likely zone of operation. Earthquake, elevation (m), geology, NDVI, lineaments, distance from the river, slope, and road distance are among the parameters taken into consideration (Biswas & Raha, 2023).

Figure 5.10 shows the five landslide susceptibility zones classified as very high, high, moderate, low, and very low. The catchment areas for the three rivers in

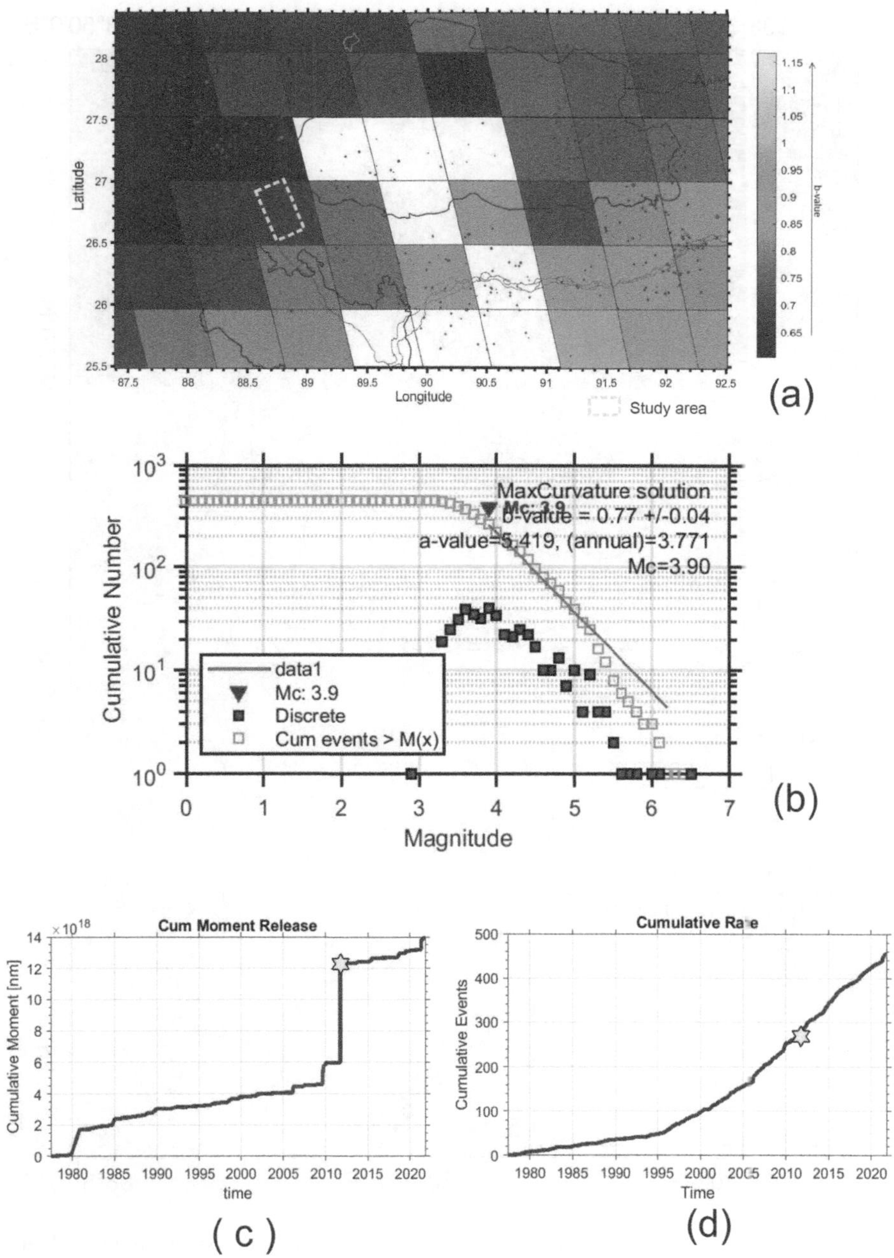

FIGURE 5.9 (a) Depicting the b-value's spatial variation throughout the study area and its surrounding zones. (b) Plotting of frequency magnitude distribution (Gutenberg-Richter Relation). (c) Temporal variation in cumulative moment (in nm.) and its relation with earthquakes of greater magnitudes. (d) Temporal variation in cumulative number of seismic events (Biswas et al., 2024). Data available in https://earthquake.usgs.gov/earthquakes/search/

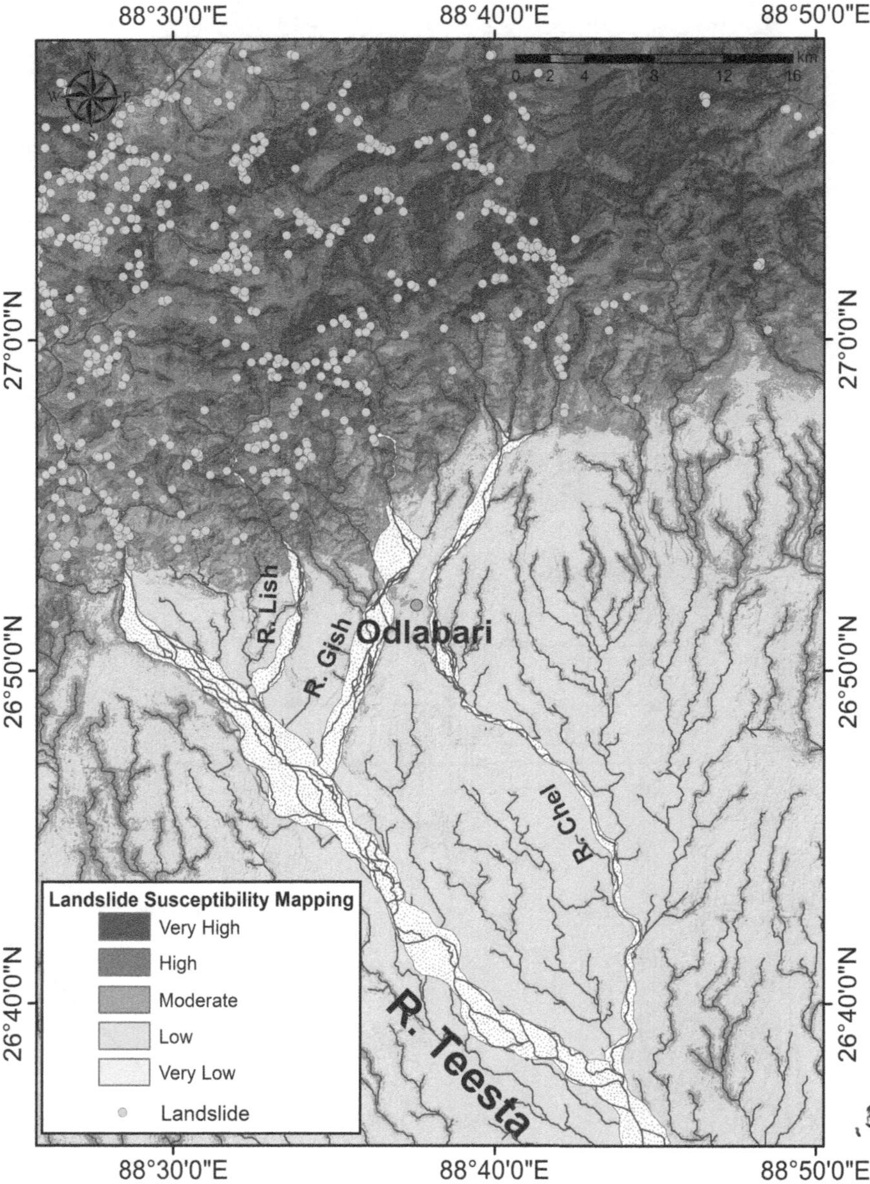

FIGURE 5.10 Landslide susceptibility map of the area with landslide events with five distinct classifications as very high, high, moderate, low, and very low (Biswas et al., 2024). https://bhukosh.gsi.gov.in/Bhukosh/MapViewer.aspx\

question—the Gish, Chel, and Lish—fall within the high and very high vulnerable zones, which should be noted. Because of the steep gradient and higher elevation, landslides are more likely to occur. Furthermore, it has been determined that the catchment area contains specific lineaments that have made the area vulnerable to erosion.

Highway construction along hillslopes, especially in the Gish and Chel River basin, is one man-made factor that has increased the region's susceptibility to landslides.

5.3 HYDROLOGY, ECOLOGY, AND ENVIRONMENTAL LINKAGES IN RISK

Comparatively diverse climatic regimes with the comparable-scale riverscape as the case study. Biodiversity has been eroding recently as a result of human actions in the ecosystem. Natural landscape fragmentation and habitat loss have been blamed for the decline in biodiversity. Among the ecological processes that are most impacted by unplanned developmental activities are succession, normal regeneration, nutrient, and hydrological cycles. Changes in land use, landscape dynamics, and in order to comprehend the temporal changes brought about by developmental activities in the watershed, landscape characterization has been examined at the patch, river basin, and lake catchment levels in an urban setting. The indices of shape, richness, and diversity, according to landscape analysis, offered an extra assessment of the spatial distribution of land cover within the intricate mountain landscape. An outline of the extent of fragmentation and disturbance propagation from non-biotic sources has been provided by the landscape analysis. The meta-population structure, ecotones, corridors, hydrological connectivity, and loss of connectivity have all been attributed to fragmentation. Since it measures the relationship between structural and functional components at various scales, an understanding of landscape dynamics has become crucial to ecosystem management (Ramachandra, 2010).

Globally speaking, the hydrological cycle is made up of three main systems: the ocean, which is the primary source of water, the atmosphere, which transports water, and the land, which uses it. Although there is no water gain or loss during this cycle, variations at the sources or in the delivering system may cause the user's water supply to fluctuate. On a global scale, the water movement cycle is a closed system, but on a local scale, it is an open system. Local or river basin-based water movement takes into account the aquifer, subsurface, and land systems: ecology, biodiversity, and hydrology (Ramachandra, 2010).

To examine the probable hazard zone, we have considered River Chel as an example where four parameters have been considered for inter-item co-relation matrix as depth (m), velocity (m/sec), manning_n, and debrise_p. Flooding is likely to occur every two to five years due to the region's heavy rainfall, which amounts to approximately 1582 mm during rainy seasons (Tea Estate and Meteorological Department, 2010). The individual parameters are then ranked in SPSS. The velocity is ranked with assigning 1 to the highest velocity recorded, and the same is done for depth and debris factor while manning's n is ranked inversely with 1 assigned to the lowest manning's n value. Then the mean of the ranks provides the final ranking. And the lowest rank is identified as the highest risk zone within the channels (Tables 5.1 and 5.2).

The probable risk zone within the channel is identified with the help of four parameters based on the field survey being carried out from Pathorjhora to Damdim:

TABLE 5.2
Statistical Calculations of Inter-Item Correlation Matrix of Four Considered Indicators

	Item Statistics		
	Mean	**Standard Deviation**	**N**
Depth	.21156	.316594	218
Manning_n	.04738	.040872	218
Velocity	.62779	1.210065	218
Debris_p	.26651	.362534	218

- Velocity at different locations along the channel with the help of the current meter.
- Depth of water at the same location of measurement of velocity with the help of the staff.
- Manning's n through observation of the channel bed condition, riparian environment, flow condition and anthropogenic involvement, and marking based on the reading done through various literatures.
- Debris factor is marked in terms of probability of the debris to affect the bridges, human, ecology based on the sizes (Figures 5.11 and 5.12).

The observed velocity, depth, manning's n, and debris factors are recorded with the location where they are being observed and then inputted into excel sheet. To see if there exists any correlation among those parameters, Inter-Item Correlation is being carried out in SPSS and are further analyzed to see how much significant the correlation is at 0.05 significance level. The correlation reveals that velocity-depth is highly positively correlated, debris movement is more when the velocity increases, thus, they too are positively correlated, the manning's n, i.e., the roughness increase results in retardation of the flow till the water level rises to a certain extent when its effects decrease and thus show negative correlation.

According to the risk mapping, the lowest rank is identified as the highest risk zone within the channels. The four classes are then identified:

- Very High Risk Zone with value less than 30
- High Risk Zone with value 30 to 60
- Low Risk Zone with value 60 to 90
- Least Risk Zone with greater than 90

Based on the number of points taken into consideration and changes in the value of the parameters the ranking would also differ. The identification of the risk zone indicates that immediately downstream of the bridges, the combination of high velocity and debris factors, along with minimal channel vegetation, offers little to no resistance.

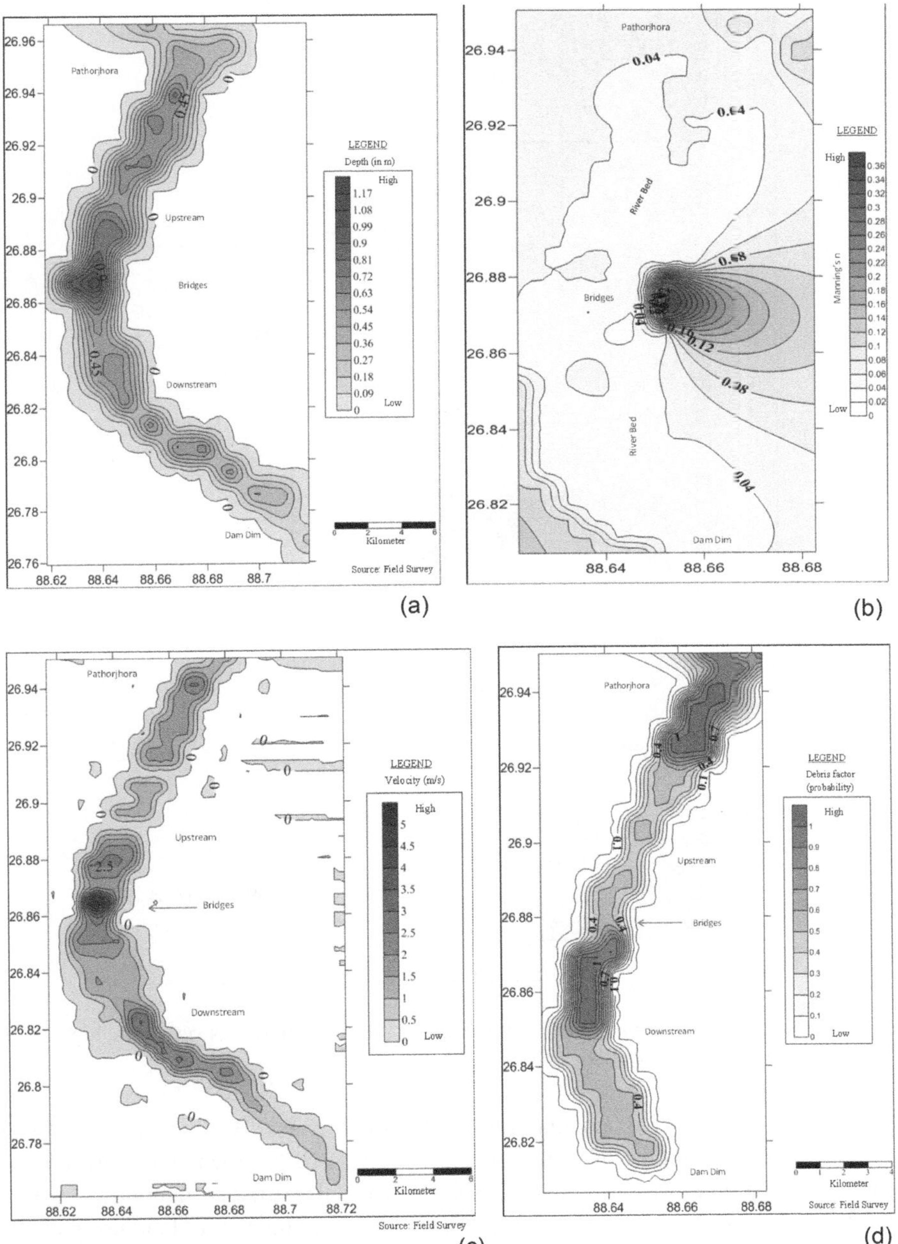

FIGURE 5.11 (a) Spatial distribution of depth along the channel near upstream and downstream of the bridges. (b) Magnitude distribution of manning_n along the channel from Pathorjhora to Damdim. (c) Spatial distribution of velocity along the channel near upstream and downstream of the bridges from Pathorjhora to Damdim. (d) Distribution of debris factor probability along the channel (Banerjee & Biswas, 2021).

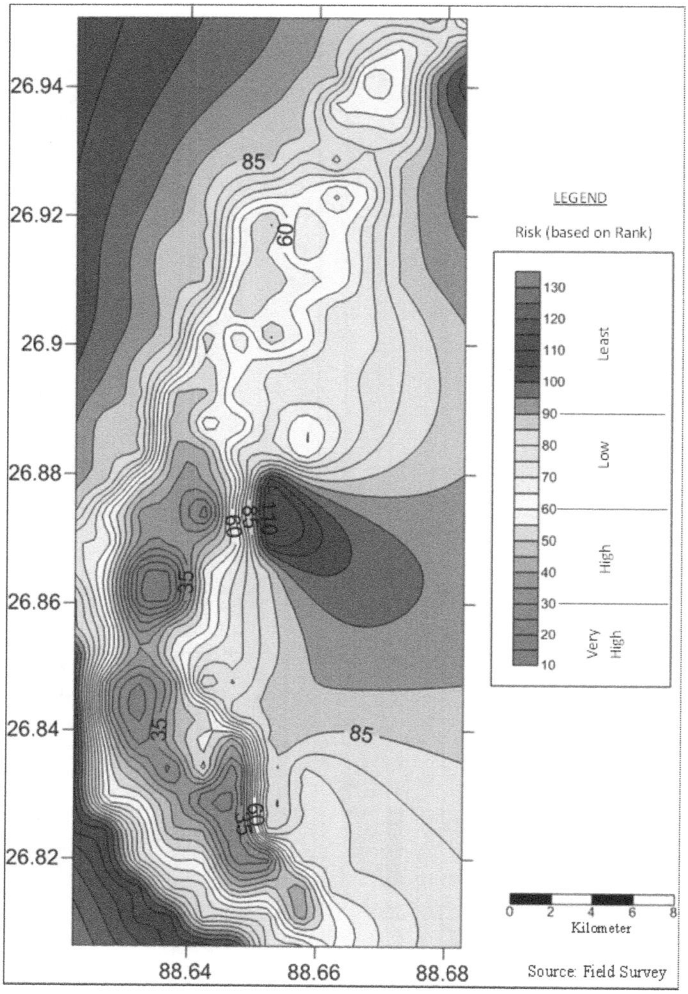

FIGURE 5.12 Risk mapping based on rank under four magnitudes as very high, high, moderate, and least (Banerjee & Biswas, 2021).

Consequently, these areas are classified as high-risk zones. However, as one moves away from the central region of the stream toward the banks, the risk diminishes due to the retardation effect of the banks, which reduces the velocity. In contrast, within the floodplain, the presence of vegetation, settlements, and overall land use patterns contributes to a reduction in flow velocity and depth, thereby categorizing this area as a lower risk zone.

REFERENCES

Adhikari, P., Y. Hong, K. Douglas, D. Kirschbaum, J. Gourley, R. Adler, and G.R. Brakenridge. 2010. A digitized global flood inventory (1998–2008): Compilation and preliminary results. *Natural Hazards*, 55(2): 405–422. https://doi.org/10.1007/s11069-010-9537-2

Arabameri, A., W. Chen, M. Loche, X. Zhao, Y. Li, L. Lombardo, … and D.T. Bui. 2020. Comparison of machine learning models for gully erosion susceptibility mapping. *Geoscience Frontiers*, 11(5): 1609–1620.

Banerjee, P., and M. Biswas. 2021. Application of habitat modification score and fluvial functioning index in discussion of eco-hydrological behavior and flood risk zonation of Himalayan foothill rivers, West Bengal, India. *Acta Geophysica*, 69: 877–893. https://doi.org/10.1007/s11600-021-00570-0

Biswas, M., and P. Banerjee. 2018. Bridge construction and river channel morphology—a comprehensive study of flow behavior and sediment size alteration of the River Chel, India. *Arabian Journal of Geosciences*, 11(16). https://doi.org/10.1007/s12517-018-3789-7

Biswas, M., and A. Raha. 2023. An overview of open channel monsoon hydraulics of Himalayan foothill rivers, West Bengal, India

Biswas, M., A. Raha, and S. Roy. 2024. Impacts of bridge constructions on channel dynamics and sediment aggradation: An integrated discussion, West Bengal, India. *Geomatica*, 76(2): 100021.

Boender, C.G.E., J.G. de Graan, and F.A. Lootsma. 1989. Multi-criteria decision analysis with fuzzy pairwise comparisons. *Fuzzy Sets and Systems*, 29(2): 133–143.

Boutkhoum, O., M. Hanine, A. Tikniouine, and T. Agouti. 2015. Multicriteria decisional approach of the OLAP analysis by fuzzy logic: Green logistics as a case study. *Arabian Journal for Science and Engineering*, 40(8): 2345–2359.

Burton, I., and R.W. Kates. 1964. The perception of natural hazards in resource management. *Natural Resources Journal*, 3: 412–441.

Chakraborty, S., and K. Datta. 2013. Causes and consequences of channel changes—A spatio-temporal analysis using remote sensing and GIS: Jaldhaka-Diana river system (Lower Course), Jalpaiguri (Duars), West Bengal, India. *J. Geogr. Nat. Disasters*, 3: 107. https://doi.org/10.4172/2167-0587.1000107

Chang, D.Y. 1992. Extent analysis and synthetic decision. *Optimization Techniques and Applications*, 1: 352–355.

Chang, D.Y. 1996. Applications of the extent analysis method on fuzzy AHP. *European Journal of Operational Research*, 95(3): 649–655.

Darabi, H., B. Choubin, O. Rahmati, A. TorabiHaghighi, B. Pradhan, and B. Kløve. 2019. Urban flood risk mapping using the GARP and QUEST models: A comparative study of machine learning techniques. *Journal of Hydrology*, 569: 142–154. https://doi.org/10.1016/j.jhydrol.2018.12.002

Das, M., A. Chattopadhyay, and R. Basu. 2017. Spatial flood potential mapping (SFPM) with flood probability and exposure indicators of flood vulnerability: Case study former Jalpaiguri District, West Bengal, India. *Journal of Geography and Natural Disasters*, 7(3). https://doi.org/10.4172/2167-0587.1000210

Foudi, S., N. Osés-Eraso, and I. Tamayo. 2015. Integrated spatial flood risk assessment: The case of Zaragoza. *Land Use Policy*, 42: 278–292. https://doi.org/10.1016/j.landusepol.2014.08.002

Ghosh, T. 2018. Floods and people, colonial north Bengal, 1871–1922. Studies in People's History, 5(1), 32–47.

Gupta, N. 2008. Geoinformatics for inter-basin water transfer assessment: A study in parts of Ganga-Brahmaputra Basin, Eastern India (M.Sc.,). *International Institute Geo-Information Science and Earth Observation*, Enschede, The Netherlands.

Gupta, S., & Biswas, M. 2023. Seismo-tectonic and morphological study of the north-east Himalaya. *Geosciences Journal*, 27(1), 1–21.

Gutenberg, B., & Richter, C. F. (1944). Frequency of earthquakes in California. *Bulletin of the Seismological society of America*, 34(4), 185–188.

Haque, C.E. 1997. *Hazards in a Fickle Environment: Bangladesh.* Kluwer Academic Publishers.

Kar, R., T. Chakraborty, C. Chakraborty, P. Ghosh, A.K. Tyagi, and A.K. Singhvi. 2014. Morphosedimentary characteristics of the Quaternary Matiali fan and associated river terraces, Jalpaiguri, India: Implications for climatic controls. *Geomorphology*, 227: 137–152.

Lee, S., and B. Pradhan. 2006. Probabilistic landslide hazards and risk mapping on Penang Island, Malaysia. *Journal of Earth System Science*, 115(6): 661–672. https://doi.org/10.1007/s12040-006-0004-0

Malik, S., S.C. Pal, I. Chowdhuri, R. Chakrabortty, P. Roy, and B. Das. 2020. Prediction of highly flood-prone areas by GIS-based heuristic and statistical model in a monsoon-dominated region of Bengal Basin. *Remote Sensing Applications: Society and Environment*, 19: 100343. https://doi.org/10.1016/j.rsase.2020.100343

Mandal, B., and S. Mandal. 2017. Landslide susceptibility mapping using modified information value model in the Lish river basin of Darjiling Himalaya. *Spatial Information Research*, 25(2): 205–218. https://doi.org/10.1007/s41324-017-0096-4

Mandal, S., S. Mandal. 2019. *Statistical Approaches for Landslide Susceptibility Assessment and Prediction*. Springer International Publishing. https://doi.org/10.1007/978-3-319-93897-4

Marshak, S. 2004. Arcs, oroclines, salients, and syntaxes—the origin of map-view curvature in fold-thrust belts. *Thrust Tectonics and Petroleum Systems: AAPG Memoir*, 82: 131–156.

Meshram, S. G., Alvandi, E., Singh, V. P., & Meshram, C. 2019. Comparison of AHP and fuzzy AHP models for prioritization of watersheds. *Soft Computing*, 23(24), 13615–13625.

Nakata, T. 1989. Active faults of the Himalaya of India and Nepal. *Geological Society of America Special Papers*, 232: 243–264.

Paul, A., M. Biswas. 2019. Changes in river bed terrain and its impact on flood propagation—A case study of River Jayanti, West Bengal, India. *Geomatics, Natural Hazards and Risk*, 10(1). https://doi.org/10.1080/19475705.2019.1650124

Raha, A., & Biswas, M. 2022a. Quaternary alluvial fan dynamics of the Jaldhaka basin. *Journal of Mountain Science*, 19(8), 2160–2179.

Raha, A., & Biswas, M. 2022b. Himalayan foredeep neotectonics and deformed riverscape landforms: An integrated discussion, West Bengal, India. *In Himalayan neotectonics and channel evolution (pp. 289-313). Cham: Springer International Publishing.*

Raha, A., Gupta, S., & Biswas, M. (2023). Flood risk assessment of Himalayan Foothill rivers: A study of Jaldhaka river, India. In Floods in the Ganga–Brahmaputra–Meghna Delta (pp. 63–90). Cham: Springer International Publishing.

Ramachandra, T.V. 2010. *Ecology, Biodiversity and Hydrology: Linkages.*

Rimba, A. B., Setiawati, M. D., Sambah, A. B., & Miura, F. (2017). Physical flood vulnerability mapping applying geospatial techniques in Okazaki City, Aichi Prefecture, *Japan. Urban Science*, 1(1), 7.

Roy, S. 2011. Flood hazards in Jalpaiguri District. Unpublished Ph.D. Thesis, Department of Applied Geography, University of North Bengal, Siliguri. http://ir.nbu.ac.in/handle/123456789/1335

Roy, S., A. Bose, and I. Chowdhury. 2021. Flood risk assessment using geospatial data and multi-criteria decision approach: A study from a historically active flood-prone region of the Himalayan foothill, India. *Arabian Journal of Geosciences*. https://doi.org/10.1007/s12517-021-07324-8

Singh, S., A. Joshi, A. Sahu, R. Arun Prasath, S. Sharma, and C.S. Dwivedi. 2022. Himalayan landslides—causes and evolution. In S. Kanga et al. (Eds.), *Disaster Management in the Complex Himalayan Terrains*, Geography of the Physical Environment, 33–42. https://doi.org/10.1007/978-3-030-89308-8_3

Smith, K., and R. Ward. 1998. *Mitigating and Managing Flood Losses: Physical Processes and Human Impacts*. John Wiley & Sons.

Tingsanchali, T., & Karim, M. F. 2005. Flood hazard and risk analysis in the southwest region of Bangladesh. *Hydrological Processes: An International Journal*, 19(10), 2055–2069.

UNESCO Working Party for World Landslide Inventory. 1993. *Multilingual Landslide Glossary*. The Canadian Geotechnical Society, BiTech Publishers Ltd, 32.

Van Laarhoven, P.J.M., and W. Pedrycz. 1983. A fuzzy extension of Saaty's priority theory. *Fuzzy Sets and Systems*, 11: 229–241.

Varnes, D.J. 1984. *Landslide Hazard Zonation: A Review of Principle and Practice*. vol. 3. International Association for Engineering Geology and the Environment, 63.

Wisner, B., et al. 2004. *At Risk: Natural Hazards, People's Vulnerability and Disasters*. Routledge.

World disaster report 2022. https://www.ifrc.org/document/world-disasters-report-2022

6 Flood Modeling and Vulnerability Assessment

6.1 APPLICATION OF HEC-RAS FLOOD MODELING

Floods are among the most hazardous hydro-meteorological disasters. Flooding accounts for about one-third of all natural catastrophes globally and has a more detrimental impact on humans than any other geophysical event (Smith & Ward, 1998). Flooding ranks among the most devastating natural disasters, disrupting lives, economies, and infrastructure. Each year, catastrophic floods result in loss of millions of lives and cause significant property damage. The HEC-RAS model provides a technique for examining floods together with other pertinent hydrological factors and their effects. The results from this research may significantly aid in forecasting flood risk as well as establishing future Water Surface Elevation (WSE) and flow rate to strategize the flood defines efforts for a city situated on a riverbank. The situation is particularly perilous for developing and underdeveloped nations. The incidence of flooding is on the rise globally, presenting a major issue. Unplanned urbanization, deforestation, shifting land use and land cover patterns, river engineering, and climate change are some of the major factors contributing to the increased frequency of floods (Jarrett & Tomlinson, 2000; Chang & Franczyk, 2008; Quesada Román, 2021; Raha et al., 2023). The frequency of floods can also be significantly influenced by river basin features in addition to these factors (Bhat et al., 2019; Mohamed & El-Raey, 2019; Ahmed et al., 2021). Using computer-aided approaches to identify flood-prone areas is essential since it allows for an understanding of how the area might be affected by various scenarios (Kaya et al., 2022). West Bengal's Himalayan foothill area is home to several rivers with different stream orders. The principal rivers of North Bengal, the Teesta, Jaldhaka, Torsa, Raidak, and Sankosh, are infamous for their frequent and destructive floods. It is important to note that a number of tributaries join the main rivers, and these tributaries significantly increase the total amount of hydro-sediment discharge to the main rivers.

One of the principal tributaries of the Brahmaputra, which rises in the Sikkim-Himalaya and traverses the vast foreland region of the Jalpaiguri district, is the Teesta River. Lish, Gish, and Chel are Teesta's principal left-hand tributaries, and they are significant for their own hydro-sediment discharge. These three rivers, the Lish, Gish, Chel, and others, are responsible for the flooding that occurs in the Jalpaiguri region of West Bengal, according to the West Bengal Irrigation Department's yearly flood reports. The hydraulics and channel dynamics of these rivers have been seen to alternate when rail and vehicle bridges are built over them (Biswas & Raha, 2022). The influence of bridge building on this river's flood coverage area is the main topic of the current study.

DOI: 10.1201/9781003562979-6

In this work, flood modeling is done using HEC-RAS 6.4. HEC-RAS is an open-source application created by the US Army Corps of Engineers that allows users to generate a variety of river-based models. For the purpose of creating flood models, velocity and discharge data were gathered straight from the field. The 2.5 m × 2.5 m spatial resolution Cartosat1 DEM was used to produce the landscape Flood modeling of the three rivers under consideration has been calculated in four different scenarios—with bridges, without bridges, with relocated bridges, and with reframed bridges—in order to meet the goal of the current study. The figure illustrates the procedures used to run the flood simulations.

The study focused on simulating flood coverage in the area for the Rivers Lish, Gish, and Chel with and without bridges, with relocation of the bridges and with reframed bridges. To enhance the precision and dependability of the findings, field verification was conducted employing a Leica Total Station TS 07, a high-precision surveying instrument. It is noticeable that for River Lish the extent of the flood coverage area is least if the bridge is reframed. For Rivers Gish and Chel, the flood coverage areas are least in the upstream right banks with the relocated bridges. However, for the other sections, the flood coverage areas are the least with reframed bridges. The model has been simulated for the three considered rivers, i.e., Lish, Gish, and Chel considering four circumstances—with the bridges, without bridges, reframed bridges, and relocated bridges. In case of reframed bridge condition, it is assumed that the bridge width will be increased according to the width of the river. For the relocation of the bridge, the shortest width of the riverbed with the bank of hard rock has been considered. The tensile strength of the rocks in the catchment area of the three rivers has been measured to mark the ideal location for the bridge relocation.

- River Chel—For the HEC-RAS modeling along the thalweg line, a number of cross-sections have been considered on the DEM and simulated the model on the basis of velocity as this is the most applicable parameter to assess the scouring in the bridge pier (Figure 6.1). With bridges, the velocity tends to be higher from the foothills to the 500 meter upstream of the road bridge. It is significant to note that the velocity again increases near the road and rail bridge of the River Chel during flooding which propagates bed degradation. The swirling movement of water during this time causes scouring at the bridge piers as well. In the immediate downstream portion of the bridges, the flood water expands beyond both the banks. In this reach, due to the decreasing velocity of the river, huge sediment which comes from the upstream reaches of Chel gets deposited and form a mid-channel bar. In the far downstream reaches, the river velocity again tends to increase which causes bank erosion during the flooding (Figure 6.2a). The velocity profile also varies considerably during flooding without the bridges. The velocity is much lower (0.85 meter/second) near the location of the bridges (Figure 6.2b, black box area 1). The decreasing trend of velocity interrupts the bed degradation and consequently the degree of scouring would be less. In the far downstream section of the rivers, the velocity of the flood water beyond the banks has noticeably reduced. The changes without a bridge are marked in the black box area near the bridge and far downstream. River Chel has

FIGURE 6.1 Considered cross-sections for HEC-RAS flood modeling based on velocity data.

almost a similar flood coverage area with the bridges and with relocated bridges (Figure 6.2c and d). The area decreases slightly without the intervention of bridges (Figure 6.3).

- River Gish—With bridges after emerging from the foothill, the river water expands excessively over low-gradient topography and cross-sections have been considered to simulate the model (Figure 6.4). The velocity of flood

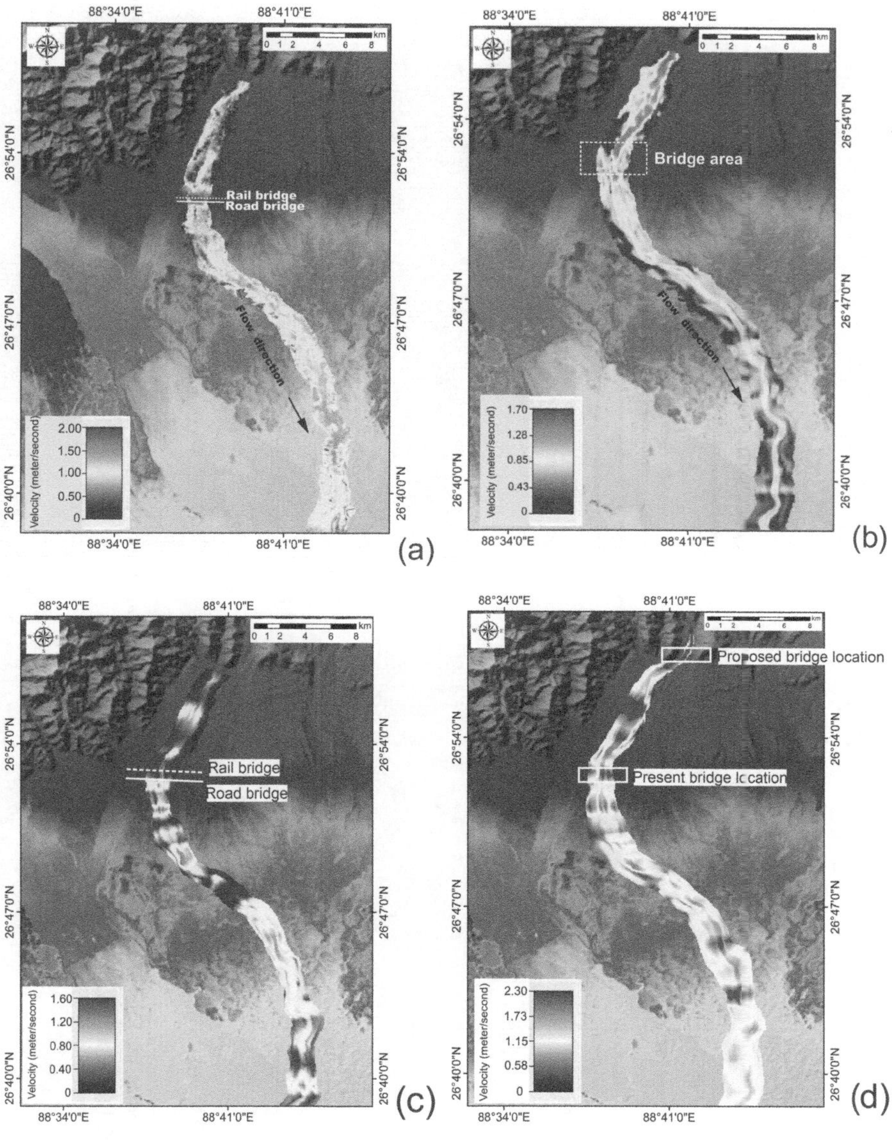

FIGURE 6.2 HEC-RAS flood simulation modeling of River Chel with bridge (a), without bridge (b), relocating bridge (c), and reframing bridge (d).

water seems to be maximum at the break of the slope near the foothill region (1.47 meter/second). It is interesting to note that the velocity decreases near the rail and road bridge, which is responsible for the bed aggradation process. In the downstream section, the flood water creates bank full stage. The velocity seems to increase just after the road ridge up to 800 meters downstream, which eventually decreases in the confluence with Teesta. With the

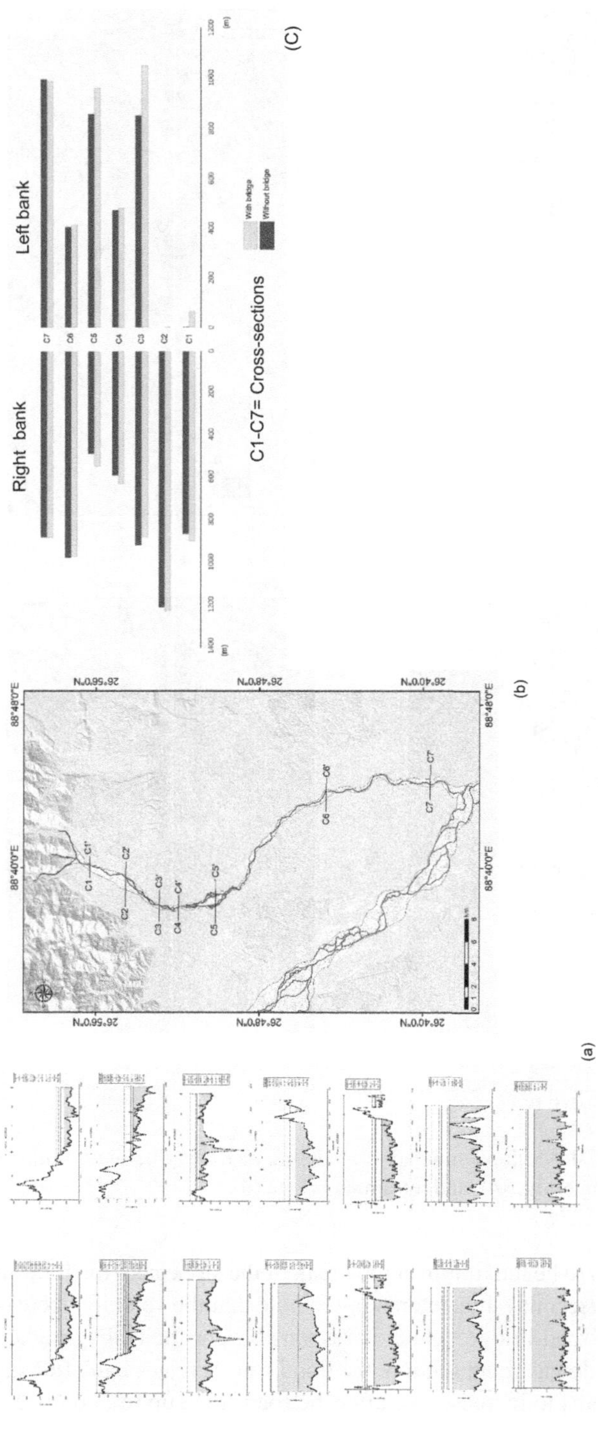

FIGURE 6.3 (a) Details of seven cross-sections of HEC-RAS flood modeling. (b) Location map of the cross-sections. (c) Flood coverage area of right and left bank of the River Chel with bridge and without bridge.

FIGURE 6.4 Considered cross-sections for HEC-RAS flood modeling based on velocity data.

absence of road and rail bridge on the River Gish, the flood water does not extend as much as in case of the intervention of bridges. It can be said that without the presence of bridges, velocity decreases in the upstream and near the bridges and the flood water gets enough space to pass toward the downstream section without any obstacles. Hence, the flood water during heavy precipitation does not expand much in the upstream section of the bridges. In the downstream section, the flood water creates bank full stage. The changes without a bridge are marked in the black box area upstream (Figure 6.4a, black box area 1) of the bridge and downstream of the bridge (Figure 6.5a and b, black box area). For River Gish, the flood coverage areas without the bridges and relocated bridges are almost similar. The velocity also increases slightly without the intervention of bridges. However, the flood model simulated with reframed bridges shows the minimum flood coverage area for River Gish. The velocity also increases slightly (1.52 m/sec) without the intervention of bridges. However, the flood model simulated with reframed bridges show the minimum flood coverage area (4.316 km^2 at the downstream left, 2.520 km^2 at the downstream right, 1.688 km^2 at the upstream left and 0.916 km^2 at the upstream right) for River Gish (Figures 6.6 and 6.7).

- It is noted that, in the case of River Lish, the flood coverage area is minimal (0.728 km^2 at downstream left, 0.677 km^2 at downstream right, 0.426 km^2 at upstream left and 0.615 km^2 at upstream right) if the bridge is reframed according to the width of the river bed. It is also noted that the velocity

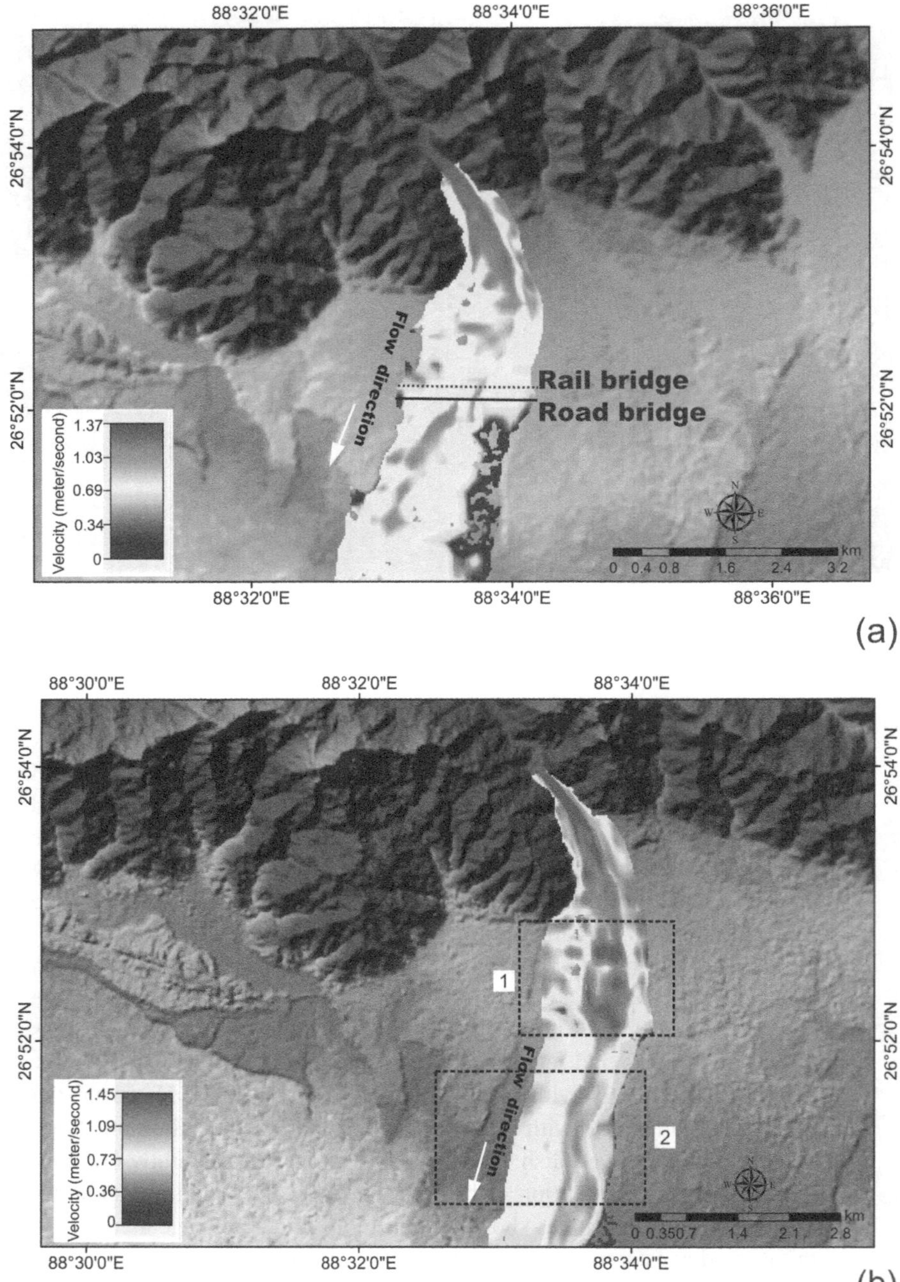

FIGURE 6.5 HEC-RAS flood simulation modeling of River Gish with bridge (a), without bridge (b), relocating bridge (c), and reframing bridge (d).

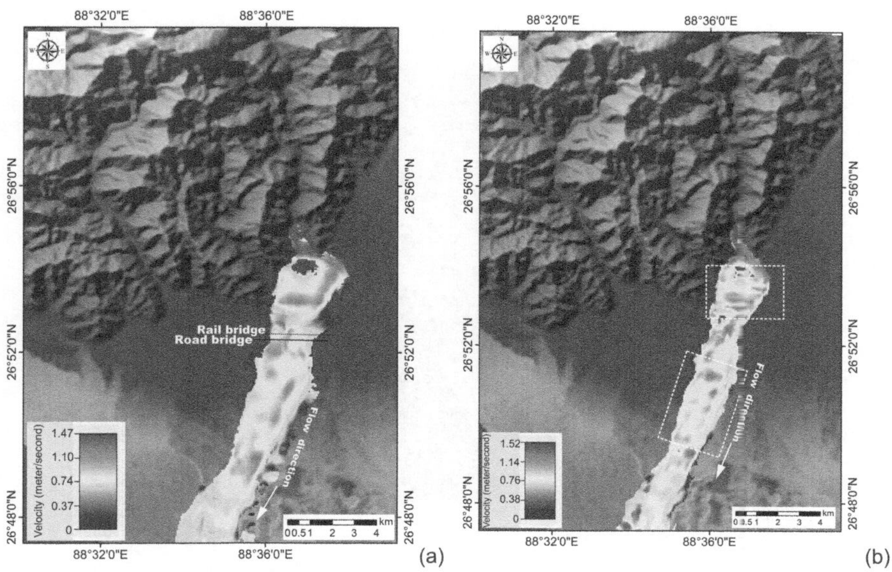

FIGURE 6.6 Location of the seven selected cross-sections of River Gish.

of the river in the present condition of the bridges is much lower (1.37 m/sec) which is causing significant bed aggradation and subsequent rise of the riverbed elevation. If the bridges are reframed, the velocity near the bridge piers will be higher (2.00 m/sec) causing the deposited materials to get washed away farther downstream. The area decreases slightly without the intervention of bridges. However, it is to be noted that the flood coverage area decreases significantly if the bridges are reframed according to the width of the river bed (19.590 km^2 at the downstream left, 18.217 km^2 at the downstream right, 6.217 km^2 at the upstream left, and 7.264 km^2 at the upstream right). It is also noted that the velocity, which is maximum near the bridge piers with the bridges, decreases considerably with reframed bridges. Thus, it is evident that huge bed degradation and scouring near the bridge piers which are the results of higher velocity (2.00 m/sec) will be decreased (1.60 m/sec) with the bridges to be reframed. It can also be said that the problem of scouring will decrease considerably after the reframing of the bridges (Figure 6.8).

It is noticed that the flood water expands beyond the banks just after the foothill region. The velocity also seems to decrease after the foothill. It is important to note that the velocity is very low (0.69 meter/second) near the rail and bridge which propagates the sedimentation process at the river bed, especially near the bridge piers. In the downstream sections of Lish, the flood water coverage is very high. Vast tract of land near the right bank of Lish gets flooded with water. The decreased velocity of river water in the downstream section helps in the sediment aggradation process (Figure 6.9a).

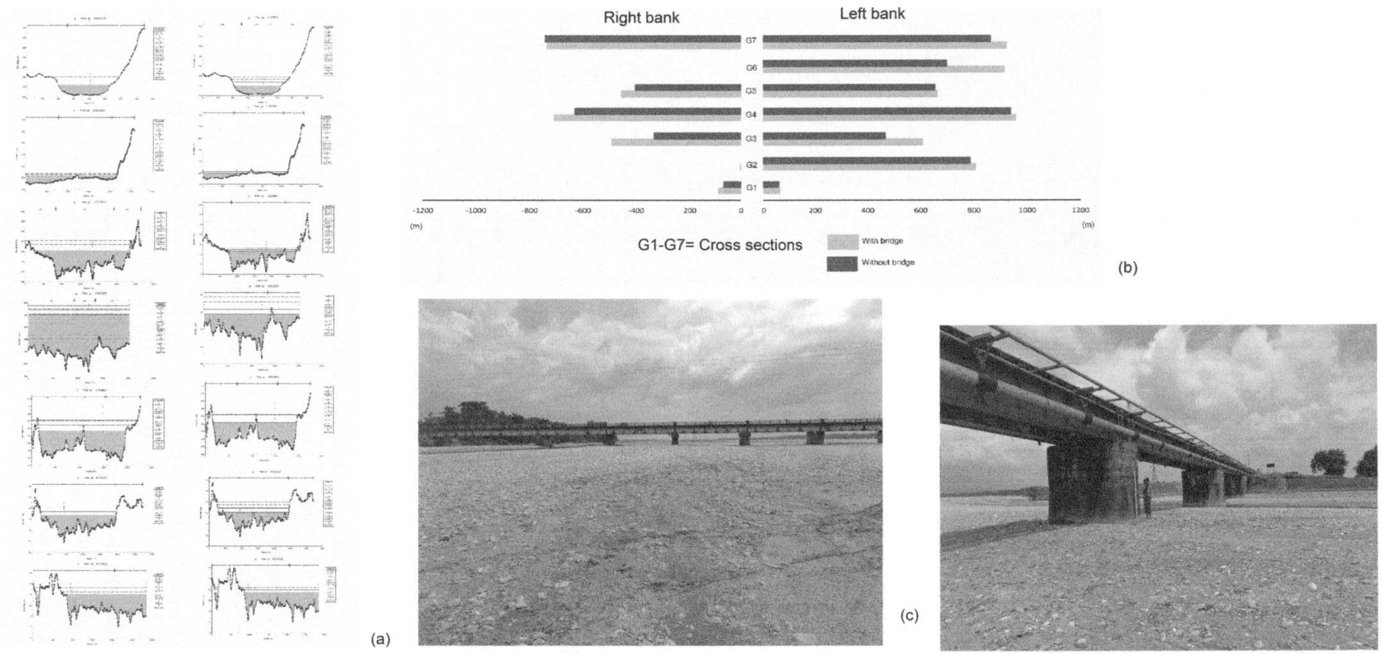

FIGURE 6.7 (a) Details of seven cross-sections of HEC-RAS flood modeling with bridge and without bridge. (b) Flood coverage area of right and left bank of the River Chel with bridge and without bridge. (c) Photograph of the road bridge over River Gish. (d) Photograph of the rail bridge over River Gish.

FIGURE 6.8 Considered cross-sections for HEC-RAS flood modeling based on velocity data.

The scenario is quite different without bridge. The extension of flood water in River Lish without the presence of road and rail bridge is much less without the intervention of bridges on the river in the upstream section of the river. The velocity of the river is higher (1.45 meter/second) which helps to pass the flood water in the downstream section of the river without being overflown beyond the banks in the upstream reaches. It is important to note that the coverage of flood water is significantly less in the far downstream reaches (Figure 6.9a). The black box area near the bridge defines that without a bridge, velocity will increase (Figure 6.9b, black box area 1) and has a probability to wash out the aggraded sediment near the rail bridge (Figure 6.9c and d).

The flood model incorporates the discharge data, velocity, and Manning's n to simulate the flood coverage area. The model has been simulated for the three considered rivers, i.e., Lish, Gish, and Chel considering four circumstances—with the bridges, without bridges, reframed bridges, and relocated bridges. In the case of reframed bridge condition, it is assumed that the bridge width will be increased according to the width of river. For the relocation of the bridge, the shortest width of the river bed with bank of hard rock has been considered. The tensile strength of the rocks in the catchment area of the three rivers has been measured to mark the ideal location for the bridge relocation. It is noted that, in case of River Lish, the flood coverage area is minimal (0.728 km^2 at downstream left, 0.677 km^2 at downstream right, 0.426 km^2 at upstream left, and 0.615 km^2 at upstream right) if the bridge is reframed according to the width of the river bed. It is also noted

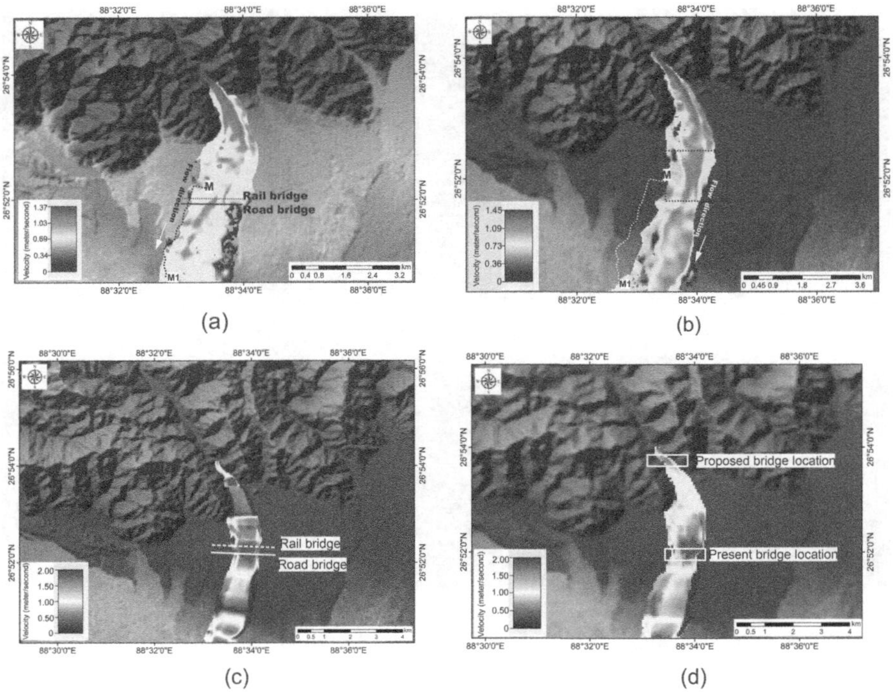

FIGURE 6.9 HEC-RAS flood simulation modeling of River Lish with bridge (a), without bridge (b), relocating bridge (c), and reframing bridge (d).

that the velocity of the river in the present condition of the bridges is much lower (1.37 m/sec) which is causing significant bed aggradation and subsequent rise of the riverbed elevation. If the bridges are reframed, the velocity near the bridge piers will be higher (2.00 m/sec), causing the deposited materials to get washed away farther downstream (Figures 6.10 and 6.11).

However, it is to be noted that the flood coverage area decreases significantly if the bridges are reframed according to the width of the river bed. It is also noted that the velocity which is maximum near the bridge piers with the bridges decreases considerably with reframed bridges. Thus, it is evident that huge bed degradation and scouring near the bridge piers which are the results of higher velocity will be decreased with the bridges to be reframed. It can also be said that the problem of scouring will decrease considerably after the reframing of the bridges.

6.2 SPATIAL DISTRIBUTION OF FLOOD WATER WITH BRIDGE, WITHOUT BRIDGE, REFRAMING, AND RELOCATING THE BRIDGES

Flood modeling using HEC-RAS: The flood hazard rating map and flood susceptibility map which were prepared during the second year of the project have shown

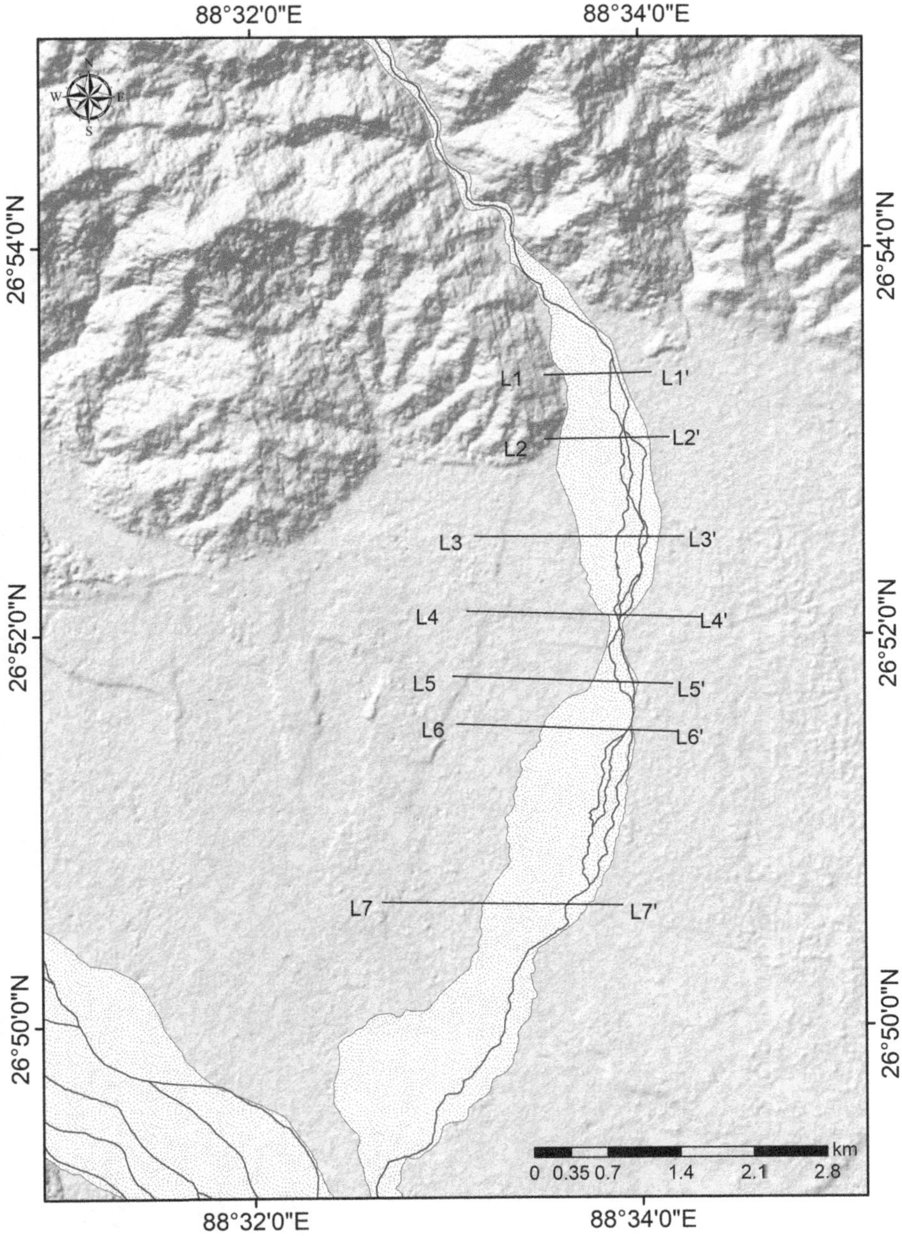

FIGURE 6.10 Location of the seven selected cross-sections of River Lish.

that at the downstream reaches, River Chel is very much susceptible to floods. River Gish throughout its reaches have shown the potentiality of flooding. To validate the maps, flood models have been simulated on the HEC-RAS 6.4 platform. The flood model incorporates the discharge data, velocity, and Manning's n to simulate the flood coverage area. The model has been simulated for the three considered rivers,

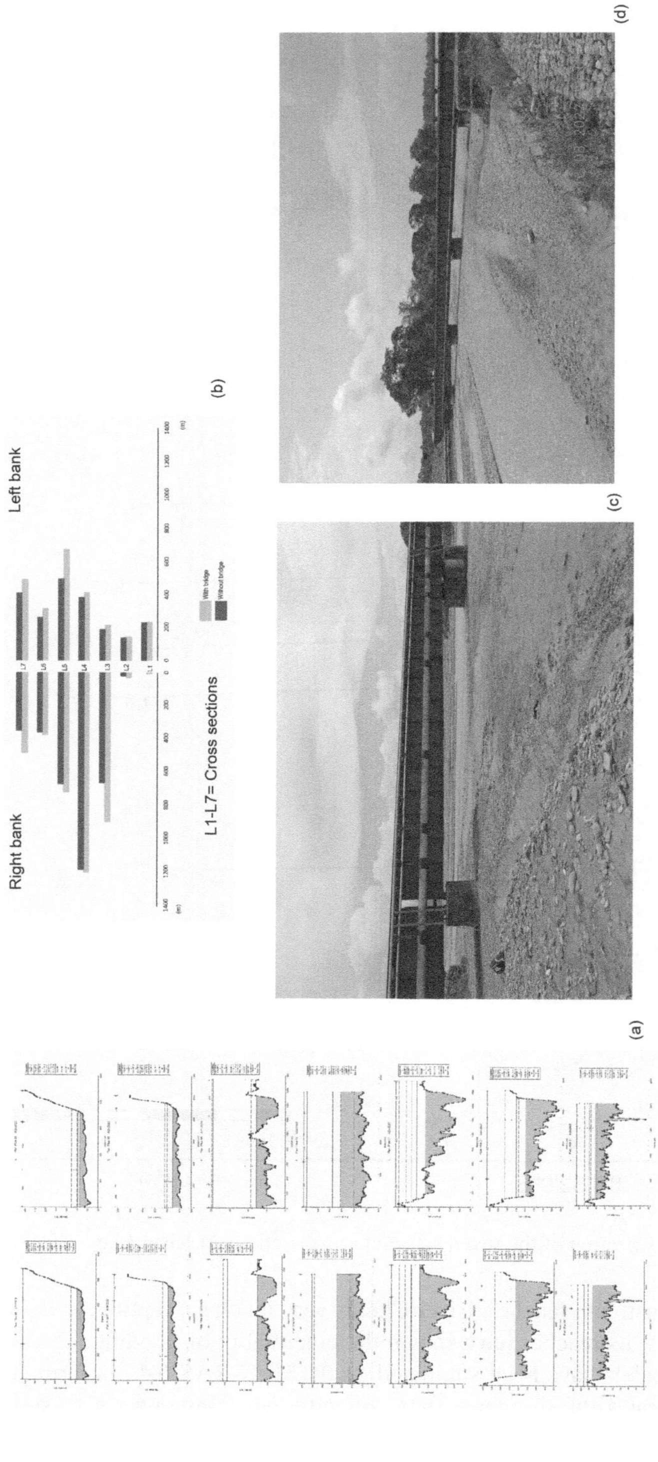

FIGURE 6.11 (a) Details of seven cross-sections of HEC-RAS flood modeling with bridge and without bridge. (b) Flood coverage area of right and left bank of the River Lish with bridge and without bridge. (c, d) Photographs of the rail bridge over River Lish.

i.e., Lish, Gish, and Chel considering four circumstances—with the bridges, without bridges, reframed bridges, and relocated bridges. In case of reframed bridge condition, it is assumed that the bridge width will be increased according to the width of the river. For the relocation of the bridge, the shortest width of the river bed with bank of hard rock has been considered. The tensile strength of the rocks in the catchment area of the three rivers has been measured to mark the ideal location for the bridge relocation.

- It is noted that, in case of River Lish, the flood coverage area is minimal (0.728 km^2 at downstream left, 0.677 km^2 at downstream right, 0.426 km^2 at upstream left, and 0.615 km^2 at upstream right) if the bridge is reframed according to the width of the riverbed. It is also noted that the velocity of the river in the present condition of the bridges is much lower (1.37 m/sec), which causes significant bed aggradation and a subsequent rise of the river-bed elevation. If the bridges are reframed, the velocity near the bridge piers will be higher (2.00 m/sec) causing the deposited materials to get washed away farther downstream.
- For River Gish, the flood coverage areas without the bridges (5.131 km^2 at the downstream left, 3.13 km^2 at the downstream right, 2.099 km^2 at the upstream left, and 0.916 km^2 at the upstream right) and relocated bridges (4.499 km^2 at the downstream left, 3.286 km^2 at the downstream right, 2.092 km^2 at the upstream left, and 0.783 km^2 at the upstream right) are almost similar. The velocity also increases slightly (1.52 m/sec) without the intervention of bridges. However, the flood model simulated with reframed bridges show the minimum flood coverage area (4.316 km^2 at the downstream left, 2.520 km^2 at the downstream right, 1.688 km^2 at the upstream left, and 0.916 km^2 at the upstream right) for River Gish.
- River Chel has almost similar flood coverage area with the bridges (24.887 km^2 at the downstream left, 27.42 km^2 at the downstream right, 6.364 km^2 at the upstream left, and 7.291 km^2 at the upstream right) and with relocated bridges (24.596 km^2 at the downstream left, 28.954 km^2 at the downstream right, 6.475 km^2 at the upstream left, and 5.714 km^2 at the upstream right). The area decreases slightly without the intervention of bridges. However, it is to be noted that the flood coverage area decreases significantly if the bridges are reframed according to the width of the river bed (19.590 km^2 at the downstream left, 18.217 km^2 at the downstream right, 6.217 km^2 at the upstream left, and 7.264 km^2 at the upstream right). It is also noted that the velocity which is maximum near the bridge piers with the bridges decreases considerably with reframed bridges. Thus, it is evident that huge bed degradation and scouring near the bridge piers which are the results of higher velocity (2.00 m/sec) will be decreased (1.60 m/sec) with the bridges to be reframed. It can also be said that the problem of scouring will decrease considerably after the reframing of the bridges (Figure 6.12).

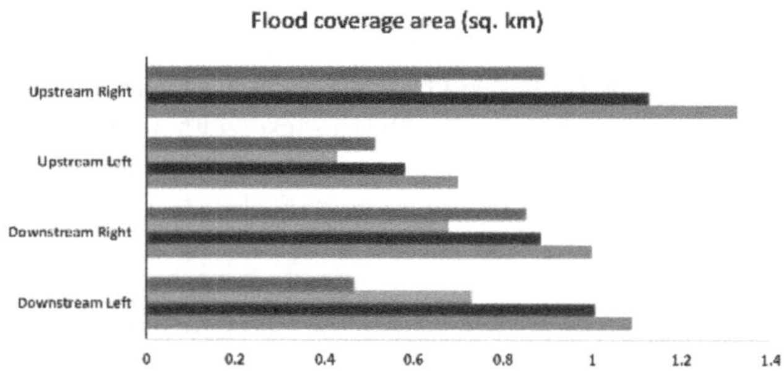

	Downstream Left	Downstream Right	Upstream Left	Upstream Right
Relocated bridge	0.465569	0.85057	0.511858	0.890091
Reframed bridge	0.728705	0.677926	0.426763	0.615052
Without bridge	1.00694	0.88523	0.5795	1.1253
With bridge	1.08914	0.99722	0.69842	1.3253

(a)

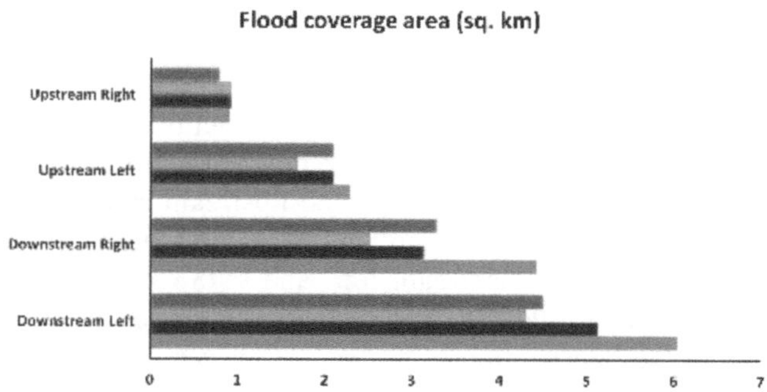

	Downstream Left	Downstream Right	Upstream Left	Upstream Right
Relocated bridge	4.499967	3.2868	2.09275	0.783769
Reframed bridge	4.31606	2.52069	1.68871	0.91663
Without bridge	5.13146	3.13	2.0998	0.916
With bridge	6.04437	4.43031	2.286	0.909947

(b)

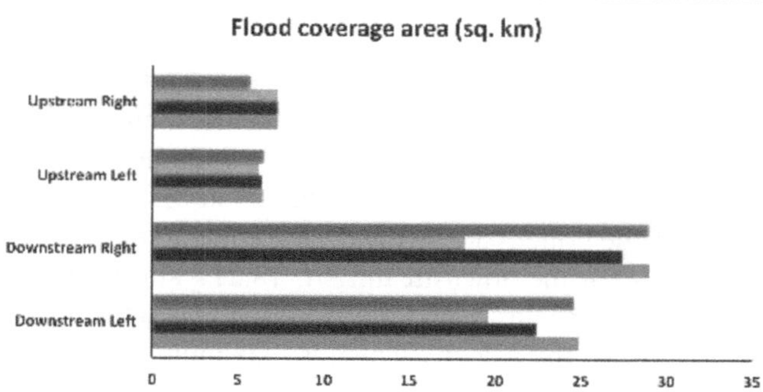

	Downstream Left	Downstream Right	Upstream Left	Upstream Right
Relocated bridge	24.5967	28.9548	6.47575	5.714
Reframed bridge	19.5908	18.2172	6.2173	7.26426
Without bridge	22.453	27.42	6.364	7.2914
With bridge	24.887	28.991	6.4231	7.3024

(c)

FIGURE 6.12 Flood coverage area with, without bridge, reframing, and relocating bridges of Rivers Chel, Gish, and Lish.

In terms of relocating the bridges, I have considered the minimum width area where hard rock is available (Figure 6.13). A Smidth hammer was used for measuring the tensile strength (MPa).

Table 6.1 shows that River Chel has the highest tensile strength and three layers with two contact zone are visible. It is followed by River Gish (42.56 MPa) and Lish (31.43 MPa). But flood propagation is not least in that case and moreover that road and rail line may hamper the forest ecosystem near Lish River. Figure 6.14 shows the location of the relocating bridge where tensile strengths have been measured.

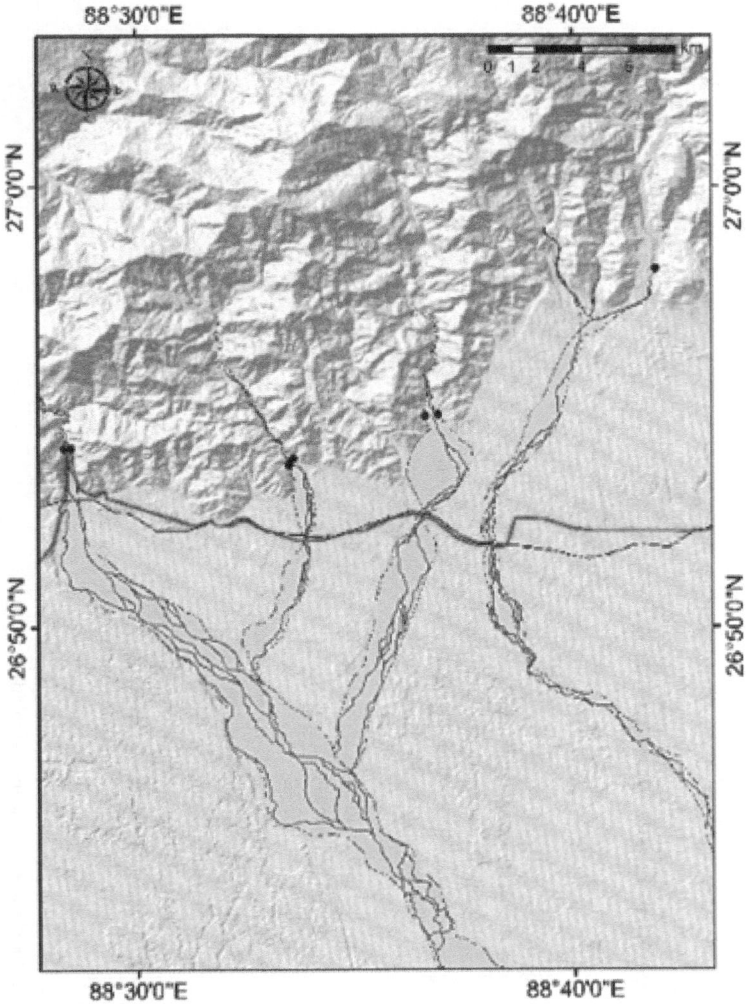

FIGURE 6.13 Locations of Smidth hammer measurements of tensile strength (MPa) for relocating bridges of Rivers Chel, Gish, and Lish.

TABLE 6.1
Measurement of Tensile Strength (MPa)

Sl. No.	Place	Latitude	Longitude	Tensile Strength (MPa)		
				Layer 1	Layer 2	Layer 3
1	Leesh (Rock Sample-Sandstone)	26.89668	88.55778	24.375		
2	Leesh (70 m downstream from point 1)	26.89612	88.55816	31.4375		
3	Chel (Beyond Foothill)-Gorubathan [Weathered Layer]	26.97027	88.69795	22		
4	Chel (Beyond Foothill)-Gorubathan [Weathered Layer]	26.97027	88.69795	**40.275**	**29.5**	**38.4375**
5	Chel (Schist-Rocky Exposure beside road)	27.0014	88.698	28.1875		
6	Chel (Schist-Rocky Exposure) Left Bank of Chel	26.98494	88.69811	**60.625**	**54.0625**	
7	Paparkheti Landslide	27.02621	88.69651	41.8125		
8	Chel Landslide –2	26.02486	88.69514	30.68		
9	Coronation Bridge (Left Bank of Teesta)	26.89718	88.47412	36.5		
10	Coronation Bridge (Right Bank of Teesta)	26.89715	88.47556	31.125		
11	Sevok—Rock sample	26.90229	88.47284	39		
12	Gish Catchment (Schist, muscovite, quartz vein)	26.97849	88.64481	42.56		
13	Gish Catchment (Rocky exposure)	26.98134	88.62575	42		
14	Gish Catchment (Landslide connecting first order stream)	26.9866	88.62031	26.5		
15	Gish Catchment (Rocky exposure)	26.98824	88.6163	39.36		

Black bold colored values are the highest in different layers.

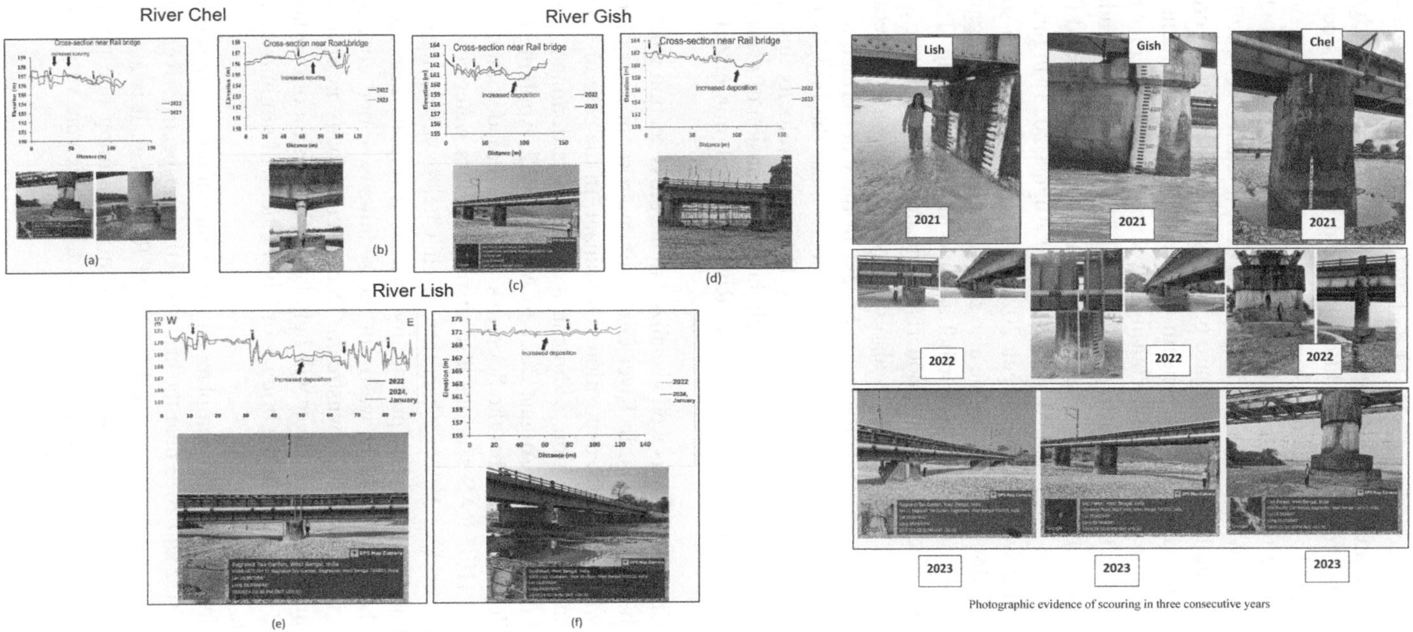

FIGURE 6.14 Cross-sections of rivers using Total Station; a, c, e near rail bridge and b, d, f near road bridge.

- The analysis encompassed both the expansion and contraction of the channel width between the years 1913 and 2021. It also involved determining the impacted area, considering scenarios with or without bridge structures or bridge relocations. Additionally, field verification was conducted, which included a comparative assessment with adjacent rivers, ensuring the accuracy of the findings. A total station survey has been scheduled for the relevant parts. We have finished the areas next to each river's rail bridge. Scouring in the River Chel (Figure 6.14a) is clearly seen in the cross-section where the river crosses beneath the piers. Whereas sediment deposition is predominating, lowest scouring happened close to the pier in River Gish (Figure 6.14b). Regarding the River Lish, sediment aggradation is predominant in most of the segment and has grown by 0.03 and 0.032 m between August 2021 and August 2022 and August 2023 (Figure 6.14c). measurements were taken between 2022 and 2023. Total station survey of the cross-sections near the bridge also shows that in River Lish aggradation is the highest and less in River Chel as scouring is increasing. The HEC-RAS model reveals that the flood risk zone decreased without a bridge. Considered HEC-RAS sections also display the less water coverage area without the bridges. Velocity becomes less in the Chel bridge area minimizing scouring probability. Increased velocity near the bridges of River Lish will help to wash out the sediments. Decreased velocity upstream and downstream of the Gish River will help lateral expansion of overland flow. Decreased velocity near the bridges of River Chel will minimize the scouring effect.
- In River Chel, scouring is well visualized in the cross-section where river is crossing below the piers. In River Gish, minimum scouring occurred near the pier, while sediment deposition is dominant. In the case of River Lish, in most part of the section, sediment aggradation is dominant and it has increased 0.03 m from August 2021 to August 2022. Applying the HEC-RAS model, it is clear that the flood risk zone decreased without a bridge. Considered HEC-RAS sections also display the less water coverage area without the bridges. Velocity becomes less in Chel downstream minimizing the flood and bank erosion probability. Increased velocity near the bridges of River Lish will help to wash out the sediments. Flood coverage area decreases in River Lish downstream. Decreased velocity in upstream and downstream of the Gish River will help lateral expansion of overland flow. Decreased velocity near the bridges of River Chel will minimize the scouring effect.
- Total station survey has been planned to conduct in respective sections. TS survey in 2022 and 2023 near the bridges of three respective rivers discloses the increased scouring near road bridge of River Chel. It is also observed that aggradation of sediments increased in an alarming rate upstream of River Gish and near the rail bridge in River Lish which are very clear in the photographic evidences from 2021 to 2023.

6.3 HUMAN INTERVENTION, CONSTRUCTIONS, LULC CHANGE, LINKAGES WITH THE HYDROLOGIC REGIME, AND CONSEQUENT VULNERABILITIES

A river is defined by a wide range of planform patterns and is considered to be a continuum of morphological diversity. The morphological conditions of the river are influenced by both independent factors (geology, climate, and human activity) that "impose conditions on the watershed" and dependent factors (which adapt to the imposed conditions) that are heavily influenced by the watershed's natural vegetation. Human changes to the landscape have been continuously emphasized as a major factor in altering the hydrology, which impacts the frequency, volume, and quality of the sediments carried by the river; discharge, which includes its frequency, duration, and variability in time and space; and channel erodibility, which is influenced by deposition and affects the erodibility of the bank. These tributaries also play an important role because they contribute to a huge amount of water at Teesta with the monsoon and the form of a dynamic channel plan due to the condition of a bottleneck has been created. Consequently, the constructions on rivers distinguish distinctly the morphological scenario downstream lately. In fact, after the construction of the road bridge in the 1970s, the flow of these rivers minimized the structural force, but during the monsoon, a huge amount of water is released by the narrow passage leading to the increase in recharge at the bottom of the bridge, which permanently weakens the pillars of the bridge creation of the probability of collapse of the bridge in the near future and downstream of the bridge due to the expansion of the area, the loss energy has induced a risk of flooding. The variety of literary engineering works has enlightened the negative impacts of the bridges on the morphology of existing streams and also its impact on the surrounding local areas if they are poorly designed. The bridge, built along the cross section of the river, was constructed to develop transportation equipment. This was an excellent planned engineering structure parallel to the railway bridges previously manufactured during the British era. Engineers who build highway road bridges have minimized the width of the bridge for structural efficiency. However, decades later, the morphology, river ground, nature of deposit damage, and channel networks are changing very frequently in flood coverage areas, particularly in rivers and in relation to these factors. And there are risks associated with them when bank erosion is developing.

Given the previous discussion, this study also focuses on the characteristics of land use changes detected from a respected two-year terrestrial images, 2005 and 2023. Classified images illustrate a wonderful change in colony and tea gardens. The coating areas of colonies and roads have increased 5–11% and agricultural land from 11 to 30%, but the alarming scenario of the tea garden is the place where it was reduced from 25.9 to 10.75%. Deposited river channels precipitation increased from 4.96 to 8.63%. The classified image, it is clear that tea gardens in the upstream area from the bridge are greatly affected. Most of the time it was noticed during my visit to the site that land is kept to grow vegetables and paddy fields. The largest land in the downstream section was converted to facilities and farmland.

During the rainy season, large amounts of water cannot flow through the constricted passage; consequently, the upstream area becomes inundated, significantly affecting nearby land uses such as tea gardens and farmland adjacent to the river. With the abrupt reduction of river energy, slope, and velocity after the road bridge helps the water to overflow the banks, impacting the agricultural lands and communities on the riverbank. However, the land on the river side is also seasonally used. Such kinds of changes in land use ultimately produce an impact on the landscape architecture in the field of environmental ecology.

North Bengal, located in the eastern foothills of the Himalayas, is famous for its remarkable biodiversity, which includes a range of ecosystems from dense tropical forests to vast tea gardens and productive plains. This area, including districts like Darjeeling, Jalpaiguri, and Cooch Behar, contains an abundance of plant and animal life that plays a vital role in its ecological wealth and cultural diversity. Climate change is altering ecological interactions in North Bengal, resulting in significant changes in species relationships and ecosystem dynamics. Here's a review of these modified interactions within 400 words: changes in temperature and precipitation patterns, induced by climate change, are modifying the distribution and behavior of predator and prey species in North Bengal. Modifications in seasonal cues, including the timings of flowering and fruiting for plants, can create mismatches between predators and their prey. Climate change influences the timing of flowering and the active periods of pollinators such as bees, butterflies, and birds. This disruption can result in mismatches where pollinators may not be present when plants need pollination, impacting plant reproductive success and diminishing biodiversity. Flood-affected areas in North Bengal, including those in the Terai-Dooars area, are vulnerable to shifts in hydrology. Changes in water levels and the drying of wetlands caused by decreased precipitation or heightened evaporation endanger the existence of aquatic plants, fish, amphibians, and migratory birds that rely on these ecosystems. Rivers such as the Teesta and Ganges are crucial for North Bengal's biodiversity, sustaining various fish species, and riverbank vegetation. Alterations in river flow dynamics impact nutrient cycling, sediment deposition, and the structure of aquatic habitats, affecting the overall riverine ecosystem.

REFERENCES

Ahmed, A., G. Hewa, and A. Alrajhi. 2021. Flood susceptibility mapping using a geomorphometric approach in South Australian basins. *Natural Hazards*, 106(1): 629–653. https://doi.org/10.1007/s11069-020-04481-z

Bhat, M., A. Alam, S. Ahmad, H. Farooq, and B. Ahmad. 2019. Flood hazard assessment of upper Jhelum basin using morphometric parameters. *Environment and Earth Science*, 78(2). https://doi.org/10.1007/s12665-019-8046-1

Biswas, M., and A. Raha. 2022. An overview of open channel monsoon hydraulics of Himalayan foothill rivers, West Bengal, India. *Modeling Earth Systems and Environment*, 9(2): 2969–2979. https://doi.org/10.1007/s40808-022-01481-9

Chang, H., and J. Franczyk. 2008. Climate change, land-use change, and floods: toward an integrated assessment. *Geography Compass*, 2(5): 1549–1579. https://doi.org/10.1111/j.1749-8198.2008.00136.x

Jarrett, R., and E. Tomlinson. 2000. Regional interdisciplinary paleoflood approach to assess extreme flood potential. *Water Resources Research*, 36(10): 2957–2984. https://doi.org/10.1029/2000wr900098

Kaya, Y.Z., F. Üneş, and M. Demirci. 2022. Flood modeling based on the precipitation data by using HEC-RAS software version (5.0.7). *Aerul Şi Apa, Componente Ale Mediului.* https://doi.org/10.24193/awc2022_05

Mohamed, S., and M. El-Raey. 2019. Vulnerability assessment for flash floods using GIS spatial machine-learning-based sensed data in El-Arish City, North Sinai, Egypt. *Natural Hazards*, 102(2): 707–728. https://doi.org/10.1007/s11069-019-03571-x

Quesada Román, A. 2021. Landslides and floods zonation using geomorphological analyses in a dynamic basin of Costa Rica. *Revista Cartográfica*, 102: 125–138. https://doi.org/10.35424/rcarto.i102.901

Raha, A., S. Gupta, and M. Biswas. 2023. Flood risk assessment of Himalayan foothill rivers: A study of Jaldhaka River, India. In A. Islam, et al. (eds.), *Floods in the Ganga–Brahmaputra–Meghna Delta*. Springer, 63–90. https://doi.org/10.1007/978-3-031-21086-0_4

Smith, K.D., and R.C. Ward. 1998. Floods: physical processes and human impacts. https://agris.fao.org/agris-search/search.do?recordID=GB1997043473

7 Assessment of Bottleneck Conditions and Concluding Remarks

7.1 HYDRO-MORPHOLOGY OF BOTTLENECK CONDITIONS

According to S. Kale (2005), the Indian subcontinent is thought to be the world's largest monsoon-dominated region. The total reversal of the monsoon wind is linked to India's seasonal variations. The south-west monsoon, which enters the country after splitting into two branches the Arabian Sea branch and the Bay of Bengal branch largely, shapes the rainfall characteristics. Therefore, the monsoon rainfall determines the hydro-sediment discharge of the Indian rivers and their yearly flow pattern, which varies accordingly. As a result, the rivers' channel shape and hydro-sediment discharge characteristics vary seasonally. It should be mentioned that the major riverine systems in India bear the greatest amount of hydro-sediment load from June to September, which is the monsoon season. Channel dynamics, which includes channel migration, erosion, and deposition, is a significant geomorphological phenomenon. Due to human intervention in the form of bridge constructions, there is a bottleneck condition due to the maximum wide extension of the Gish River upstream and downstream between 1913 and 2021. The sediments in Lish and Gish aggrade laterally, widening the upstream and downstream reaches. The increased velocity and stream power near the bridges accelerate sediment degradation at the bed near piers. The rivers get sediments from the landslides as well which are connected with the first-order streams in the catchment area. It is interesting to note that the lengths of Rivers Lish and Gish are much shorter than that of Chel which have modified the sedimentation pattern at the beds. The sediments are getting enough lengths for River Chel and transported to farther downstream sections. On the other hand, in Lish and Gish, the sediments aggrade laterally increasing the width of the upstream and downstream reaches from the bridges as they have smaller catchment, and as bridges are quite near from the point, they enter into the foothill section. This has resulted in huge deposition in the upstream of the bridge and just after the bridge in the downstream section as well. It has created a bottleneck condition of these two rivers which is most profound for River Gish. The narrowing of the river flow path near the bridges of River Gish is maximum that induced upstream-downstream deposition of sediments, widening the river width near Pathorjhora village. Large sediment deposits near the bridges are accompanied by catchment landslides that cause sediments, which are primarily carried by the first- and second-order rivers mostly

DOI: 10.1201/9781003562979-7

during the monsoon season. The sediment transport index map and landslide susceptibility map are interlinked with sediment generation, transportation, and deposition over the foothill.

In North Bengal's Himalayan foreland region, floods are common, causing extensive destruction and a great lot of misery for the local population. It is imperative to carefully investigate the most likely causes of catastrophic floods, identify the areas most susceptible to such disasters, and put targeted actions in place to lower the risk of flood dangers. A key tactic to address this challenge is the implementation of sensible technological solutions, such as the construction of levees and embankments, which are placed strategically in the most susceptible areas of the river. These constructions serve as strong barriers that efficiently contain and regulate the overflow of river water during periods of intense rainfall.

This work reveals some clear signs of engineering construction on channel hydraulics, which are caused by bed scouring and sediment aggradation. It is crucial to remember that the three rivers under consideration are affected differently by bridge construction. The bridge length was drastically reduced during construction on the Lish and Gish Rivers, resulting in a bottleneck situation in both rivers. According to the analysis, the channel hydraulics parameters are suitable for the sediment aggradation process close to the rail and road bridge of both rivers, which greatly aids in the rising of the riverbed. Large amounts of water and silt have been deposited upstream due to the inability of the tight path between the bridges to carry them, thus enlarging the channel in the upstream reach. The Gish River is where this situation is most noticeable. Since the parameters have caused a state of degradation, particularly scouring at bridge piers and the subsequent lowering of the bed, the channel hydraulics parameters have shown a variety of conditions in the Chel River. Therefore, the entire study emphasizes the main issues with bridge construction on rivers, which changes the amount of suspended sediment and the spatial distribution of hydrological parameters at various river reaches, particularly before, between, and after the bridges.

7.2 CONCLUDING OUTLINE

The outcomes discussed in the book address the problems associated with the construction of bridges across rivers in connection with planform dynamics and sediment aggradation/scouring mechanisms. It is noticeable that the three considered rivers, i.e., Lish, Gish, and Chel, have different impacts of bridge construction on them. River Lish experiences significant sediment aggradation near the rail and road bridges which has resulted in the rise of bed elevation. It is noticed that huge amounts of fine sediments (mostly less than 40 micron) get deposited near the bridge pier under hydraulic conditions of lesser velocity and decreased stream energy. River Gish, in spite of the huge suspended sediment load, deposits a lesser amount of sediment near the bridge which is attributed to the higher velocity and stream power than that of River Lish. Scouring is also observed in Gish. However, the problem of scouring is massive in River Chel. Flood vulnerability of the region is further accelerated by the construction of bridges. Due to shortening of the width of the rivers near the bridges, huge discharge in the monsoon gets overflowed causing floods.

The HEC-RAS models portray a lesser extent of flood coverage areas if the bridges are reframed maintaining the actual width of the rivers. This study highlights the future scope of research in the Himalayan foothill regions where numerous tributaries of Jaldhaka and Torsa traverse. All these tributaries are connected with road and rail bridges which have a contrasting impact on the hydro-sediment interaction and planform dynamics.

Sediment degradation at the bed near piers is accelerated by the increased stream strength and velocity close to the bridges. Landslides that are connected to the catchment area's first-order streams also contribute sediments to the rivers. Interestingly, the lengths of the Gish and Lish Rivers are significantly shorter than those of the Chel, which has changed the pattern of sedimentation at the beds. Sediments that reach sufficient length for Chel are carried to areas further downstream. Bridge construction further increases the region's susceptibility to flooding. Floods are caused by massive monsoon discharges that overrun because the rivers surrounding the bridges have narrowed. If the bridges are reframed while maintaining the real width of the rivers, the HEC-RAS models show a smaller extent of flood coverage areas. Due to its lower height and extremely gentle slope, which allow flood water to stagnate for an extended amount of time, the southeast portion of the study area under consideration shows significant susceptibility. The likelihood of flooding is further increased by the land's bareness and lack of vegetation cover. Geomorphologically, the high altitudinal profile and moderate to steep slope from the dissected hills to the distal half of the alluvial fans allow river water to flow down quickly, making the areas extremely vulnerable to flooding. In their various catchment areas, the three foothill rivers under consideration receive enormous amounts of rainfall. This study highlights the need for more research in the foothill parts of the Himalayas, where many of the Jaldhaka and Torsa tributaries pass through. Road and rail bridges connect all of these streams, which have differing effects on the dynamics of the planform and the hydro-sediment interaction. Sediment aggradation has increased the elevation of the riverbed near the rail bridge of 0.03 m from 2021 to 2022. Scouring rate is high and positive in River Chel. First- and second-order streams connected with landslides near catchment of Chel, Lish, and Gish added huge sediments mainly in the monsoon. The distance between the rail bridge and riverbed mostly decreased in River Lish (~2.27 m in 2022 that was ~3.00 m in 2021). Near the bridge, River Gish is mostly under the bottleneck condition. Applying the HEC-RAS model, it is clear that the flood-risk zone decreased without a bridge. Considered HEC-RAS sections also display the less water coverage area without the bridges. Velocity becomes less in Chel downstream minimizing the flood and bank erosion probability. Increased velocity near the bridges of River Lish will help to wash out the sediments and it will help to minimize the flood coverage area further downstream. Decreased velocity upstream and downstream of the Gish River will help to minimize the lateral expansion of overland flow. Decreased velocity near the bridges of River Chel will minimize the scouring effect that will protect the bridge piers.

The entire study reveals that between 1913 and 2021, there was a significant east-west channel change, yet the maximum width of the Gish River increases upstream, resulting in a classic bottleneck condition close to the bridge. The likelihood of landslide events in the watersheds' catchment area is determined by landslide susceptibility zones and occurrences. The mineral composition of the landslide origin base rocks, which are primarily composed of chlorite, phyllite, muscovite, recrystallized mica, and quartz, is supported by microscopic analysis. These minerals are carried by channels and deposits in the bed close to bridges, which causes the topographic planform to change over time. This additional surge water can be restored and utilized for drinking after filtration, and during the Rabi period, this water may be used for farming purposes. As this foot slope segment is beneath the "BHABAR" zone, there's water shortage that emerge during the winters. In this connection, it may be suggested that some micro-watershed management policies in the catchment area of the tributaries may be consider for multi development and hazard reduction.

Index

Pages in *italics* refer to figures and pages in **bold** refer to tables.

F

G

H

I

J

K

L

M

N

O

P

R

S

For Product Safety Concerns and Information please contact our EU representative GPSR@taylorandfrancis.com Taylor & Francis Verlag GmbH, Kaufingerstraße 24, 80331 München, Germany

Responsible Technology and Intelligence: A Review Series

1. A Compendium of Responsible Artificial Intelligence
   Deepak Kumar Jain, Kirti Raj Bhatele, Deepak Garg and *Gaurav Dhiman (Editors), 2025*
2. Ambient Assisted Living: Technological Advances
   Rashmi Gupta, Shashi Shekhar, Jeetendra Kumar and *Gaurav Dhiman (Editors), 2025*
3. Industrial Internet of Things for Responsible Technology
   B.K. Mishra, Gaurav Dhiman, Hemant Kasturiwale, Sujata Alegavi and *Kusum Yadav (Editors), 2026*

Responsible Technology and Intelligence: A Review Series

Industrial Internet of Things for Responsible Technology

Editors:

B.K. Mishra
Principal & Professor
Department of Electronics and Telecommunication Engineering
Thakur College of Engineering and Technology
Mumbai, India

Gaurav Dhiman
Assistant Professor
School of Sciences and Emerging Technologies
Jagat Guru Nanak Dev Punjab State Open University
Punjab, India

Hemant Kasturiwale
Associate Professor & HOD
Electronics and Computer Science
Thakur College of Engineering and Technology
Mumbai, India

Sujata Alegavi
Associate Professor & HOD
Internet of Things
Thakur College of Engineering and Technology
Mumbai, India

Kusum Yadav
University of Ha'il
Saudi Arabia

CRC Press
Taylor & Francis Group
Boca Raton London New York

CRC Press is an imprint of the
Taylor & Francis Group, an **informa** business
A SCIENCE PUBLISHERS BOOK

First edition published 2026
by CRC Press
2385 NW Executive Center Drive, Suite 320, Boca Raton FL 33431

and by CRC Press
4 Park Square, Milton Park, Abingdon, Oxon, OX14 4RN

CRC Press is an imprint of Taylor & Francis Group, LLC

Library of Congress Cataloging-in-Publication Data (applied for)

ISBN: 978-1-032-82940-1 (hbk)
ISBN: 978-1-032-96083-8 (pbk)
ISBN: 978-1-003-58790-3 (ebk)

DOI: 10.1201/9781003587903

Typeset in Times New Roman
by Prime Publishing Services

Preface

The Industrial Internet of Things (IIoT) is rapidly transforming the landscape of modern industry, fostering efficiencies, insights, and automation at an unprecedented scale. As we advance deeper into this era of hyperconnectivity and intelligent systems, it is essential to pause and reflect on the broader implications of such technologies. This book, Industrial Internet of Things for Responsible Technology, is a step in that direction—an effort to explore not only the technical dimensions of IIoT but also the ethical, social, and sustainable responsibilities that must accompany innovation.

Our primary motivation in compiling this volume is to provide a comprehensive resource that bridges the gap between IIoT advancements and the pressing need for responsible technology practices. The chapters in this book reflect a multidisciplinary perspective, encompassing foundational knowledge, regulatory insights, design ethics, and real-world implementations. From sustainable design principles and human-centric approaches to security, data governance, and social impact, each chapter builds a narrative that encourages mindful development and deployment of IIoT systems.

The book opens with an introduction to the concept of responsible technology within the IIoT context, followed by foundational discussions that set the technical stage. We then delve into ethical considerations, sustainability, and governance, presenting a balanced mix of theory, application, and case studies. Special emphasis is placed on human factors, resilience, and collaborative innovation to highlight how IIoT can benefit not only industries but also communities and the environment. Notably, this volume includes a detailed technical case on battery management systems, demonstrating how IIoT frameworks can be tailored for high-impact, responsible solutions.

This book is intended for engineers, researchers, policymakers, educators, and industry professionals who are committed to shaping the future of IIoT with integrity and foresight. As the field continues to evolve, we hope this work serves as both a guide and an inspiration for cultivating technologies that are not only intelligent and efficient but also just, inclusive, and sustainable.

We thank the contributors for their valuable insights and the readers for their commitment to advancing technology responsibly. Let this book be a catalyst for thoughtful dialogue and impactful action in the ongoing journey of industrial transformation.

Series Preface

The rapid pace of technological advancement in the 21st century has brought with it transformative capabilities and unprecedented possibilities across sectors. From artificial intelligence and ambient intelligence to industrial automation, technology is deeply embedded in our social, economic, and personal lives. While these innovations offer immense benefits, they also raise critical questions about ethics, equity, sustainability, and human well-being. It is within this context that the "Responsible Technology Series" has been conceived—a multidisciplinary initiative aimed at fostering technological progress that is not only intelligent and efficient but also inclusive, ethical, and accountable.

This series brings together thought leaders, researchers, and practitioners to explore how emerging technologies can be designed and deployed responsibly to serve both individual and collective good. Each volume in the series addresses a specific domain of technological innovation while upholding a common commitment: to highlight frameworks, practices, and case studies that emphasize responsible development, deployment, and governance.

The first volume, A Compendium of Responsible Artificial Intelligence, offers a comprehensive overview of the ethical, social, and regulatory challenges associated with AI systems. It explores fairness, transparency, accountability, and the societal impacts of machine learning and decision-making algorithms, laying the groundwork for sustainable AI applications.

The second volume, Ambient Assisted Living: Technological Advances, focuses on the intersection of healthcare, aging populations, and intelligent environments. It presents innovations in assistive technologies and smart living spaces, while foregrounding the need for privacy, user dignity, and human-centric design in care ecosystems.

The third volume, Industrial Internet of Things for Responsible Technology, examines how industrial automation and IoT can be harnessed responsibly. It covers foundational technologies, ethical considerations, sustainability, human-centered design, and resilience—demonstrating how IIoT can drive industrial efficiency without compromising societal and environmental values.

Together, these volumes reflect the core mission of the series: to enable a balanced narrative in which innovation and responsibility go hand in hand. By bringing diverse voices and insights to the forefront, this series aims to equip readers with the tools, perspectives, and inspiration needed to build a more equitable and conscientious technological future.

We invite scholars, professionals, students, and policymakers to engage with these works, contribute to this growing dialogue, and join us in shaping a future where technology serves humanity with foresight and integrity.

Contents

List of Contributors

Abhishek Patil

Department of Mechanical & Mechatronics Engineering, Thakur College of Engineering & Technology, Mumbai, Maharashtra, India.

Achala Khandelwal

Department of Engineering Science and Humanities, Thakur College of Engineering & Technology, Mumbai, Maharashtra, India.

Aditeya Prajapati

Department of Information Technology, Thakur College of Engineering & Technology, Mumbai, Maharashtra, India.

Akanksha Manoj Singh

Internet of Things, Thakur College of Engineering & Technology, Mumbai, Maharashtra, India.

Aman Verma

Department of Electronics Engineering, Thakur College of Engineering & Technology, Mumbai, Maharashtra, India.

Amit Rakesh Maurya

Department of Internet of Things, Thakur College of Engineering & Technology, Mumbai, Maharashtra, India.

Anjali Priya

Department of Electronics Engineering, Thakur College of Engineering & Technology, Mumbai, Maharashtra, India.

Ansh Santosh Mishra

Internet of Things Thakur College of Engineering & Technology, Mumbai, Maharashtra, India.

Archana Belge

Department of Mechanical and Mechatronics Engineering (Additive Manufacturing)

Archana Kulkarni

Department of Computer Science and Engineering (Cybersecurity), Thakur College of Engineering & Technology, Mumbai, Maharashtra, India.

Archana Talhar Belge
Department of Mechanical & Mechatronics Engineering, Thakur College of Engineering & Technology, Mumbai, Maharashtra, India.

Bhushankumar Nemade
Shree L.R. Tiwari College of Engineering, Mumbai, Maharashtra, India.

Deven Shah
Shree L.R. Tiwari College of Engineering, Mumbai, Maharashtra, India.

Dipesh Tare
Department of Mechanical & Mechatronics Engineering, Thakur College of Engineering & Technology, Mumbai, Maharashtra, India.

Disha Garje
Department of Electronics Engineering, Thakur College of Engineering & Technology, Mumbai, Maharashtra, India.

Ekta Ronak Desai
Thakur College of Engineering & Technology, Mumbai, Maharashtra, India.

Foram Shah
Department of Computer Engineering, Thakur College of Engineering & Technology, Mumbai, Maharashtra, India.

Gauravi Suraj Naik
Internet of Things Thakur College of Engineering & Technology, Mumbai, Maharashtra, India.

Jalpaben Pandya
Faculty of Electronics Engineering, Thakur College of Engineering & Technology, Mumbai, Maharashtra, India.

Janhvi Vinod Kushwaha
Internet of Things Thakur College of Engineering & Technology, Mumbai, Maharashtra, India.

Jignesh Patel
Assistant Professor, Thakur College of Engineering & Technology, Mumbai, Maharashtra, India.

Jitendra Chavan
Department of Mechanical & Mechatronics Engineering, Thakur College of Engineering & Technology, Mumbai, Maharashtra, India.

Kavita Maskar
Assistant Professor, Department of Engineering Science and Humanities, Thakur College of Engineering & Technology, Mumbai, Maharashtra, India.

Keyur Amit Chauhan
Internet of Things, Thakur College of Engineering & Technology, Mumbai, Maharashtra, India.

Kiran Maharana
Researcher, ICICI Lombard GIC Ltd, Mumbai, Maharashtra, India.

Kshama Shukla
Department of Mechanical & Mechatronics Engineering, Thakur College of Engineering & Technology, Mumbai, Maharashtra, India.

Loukik Salvi
Department of Computer Engineering, Thakur College of Engineering & Technology, Mumbai, Maharashtra, India.

Minal Bhoir
Assistant Professor, Department of Engineering Science and Humanities, Thakur College of Engineering & Technology, Mumbai, Maharashtra, India.

Namit Singh
MSc Business Analytics, University of Warwick - Warwick Business School in Coventry, England.

Nandini Seth
Department of Electronics Engineering, Thakur College of Engineering & Technology, Mumbai, Maharashtra, India.

Neha Kapadia
Department of Information Technology, Thakur College of Engineering & Technology, Mumbai, Maharashtra, India.

Noopur Divekar
Department of Electronics Engineering, Thakur College of Engineering & Technology, Mumbai, Maharashtra, India.

Pallavi Manoj Jha
Internet of Things, Thakur College of Engineering & Technology, Mumbai, Maharashtra, India.

Pankaj Shelar
Department of Mechanical & Mechatronics Engineering, Thakur College of Engineering & Technology, Mumbai, Maharashtra, India.

Pratibha Sajwan
Department of Artificial Intelligence and Data Science, Thakur College of Engineering & Technology, Mumbai, Maharashtra, India.

Priyanka Musale

Department of Internet of Things, Thakur College of Engineering & Technology, Mumbai, Maharashtra, India.

Prutha Trivedi

Department of Electronics Engineering, Thakur College of Engineering & Technology, Mumbai, Maharashtra, India.

R.G. Mehta

Department of Computer Science & Engineering, SVNIT, Surat, India.

Rupinder Kaur

Faculty of Electronics Engineering, Thakur College of Engineering & Technology, Mumbai, Maharashtra, India.

Seema Jamal

Assistant Professor, Thakur College of Engineering & Technology, Mumbai, Maharashtra, India.

Shaurya Chaturvedi

MS - Astronomy and Astrophysics, The Australian National University, Bruce, Australian Capital Territory, Australia.

Shiwani Gupta

Department of AI ML, Thakur College of Engineering & Technology, Mumbai, Maharashtra, India.

Soham Sharad Kadam

Internet of Things, Thakur College of Engineering & Technology, Mumbai, Maharashtra, India.

Sonal Dubal

Department of Internet of Things, Thakur College of Engineering & Technology, Mumbai, Maharashtra, India.

Sony Jha

Thakur College of Engineering and Technology, Mumbai, Maharashtra, India.

Sujata Alegavi

Department of Internet of Things, Thakur College of Engineering & Technology, Mumbai, Maharashtra, India.

Sukruti Kaulgud

Department of E&TC, Thakur College of Engineering & Technology, Mumbai, Maharashtra, India.

Sumit Kumar

Department of Electronics and Computer Science, Thakur College of Engineering & Technology, Mumbai, Maharashtra, India.

Sunil Khatri

Department of Internet of Things, Thakur College of Engineering & Technology, Mumbai, Maharashtra, India.

Sushan Uchil

Department of Information Technology, Thakur College of Engineering & Technology, Mumbai, Maharashtra, India.

Swati Mahalle

Department of Artificial Intelligence and Data Science, Thakur College of Engineering & Technology, Mumbai, Maharashtra, India.

Swati Mude

Department of Artificial Intelligence and Data Science, Thakur College of Engineering & Technology, Mumbai, Maharashtra, India.

Usha Gupta

Department of Internet of Things, Thakur College of Engineering & Technology, Mumbai, Maharashtra, India.

Vidyadhari Singh

Department of Computer Science and Engineering (Cybersecurity), Thakur College of Engineering & Technology, Mumbai, Maharashtra, India.

Vinayak Bharadi

Finolex Academy of Management and Technology, Mumbai University, India.

Vishal Patwa

Department of Information Technology, Thakur College of Engineering & Technology, Mumbai, Maharashtra, India.

1 An Introduction to Responsible Technology in Industrial Internet of Things

Ekta Ronak Desai[1] and Amit Rakesh Maurya[2*]

The concept of Responsible Technology is introduced in this chapter with reference to the Industrial Internet of Things (IIoT). It distinguishes IIoT from conventional IoT by emphasizing its concentration on industrial applications and the elevated risks linked to system breakdowns. The chapter subsequently examines multiple facets of responsible technology in IIoT, encompassing ethical data utilization, security and privacy, sustainability, workforce implications, and regulatory adherence. The chapter examines the problems and potential solutions for each component, highlighting the necessity of reconciling innovation with ethical issues. The chapter emphasizes the necessity for responsible development and implementation of IIoT to guarantee that its advantages for industries correspond with long-term societal and environmental welfare.

1. Overview

The Industrial Internet of Things (IIoT) is a subset of the Internet of Things. The IIoT denotes a network of intelligent devices networked to create systems that monitor, collect, exchange, and analyze data. An IIoT ecosystem generally consists of the following elements: interconnected devices that can sense, communicate, and store data; public and private data communication infrastructures [6] analytics and applications that convert raw data into actionable business insights; data storage for the information produced by IIoT devices; and human operators. These edge devices and intelligent assets relay data directly to the communication infrastructure, where it is analyzed into actionable information regarding equipment operation. This information can be utilized for predictive maintenance and to enhance business processes [1].

The Internet of Things (IoT) and IIoT utilize similar technologies, including cloud platforms, sensors, connectivity, machine-to-machine communications,

[1] Thakur College of Engineering and Technology, Mumbai, India.
[2] Department of Internet of Things, Thakur College of Engineering and Technology Mumbai, India.
Email: ektadesai1904@gmail.com[1]
* Corresponding author: amit.maurya021@gmail.com

DOI: 10.1201/9781003587903-1

and data analytics; however, their applications are distinct [2]. IoT interconnects devices across diverse sectors such as agriculture, healthcare, enterprise, consumer, utilities, government, and smart cities. Instances of IoT technology encompass smart devices, fitness trackers, and many consumer-oriented applications, wherein malfunctions generally do not result in catastrophic emergencies. Conversely, IIoT emphasizes the interconnection of equipment and devices inside sectors such as oil and gas, utilities, and manufacturing. System malfunctions or outages in IIoT can result in high-risk, perhaps life-threatening scenarios. IIoT applications emphasize the improvement of efficiency, health, and safety, whereas IoT applications are predominantly user-focused. The objective of IIoT is to enhance industrial processes, improve productivity, and minimize downtime through the utilization of real-time data and sophisticated analytics. The network of devices networked via the internet encompasses industrial and commercial systems, including transportation networks, healthcare facilities, and the energy grid, as well as devices such as cell phones and laptops. The deployment of the IoT in industries, including manufacturing, oil and gas, and utilities, has increased. It is essential to design and deploy these technologies with a focus on societal welfare, environmental sustainability, data protection, and system security.

2. Ethical Technology in the Industrial Internet of Things

Responsible Technology pertains to the deliberate creation and utilization of technologies aimed at reducing their adverse effects on the environment [3].

2.1 Ethical Data Utilization

Industrial Internet of Things (IIoT) systems produce substantial volumes of data from industrial machinery and devices. Responsible data management guarantees that data is acquired, kept, and utilized in accordance with privacy standards and without exploitation.

Data exchange is an essential component of IIoT systems. Real-time sharing and processing of temperature and humidity data in a workshop are essential for the successful operation of control systems. Likewise, smart meter data ought to be available to electricity system operators and public utilities for subsequent analysis of power consumption [8].

The IIoT prioritizes the security of data transmission, rendering data storage protection paramount. Ensuring data security in edge-based IIoT contexts poses problems due to numerous vulnerabilities. Four primary difficulties emerge in this context: ensuring dependable data storage, permitting seamless data outsourcing, enabling effective data retrieval, and guaranteeing secure data erasure. All these elements are essential for preserving the integrity and security of IIoT systems.

2.2 Dependable Storage

The exponential increase in data from IIoT devices is beyond conventional local storage, necessitating a transition to cloud storage options. Numerous leading corporations, including Google Drive, OneDrive, and AWS, have established their own cloud platforms to fulfill this requirement. Nonetheless, cloud storage is not wholly dependable or secure, and occurrences of data loss are prevalent, rendering it a significant security issue. In August 2015, a lightning strike impacted Google's cloud data center, causing damage to disks and leading to irreversible data loss.

Fog nodes have been offered as an alternative to cloud storage to facilitate data storage nearer to the edge. Nevertheless, while fog storage has developed from cloud storage, it also retains numerous identical security problems. Fog nodes, similar to cloud services, may lack complete reliability and could erase data to optimize storage or manipulate it without users' awareness. Moreover, users frequently eliminate local copies post-uploading data to fog nodes, hence heightening the danger of irreversible data loss.

2.3 Practical Utilization

In the IIoT, devices incessantly monitor and generate substantial quantities of data. This data is generally uploaded and stored in fog nodes as ciphertext to protect sensitive information, hence complicating data sharing. Nonetheless, data sharing is an essential component of IIoT systems.

Consequently, encrypted data housed in fog nodes cannot be processed by other parties without possession of the decryption key. A straightforward method involves supplying the decryption key to individuals requesting the data; nevertheless, this heightens the danger of key exposure. Once obtained, the recipient has access to additional sensitive material that was encrypted using the same key in addition to the shared data [8]. Moreover, the key might be shared with third parties without the data owner's permission. As a result, stricter controls over data sharing within encrypted storage are needed.

2.4 Efficient Search

IoT devices commonly store data in encrypted form on fog nodes to boost security. However, this encryption makes it impossible for data owners or others to seek for specific information within the encrypted data. Nonetheless, the capacity to query outsourced data is a crucial element of data storage. A data owner may need to extract information containing the term 'urgent' from their encrypted, outsourced data to respond to emergencies. Consequently, it is imperative that encrypted data residing on fog nodes is searchable. Searching plaintext data is uncomplicated; however, performing secure and efficient searches on encrypted data poses a considerable challenge in IIoT systems [8].

2.5 Permanent Erasure

Data owners frequently need to erase their data from cloud storage service providers after it is no longer necessary. Although a reliable fog node would adhere to the request and delete the data, certain providers may neglect this obligation in favor of business interests, potentially resulting in privacy infringements or data exploitation. This presents a significant risk to users. Thus, a major problem for service providers in ensuring secure data storage is the reliable deletion of data. Ensuring that fog nodes efficiently remove data upon request is a critical issue in this field [8].

2.6 Safeguarding and Confidentiality

The IIoT presents opportunities for novel business models in various industrial sectors by providing extensive connectivity, robust data analytics tools, and improved decision support systems to enhance market competitiveness. Nonetheless, IIoT implementations encounter various security threats across multiple tiers of the connectivity and communication framework. The intricacy of IIoT systems hinders the assurance of availability, confidentiality, and integrity, potentially leading to skepticism regarding network operations. This raises worries about the potential loss of critical infrastructure, the safety of end-users, and the danger of privacy breaches involving sensitive information [10].

The following crucial areas need to be the focus of developers creating IIoT privacy solutions:

- **Assessing Privacy Needs in Business Operations:** Determining which business tasks need privacy protection is crucial. Declarations of privacy policies must be unambiguously stated and readily enforceable upon identification. This is especially difficult in IIoT, where data traverses multiple subsystems consisting of different software and hardware, frequently managed by external entities such as cloud storage providers. In such instances, discrepancies in privacy policies may arise, necessitating the establishment of mechanisms for conflict resolution [9].
- **Employing Cryptographic Techniques:** Cryptographic methods are frequently utilized to uphold privacy policies. However, the problem lies in building a privacy-enhancing cryptosystem that is lightweight enough for resource-constrained IIoT devices. Using complicated cryptography might slow down computations, making it impossible to meet time-sensitive needs. Non-cryptographic measures, such as anonymizing data or developing tools to evaluate aggregate data, should also be studied. Offloading large-scale data processing to third-party cloud services offers speed and efficiency but raises privacy concerns, as it's vital to ensure that sensitive data is processed without exposing privacy-sensitive information [10].

- **Ensuring Transparency:** It is essential to furnish end-users with clarity regarding data gathering, management, and processing. This involves notifying users about the dangers involved and the mechanisms in place to mitigate them, as well as clarifying what data is being collected and its intended purpose.

This openness makes it easier to comply with global data protection laws and improves responsibility inside the network [8].

- **Minimization of Data Collection:** Limit the amount of data that edge devices collect to that which is necessary and advantageous for network operation. Additionally, if feasible, anonymization techniques must be applied to personally identifiable information.
- **Establishing Clear Access Control:** Formulating precise data access policies is essential, detailing who is permitted to view edge node data and the conditions under which such access is authorized. Effective access control methods must be implemented to enforce these regulations [7].
- **Sustainability:** The growing use of IIoT devices may lead to heightened electrical waste and energy consumption. Responsible technology evaluates the environmental consequences of these systems, promoting energy-efficient designs and the utilization of recyclable or sustainable materials.

The increasing implementation of IoT technology is anticipated to elevate fossil fuel consumption to satisfy the energy requirements of IoT production lines.

The production of electronic devices is contributing to the excessive consumption of finite metals and resources, potentially resulting in considerable long-term difficulties. The recycling rate for electronic waste is critically low, currently at approximately 22.3% according to the World Health Organization, which raises concerns regarding the availability of resources required to satisfy the increasing market demand for IoT products. The manufacturing of electronic devices necessitates the utilization of chemicals, water, and fossil fuels, each of which has ecological repercussions. Copper, silver, gold, and palladium are essential metals in electronics manufacturing. The presence of lead in e-waste is a significant concern due to its detrimental environmental effects.

The Global e-Waste Monitor 2024 indicates that electronic waste is increasing at a rate fivefold that of documented e-waste recycling efforts. In 2022, a historic 62 million tons (Mt) of e-waste were generated, signifying an 82% rise since 2010.

Forecasts suggest that by 2030, this amount will increase by an additional 32%, totaling 82 million tons. The swift increase in e-waste has resulted in the waste and disposal of billions of dollars in strategically vital resources [5]. IIoT technologies have resulted in heightened usage of finite resources and raw materials, some of which are becoming scarce, such as specific precious metals utilized in electronics. The more affordable cost of electronic products has spurred increased manufacturing volumes, which further limits resource availability, perhaps producing a rebound impact.

The long-term environmental consequences of IoT technology remain undetermined, particularly as considerable amounts of energy are required to build and operate IoT devices.

A large surge in electronic waste is foreseen due to the planned spread of IoT-based devices in the near future.

The rapid rise of IIoT devices, which are utilized to boost automation, connectivity, and data-driven decision-making in industrial settings, can drastically increase energy usage. This is due to the fact that IIoT devices frequently necessitate uninterrupted operation, data transfer, and processing, all of which require substantial power consumption. As enterprises increasingly adopt IIoT to enhance operations, the aggregate energy consumption escalates, potentially straining energy infrastructures and amplifying the environmental impact of industries.

Energy-efficient designs are promoted by ethical technology in the IIoT space. This means designing IIoT systems and devices to use as little energy as possible during the manufacturing and operating stages. IIoT systems can operate more effectively and reduce overall energy consumption by improving power management, utilizing low-power CPUs, and including energy-saving protocols in communications.

Moreover, responsible technology prioritizes the utilization of recyclable or sustainable materials in the production of IIoT devices. Numerous IIoT devices are constructed from non-renewable materials, including certain metals and polymers, which exacerbate resource depletion and environmental deterioration [4].

By transitioning to recyclable materials or those derived from renewable, sustainable sources, industries can diminish their dependence on finite natural resources and mitigate the trash produced by outdated technologies.

In conclusion, although IIoT adoption may increase energy consumption and exacerbate environmental issues, responsible technology offers a framework for alleviating these impacts by fostering energy efficiency and sustainable material utilization. This helps guarantee that technological improvements in IIoT correspond with broader environmental sustainability goals.

The objective of expressing these concerns is not to generate a negative perspective of IoT technology, but rather to promote a full review of all aspects to ensure wise and sustainable development. IoT presents a significant opportunity for humanity, and careful study is needed to maximize its potential while reducing its threats.

Workforce Impact: Automation driven by IIoT may result in the displacement of specific jobs within various industries. An accountable strategy entails retraining employees, facilitating a seamless transition to new positions, and reducing job displacement.

The use of IoT technologies in specific industries or enterprises can result in substantial social effects. A significant consequence is the possible decrease in the necessity for human labor, since automation and intelligent systems may assume responsibilities formerly executed by employees. This could lead to people losing their jobs or the way they work changing. IoT can also make it harder for people to interact with each other directly because automated systems make human participation in many tasks unnecessary. People need to interact with others and make friends in order to be healthy, so the widespread use of IoT could have unexpected effects in this area.

In the light of these issues, the use of IoT technology should be carefully looked over again. It's important to think about the possible human and social effects in fields where automation could lessen the need for workers and social contact. IoT makes things more efficient and opens up new possibilities, but it's important to make sure that people can still interact with each other and jobs are kept. This way, IoT can serve society in a good, long-term way [5, 11].

Regulatory Compliance: IIoT solutions must follow all local and international laws and rules. This makes sure that industrial processes meet safety, data protection, and environmental standards.

When it comes to the IIoT, regulatory compliance means that IIoT systems have to follow all local and foreign laws, rules, and standards.

These rules involve things like safety, data security, and taking care of the environment. Compliance is very important to make sure that IIoT technologies not only help industries but also work properly and don't hurt people, companies, or the environment. Let's look at this idea in more depth [10].

3. Safety Regulations

IIoT systems are frequently employed in high-risk sectors, including manufacturing, oil and gas, transportation, and healthcare. Prioritizing the safety of personnel, machinery, and the general public is essential. Adherence to safety regulations necessitates the design and operation of IIoT systems that conform to industry-specific safety protocols. This may encompass:

- **Occupational Safety and Health:** Numerous nations implement regulations (e.g., OSHA in the U.S., ISO standards) that require safe working environments. IIoT devices deployed in industrial contexts must not provide dangers to workers, including equipment failures, inadvertent exposure to dangerous substances, or unsafe working conditions.
- **Equipment and Process Safety:** IIoT sensors and automation systems must adhere to safety standards that guarantee industrial equipment functions within safe parameters. Smart sensors employed to monitor machinery in factories must comply with regulations designed to avert equipment failures or explosions by facilitating early issue detection.

Noncompliance with safety standards may result in accidents, injuries, or fatalities, and it subjects businesses to fines, legal liabilities, and reputational harm.

3.1 Regulations on Data Protection and Privacy

IIoT systems aggregate substantial volumes of data, encompassing both sensitive operational and personal information. Adherence to data protection regulations is essential for responsible technology in the IIoT. This entails safeguarding data from unauthorized access, misuse, or breaches. Principal data protection regulations encompass:

The General Data Protection Regulation (GDPR) mandates that enterprises handle and process personal data with due diligence. In an IIoT setting, GDPR compliance guarantees that data gathered from employees, consumers, or other stakeholders is managed with appropriate privacy protections.

The California Consumer Privacy Act (CCPA) parallels the GDPR by safeguarding consumer data in California, requiring transparency and robust data security protocols.

IIoT solutions must implement encryption, secure communication channels, and access control mechanisms to safeguard sensitive data. Moreover, companies must guarantee that their data management policies are transparent and that individuals from whom data is collected are informed about its intended use.

3.2 Environmental Legislation

Numerous nations and international organizations have instituted environmental restrictions to alleviate the adverse effects of industrial operations on the environment. As IIoT solutions are progressively utilized to oversee and enhance industrial operations, they must be created and managed in accordance with these requirements. Adherence in this domain entails:

- **Garbage Management:** IIoT gadgets frequently possess restricted lifespans, hence exacerbating electronic garbage (e-waste). Adherence to e-waste legislation necessitates the responsible disposal or recycling of IIoT devices at the conclusion of their lives. The EU's Waste Electrical and Electronic Equipment (WEEE) Directive mandates that producers oversee the disposal of electronic devices, encompassing IIoT hardware.
- **Energy Consumption and Emissions:** Industries face pressure to diminish their carbon footprints. IIoT systems that oversee energy use and industrial operations must adhere to environmental standards designed to diminish energy consumption and greenhouse gas emissions. For example, enterprises may need to achieve the criteria set out in the Kyoto Protocol or municipal energy-efficiency laws.
- **Sustainable Use of Resources:** Using IIoT technology responsibly means following rules that encourage the long-term use of natural resources. IIoT systems should be designed to keep an eye on how well resources are being used. This way, companies can get rid of waste and improve processes while

still meeting standards like ISO 14001, which is all about environmental management systems.

4. Compliance Code for a Certain Industry

There may be different rules that IIoT systems must follow in different businesses. These rules make sure that IIoT technologies are used in businesses in a way that is safe, dependable, and trustworthy. Some examples are:

- **Health Care:** IIoT devices used in health care must meet standards like the Health Insurance Portability and Accountability Act (HIPAA) in the U.S., which protects user data rights and privacy. IIoT solutions in healthcare may include medical gadgets that monitor patients remotely or hospital equipment management systems.
- **Automotive:** IIoT systems employed in connected automobiles or smart vehicle manufacturing must comply with safety regulations such as ISO 26262, the worldwide standard for functional safety in automotive systems.

The IIoT in the aerospace and defence sectors must adhere to stringent security and safety rules, encompassing the protection of sensitive data, secure communication, and operational safety.

4.1 Ethical and Social Accountability

Beyond legal requirements, responsible technology in IIoT also requires adhering to ethical norms that enhance societal welfare and avoid harm. This includes:

- **Ensuring Fair Labor Practices:** IIoT technologies shouldn't be used to undermine labour rights or take advantage of workers. Fair labor laws must be followed, and it is crucial to prevent automation from causing a large loss of jobs without providing opportunities for reskilling or other forms of employment.
- **Preventing Discrimination:** Whether through automation or data analysis, IIoT solutions shouldn't add prejudices or discrimination. For instance, it's critical to stop data from being used in a way that discriminates or punishes workers in industries that are implementing IIoT systems for worker monitoring.

5. Reporting and Audits

Regulatory agencies typically conduct routine audits of IIoT-using industries to ensure compliance. These audits assess whether IIoT systems meet the applicable safety, environmental impact, and data protection regulatory requirements. Reputable businesses take steps to make their operations more transparent and to make compliance reporting easier.

A key component of responsible technology in the IIoT space is regulatory compliance, which guarantees the safety, security, and environmental sustainability of IIoT solutions. IIoT technology can be applied in ways that increase industrial production while preserving societal and environmental well-being by abiding by laws governing safety, data privacy, environmental protection, and industry-specific restrictions [10].

In conclusion, it is critical to emphasize the indisputable advantages that IIoT technologies provide, including increased productivity, lower costs, better product quality, increased safety, and predictive maintenance. IoT does, however, have its own set of problems and restrictions that need to be recognized and carefully investigated, just like any other technology. Given IIoT's potential to significantly impact industry, it is imperative to address a number of crucial issues in order to understand the long-term consequences of its rapid development [4].

A major component of industrial operations, such as factories and safety-critical control systems, is the IIoT. Its intrinsic complexity arises from the numerous hardware and software components, extensive subsystem relationships, and the rigorous standards for security, safety, and privacy. Responsible technology in the IIoT is vital for combining innovation and productivity with ethical, social, and environmental considerations. Regulatory compliance plays a critical role in ensuring that IIoT systems adhere to local and international norms for safety, data protection, and environmental stewardship. By addressing essential concerns such as dependable data storage, safe data management, privacy, sustainability, and workforce effect, IIoT technologies may be used responsibly. This entails adopting energy-efficient designs, employing recyclable materials, securing sensitive data, and maintaining compliance with industry-specific safety and environmental laws. Furthermore, IIoT technology must support fair labor practices, observe ethical norms, and adopt transparent regulations to preserve public trust and minimize negative outcomes like job displacement or environmental impact. In conclusion, the IIoT has the potential to revolutionize industries and promote long-term societal and environmental well-being by fusing technological advancements with moral, legal, and ecological principles.

References

[1] Khan, W.Z., Rehman, M.H., Zangoti, H.M., Afzal, M.K. and Armi, N., K. Salah. (2020). Industrial Internet of Things: Recent Developments, Enabling Technologies, and Outstanding Challenges. *Computers and Electrical Engineering, 81,* 106522. www.elsevier.com/locate/compeleceng.

[2] Farhan, A., Wang, A., Mingze, L. and Khaliq, R. (2020). *Innovation in the Context of IoT and Industry 5.0: Absolute Innovation Management (AIM) Framework.* MDPI, February.

[3] Bourgeois, J. and Kortuem, G. (2019). Towards Responsible Design with Internet of Things Data, In: *22nd International Conference on Engineering Design* (ICED19), Delft, The Netherlands, August 5–8, DOI: 10.1017/dsi.2019.349 ICED19.

[4] Sandro Nižetić, Petar Soljić, Diego Lopez-de-Ipiña, Gonzalez-de-Artaza and Luigi Patrono. Internet of Things (IoT): Opportunities, Issues, and Challenges Towards a Smart and Sustainable Future. *Journal of Cleaner Production*. www.elsevier.com/locate/jclepro.

[5] Dorian Oswaldo and Mora Sanchez. (2019). *Sustainable Development Challenges and Risks of Industry 4.0: A Literature Review*. IEEE. 978-1-7281-2171-0/19/$31.00.

[6] Jiafu, W., Shenglong, T., Zhaogang, Sh., Di, Li., Shiyong, W., Muhammad, I. and Athanasios, V. (2016). Software-Defined Industrial Internet of Things in the Context of Industry 4.0, *IEEE Sensors Journal*, *16* (20), October 15.

[7] Joaquín Ordieres, M., Javier Villalba, D. and Xiaochen, Zh. (2019). Challenges and Opportunities for Publishing IIoT Data in Manufacturing as a Service Business. In: *25th International Conference on Production Research, Manufacturing Innovation: Cyber Physical Manufacturing*, August 9–14, Chicago, Illinois (USA).

[8] Yong Yu, Ruonan Ch, Huilin Li, Yannan Li and Aikui T. Toward Data Security in Edge Intelligent IIoT. In: *Edge Intelligence for the Industrial Internet of Things*, IEEE Network, September/October.

[9] Gebremichael, T., Ledwaba, L., Eldefrawy, M. and Gerhard, H., Nuno, P., Mikael, G. (2020). Security and Privacy in the Industrial Internet of Things: Current Standards and Future Challenges. *IEEE Access*, Volume 8.

[10] Jinying Li, Ananda Maiti and Jiangang Fei. (2023). Features and Scope of Regulatory Technologies: Challenges and Opportunities with Industrial Internet of Things, *MDPI Future Internet*, *15* 256. https://doi.org/10.3390/fi15080256.

[11] Bandar, A. (2023). A Survey on Industrial Internet of Things Security: Requirements, Attacks, AI-Based Solutions, and Edge Computing Opportunities. MDPI, *Sensors*, *23*, 7470, https://doi.org/10.3390/s23177470.

[12] UNITAR, the United Nations Institute for Training and Research. (2024). *Sustainable Cycles (SCYCLE) Programme, The Global E-waste Monitor 2024*. Electronic Waste Increasing Fivefold Faster than Recorded E-waste Recycling: UN. ewastemonitor.info/the-global-e-waste-monitor-2024.

2 Foundations of Industrial Internet of Things

Shiwani Gupta[1*], Foram Shah[2] and Loukik Salvi[3]

The coming together of Artificial Intelligence and the discourse surrounding the Internet of Things (IoT) altered the landscape of technology in a critical way. Smart connectivity; the focus turned from the making of data-centric decisions. This part looks more closely at the intimacy of the worlds of machine learning (ML) models and IoT applications in interaction with one another. More profoundly, the section will dwell on the analysis of the mutually beneficial implications of linking the two, while seeking change and transformation across different industrial fronts. With an introductory discussion that covers the main subject under both ML and IoT, the whole section will serve to deepen the reader's understanding of the base technologies. It therefore traverses the taxonomy from the happenings during integration of ML models in IoT ecosystems to shed light on the problems and opportunities arising from this integration. Quite a large segment of it is on different machine learning models suitable for IoT scenarios, again, features quite a range from the most conventional supervised learning algorithms to the very advanced deep learning architectures. The section sheds light on the merits, restrictions, and practical applicability within the IoT framework. In the face of tremendous importance, we gave special attention to these challenges: issues such as heterogeneity, high volume, and the real-time nature of IoT data Additionally, this section will broach the practical aspects, laying out definite guidelines for selecting, training, and deploying ML models in IoT environments. Practical concepts will be discussed using real-world business examples of lessons learned and technical case studies with concrete approaches that led to significant benefits through the use of ML in extracting intelligent insights from large amounts of data generated by an IoT device.

[1] Department of AI ML, Thakur College of Engineering and Technology, Mumbai, India.
[2] Department of Computer Engineering, Thakur College of Engineering and Technology, Mumbai, India.
[3] Department of Computer Engineering, Thakur College of Engineering and Technology, Mumbai, India
Email: foram.shah@thakureducation.org[2]; loukik.salvi@thakureducation.org[3]
* Corresponding author: shiwani.gupta@thakureducation.org

DOI: 10.1201/9781003587903-2

1. Introduction

1.1 Foundations of Industrial Internet of Things

Technology is continually evolving, and so is the definition of the Internet of Things (IoT), which brings a shift in a way not imaginable, even earlier; dumb objects can now become smart with the touch of communication and intelligence. IoT as an emerging concept creates a weaving mesh of devices, sensors, and systems and, as such, it forms sophisticated structures for connecting and communicating, exceeding the system conventions in traditional human-computer communication. The most literal interpretation of what IoT can mean for human beings is the capability of everything—from home appliances to industrial devices—to collect, transmit, and act on real-time data. This vast connectedness offers numerous possibilities, ranging from routine tasks in a smart house to managing massive production processes in the most efficient manner possible.

IoT has its historical roots in the very early years of the 20th century and was originally founded upon visions of machine-to-machine communication. The convergence of advances in sensor technology, wireless communications, and the ubiquity of the internet spurred mass adoption of the IoT in the modern world. The next generation of IoT will be smart because it finds itself spreading into every corner of living, working, and experiencing the world. Some smart applications are found in smart cities where management is honed to make cities perform better through integrated, efficient traffic flow and waste management, as well as the use of energy effectively and efficiently. In the field of healthcare, the next-generation IoT will make possible various wearable devices with smart sensors facilitating continuous patient monitoring and personalized intervention in health management. Agriculture has applied IoT to unleash its own revolution—in precision farming that optimizes crop yields and resource usage, based on readings collected from sensors. The IoT seed-boosted potential carries a suite of issues with it. The real strength of the latest potential IoT is likely to be built on security issues. Huge amounts of information are exchanged within such an infrastructure—in fact, an unprecedented plethora of information hence a need for privacy and integrity encryption that must be addressed through robust solutions. Standardizing IoT will be necessary to ensure seamless interoperability among the myriad of devices and to ensure ethical deployment of these technologies.

These are the very target issues for a thorough exploration and innovation. These are what can bend and stretch technology to the maximum: security, privacy, and one pass to help make things function—standardization. From this putative tone, not only tech cracking but also revolutionizing forces that shape the exit personnel which future downloads were to the crucible of opportunity and challenge, but that they have become most engaged research projects to understand the IoT.

What IoT represents with the advent of data-processing modernity by everyday articles is another revolution using smart objects, which provide devices and rendering them smart in everyday life. It creates a network, interacting in a complicated way, with things, sensors, and systems that work into one another outside traditional human-computer interaction. This new empowerment means objects, from household items to huge things in industrial zones, taking data from what's happening now and measuring, sharing, and in many cases acting on that information in real time. It is truly very promising—underscoring the effectiveness of daily tasks in smart homes or boosting enterprises with new ones with unprecedented efficiency. Therefore, sensor technology advancements, wireless communications, and internet ubiquity pave ways for the establishment of the first stepping stone to IoT technology. As an influential power, IoT acts at full range in such sectors as smart cities, healthcare, agriculture, and manufacturing and promises to shift societal dynamics. However, the challenges IoT faces in terms of security, privacy, and standards are immense for its potential to be realized fully. This journey steps out of techie issues, venturing further into the exploration of the changing faces and forces of our digital future. It is an invitation to researchers and practitioners to delve into the inquiry, dissect the intricacies, and harness the transformative potentials within the domain of the IoT [3].

The foundational elements include:

- Connectivity through sensors, actuators and communication protocols like MQTT, OPC UA, and CoAP that facilitate data transfer between devices and systems.
- Data management through data acquisition, storage, and analytics.
- Edge computing to process data closer to where it is generated to reduce latency, bandwidth usage, and enhance real-time decision-making.
- Ensuring that various devices and systems are interoperable and work together seamlessly through standardized protocols and APIs (application programming interface).
- Implementing measures to protect data integrity, confidentiality, and availability. This includes securing devices, networks, and data storage against cyber threats.
- Utilizing cloud services for scalable computing resources, data storage, and analytics capabilities, which enable organizations to handle vast amounts of data generated by IIoT (industrial internet of things) devices.
- Leveraging AI (artificial intelligence) and ML (machine learning) algorithms to analyze data, predict failures, optimize operations, and support decision-making processes.

1.1.1 Connecting Devices Using IoT

Within the context of the IoT, the core concept of connecting devices serves as a foundational element essential for enabling seamless communication and collaboration within the network. Device connectivity involves establishing a robust and efficient linkage between physical entities, allowing them to collectively share information and execute tasks. This interconnectedness is pivotal for unlocking the potential of IoT, facilitating coordinated actions among devices in smart homes, industrial environments, and other application domains [6].

Connecting devices is a foundational principle for the IoT, and it is highly essential that devices are able to connect in order to be able to communicate effectively and efficiently within the network. Typically, it means building a solid and efficient connection that allows physical entities to work together in the sharing of information and pursuing activities. It opens up an environment where devices can work together in smart homes, industrial environments, as well as other application domains for the optimal exploitation of the IoT.

Sensors act as an essential component with respect to establishing communication among the devices located in any IoT environment. Be they embedded modules in this or that entity, all these little gadgets are quite a lot of sensors, which are capable of gathering physical values around themselves to process in the machine. These sensors give another dimension to devices, and then they collect data, the analysis of which is generally an essential process in decision-making. Thus, a thermostat having a temperature sensor or a security camera with one or two motion sensors; the application of the sensors itself serves as the basic building block on which any interconnected devices can work.

1.1.2 Data Collection and Processing

Data drives most processes within IoT resulting in life in the IoT; in the context within which data may be useful in making decisions as well as doing things. Data primarily captures the state and performance details of all connected devices. Data forms when sensors make it possible to make informed decisions about process optimization people and buildings, and application development that uses intelligence. Most IoT benefits derive their realization from data.

Some of the tremendous amounts of data in IoT often come from sensors placed inside devices. The sensor embeds itself within the device to record all physical phenomena, meaning body temperature, humidity, motion, and many others. So, the data that is collected through sensors is varied and dynamic and becomes the root of understanding everything in the organization. Analyzing such data makes better informed decisions in agriculture, e.g., monitoring crop condition; or health, e.g., keeping track of vital signs. Sensor-gathered data becomes an imperative element for the successful functioning of numerous IoT applications. It is mandatory to manage the vast quantity of data that interconnected devices generate by processing it efficiently. In this context, cloud service makes the biggest deal of noise in IoT. Through cloud service, the infrastructure and computational resources necessary for storage, analysis, and knowledge extraction from very large amounts of data. This cloud computing phenomenon arises when the infrastructure starts utilizing the existing IT through which some hardware may be implemented, while the rest gets driven by cloud service; IoT exploits the same beneficial technology implementation for mining value from archived data. This would allow industrial utilization of intricate calculating tasks along with machine learning algorithms and would offer scalable connectivity, which could thereby enhance the overall efficiency in the ecosystem of IoT.

1.1.3 Steps involved in Data Collection and Processing

1. **Sensor Deployment**
   - *Selecting Sensors:* Choose suitable sensors based on the parameters to monitor.
   - *Deploy Sensors:* Place sensors strategically in the environment, considering coverage area and environmental conditions.
2. **Data Acquisition**
   - *Calibrate Sensors:* Ensure sensors provide accurate measurements.
   - *Sample Data:* Determine the frequency of data collection.
   - *Transmit Data:* Use wireless technologies for data transmission to the central hub.
3. **Data Aggregation**
   - *Central Hub:* Aggregate data at a central hub.
   - *Integrate Data:* Unify data from different sensors.
   - *Timestamp Data:* Maintain chronological order.
4. **Pre-processing**
   - *Noise Reduction:* Eliminate outliers for improved accuracy.
   - *Clean Data:* Address missing or erroneous data.
   - *Normalize Data:* Standardize data scales.
5. **Data Storage**
   - *Manage Database:* Store processed data, considering scalability and reliability.
   - *Secure Data:* Implement measures for data security.
6. **Real-time Processing**
   - *Edge Computing:* Process data in real-time at the source.
   - *Complex Event Processing:* Identify patterns in real-time data streams.
7. **Cloud Integration**
   - *Utilize Cloud Storage:* Optionally transfer data for scalability.
   - *Leverage Cloud Analytics:* Use cloud-based tools for in-depth analysis.

8. Data Analysis

- *Descriptive Analysis:* Summarize data characteristics visually.
- *Diagnostic Analysis:* Identify causes of events or anomalies.
- *Predictive Analysis:* Use ML for future trend prediction.
- *Prescriptive Analysis:* Suggest actions based on analysis.

9. Decision-making

- *Visualize Data:* Present insights through dashboards.
- *Implement Alerts:* Set up automated alerts based on predefined thresholds.
- *Integrate with Systems:* Connect data insights with decision-making systems.

10. Continuous Monitoring and Improvement

- *Establish Feedback Loop:* Continuously monitor and enhance processes.
- *Adapt System:* Adjust the system based on changing requirements or technological advancements.

1.1.4 Some Applications of IIoT

Given the above small number of potential incidents, the IoT will soon surround the majority of people, playing a very essential role in the interaction between equipment in everyday living activities. Thus very small and mobile devices, harnessing surplus energy, will be able to easily merge into their internet-connected neighbors. Whenever you pick up a probe that is linked to the reader, for example, that very instant the larger unit's knowledge, combined with its own, becomes accessible to you. With the smallest available units, we are closer to living space knowing more about activities that will help to estimate the real situation. Sites will be seen as entities collaborating and generating solutions.

In the realm of manufacturing, IIoT stands as a transformative force, particularly through predictive maintenance. By embedding sensors in machinery IIoT enables real-time monitoring of performance metrics, and predictive modeling algorithms analyze this data to forecast potential equipment failures. This predictive approach not only enhances operational efficiency but also translates into substantial cost savings and resource optimization. Precision agriculture benefits significantly from IoT applications, where data from soil sensors, drones, and other devices are analyzed through predictive modeling algorithms. This data-driven approach guides farmers in optimal planting times, irrigation schedules, and nutrient management, fostering sustainable agricultural practices.

In the retail sector, the IoT contributes to a more personalized and immersive customer experience. Smart shelves equipped with RFID (Radio Frequency IDentification)sensors, along with predictive modeling, monitor inventory levels, ensuring optimal stock and reducing instances of product shortages. Wearable devices, the quintessential representation of IoT in personal well-being, have been seamlessly integrated into daily routines. These devices, including fitness trackers and smartwatches, collect data on physical activity, heart rate, and sleep patterns, not only promoting individual well-being but also contributing valuable data to broader health initiatives and medical research. Within homes, the concept of smart homes, enabled by the IoT, has become a reality. Devices such as thermostats, lighting systems, and security cameras are interconnected, allowing for remote control and automation. Predictive modeling in smart homes anticipates user preferences and behavior patterns, adjusting settings to enhance energy efficiency, security, and overall comfort. While the benefits of IoT are evident, challenges such as security, privacy, and interoperability require ongoing research and innovation. Addressing these challenges is crucial to unlocking the full potential of IoT, ensuring a future where interconnected technologies seamlessly integrate into our lives, workspaces, and urban environments, reshaping our collective experience.

2. ML/DL Models for IoT

2.1 Introduction to ML and DL

One way to think of ML is as an application of AI that gives systems the capacity to automatically learn from their experiences and get better without explicit programming. When a computer uses deep learning (DL), it can learn from inputs such as text, images, or sound and respond appropriately.

Both ML and DL provide algorithms for classifying data and training models. Let's begin by talking about a cat versus dog scenario. Which animal do you identify in a certain picture—a dog or a cat? How do you respond to that? You probably know. Recognizing cats and dogs will probably be more comfortable as you have been seeing many of them over the years. We may state that recognition and extrapolation of instances are among the things we wish to teach computers. Training the ML model so that the computer does the categorization would entail selecting manually which features of an image are significant in considering it, such as edges and corners. The former is then the model's template as well as serves as their means of studying and recognizing new objects. This is like how we recognize an object. But it can be used for object detection as well as scene recognition. There is a process you follow when challenged with an ML task. An image, viewed at the beginning, is the source from which an appropriate summary of properties is obtained. This summary may contain work like building a classification of an object. But deep learning extricates man from the tedious task of extracting features from images. Rather, it takes the image directly from the user and then predicts what the object is about. This is DL, a kind of ML that processes images displayed to it. It will be considerably more difficult to solve and map out visually.

2.2 AI and ML Algorithms for IIoT

In the landscape of technology today, ML, DL, and the IoT are poised to revolutionize how we interface with and come to understand the world. Indeed, the fusion of these three elements really opens up the possibility of new knowledge and efficiency in all aspects, from smart homes up to wearables and production processes. How can they fit in and complement each other in this new field? Most importantly, how can each learn from the other? And even more profoundly, how will they transform each other? That is our focus here concerning the visually impaired and wearable devices.

By using a number of smart devices, including the latest wearables, smartphones, eyewear, and sensors, efficiency and simplicity of human life have been ensured [9]. Therefore, input data to machines is now available over the Internet. ML algorithms made it possible to transform raw data into useful insights. Mathematical models can learn valuable patterns in the data that have been observed and can be used to predict future growth [13]. For example, object detection is the task in the field of computer vision: to know what objects appear in a particular image and acquire knowledge of where they are in that image. Object detection draws bounding boxes around those discovered objects and tags it with the name of the object. The model predicts the location of each object in the image and defines the label that should exist in that identified object.

Before the advent of DL in 2013, traditional ML methods dominated the techniques. Examples are Histogram of Oriented Gradients, Scale-Invariant Feature Transformations (SIFT), and the object detection technique of Viola-Jones. They could be used for Logistic Regression, color histograms, or Random Forests, which detect the properties of images by using common methods to classify them into clusters. These systems were essentially negative compared to the enormous successes obtained by the existing DL approaches. Extensive research on AI and the IoT is leading people to think about IIoT as one of their most powerful associations as they think as they incorporate their network of devices into a more complex web of human-centered technology.

From automation to the personal use of technology, we see how everything is getting smarter-adapting to nullify human intervention in operations and logic emulation thereof to a great degree.

- **Data Acquisition:** IIoT relies on a wide array of sensors that collect real-time data from machinery, production lines, and environmental conditions. These can include operational data (temperature, pressure, vibration), machine logs, and external data (weather, supply chain information).
- **Data Preprocessing:** Raw data from IIoT devices can be noisy or incomplete. Preprocessing techniques are used to clean and filter this data to improve quality. Techniques such as normalization help standardize the data, making it suitable for ML algorithms.

- **Data Storage and Management:** Data can be stored in cloud solutions or on local edge devices, depending on requirements for speed, security, and access. Tools like Hadoop and Apache Spark manage and analyze large data sets generated by IIoT systems.
- **Feature Engineering:** Identifying key features from raw data that enhance model performance. For instance, deriving operational metrics from time-series data. Techniques like Principal Component Analysis (PCA) help in reducing the feature space to focus on the most impactful variables.
- **Machine Learning Model Development:** Utilizing labeled data to train models for specific tasks, such as predicting equipment failures or quality outcomes. Applying clustering for anomaly detection methods to identify patterns or outliers in operational data without predefined labels. Using algorithms that learn to make decisions based on the outcomes of previous actions, applicable in optimizing manufacturing processes.
- **Model Deployment:** Deploying ML models on edge devices allows for real-time data processing and immediate decision-making, crucial in industrial environments. ML models can be integrated into existing IIoT systems via APIs, enabling seamless communication and data flow.
- **Real-Time Analytics and Insights:** ML models analyze historical and real-time data to predict when equipment is likely to fail, allowing for proactive maintenance. Using ML to monitor production quality and identify defects in real time, reducing waste and ensuring product consistency.
- **Feedback Mechanisms:** Implementing feedback loops that allow ML models to learn from new data and adapt over time, improving accuracy and relevance. Systems that can modify operations based on real-time analytics and ML predictions to optimize performance.
- **Visualization and Reporting:** Providing intuitive visualizations of ML outputs, helping stakeholders understand insights and make informed decisions quickly. Generating reports based on ML analysis to track performance metrics, maintenance needs, and operational efficiency.
- **Cybersecurity and Risk Management:** ML algorithms can identify unusual patterns in network traffic or device behavior, alerting organizations to potential cyber threats. Utilizing ML to anticipate and mitigate risks by analyzing trends and vulnerabilities in the IIoT ecosystem.

Let's take a quick look at the object detection techniques mentioned in this chapter: [1]

- **R-CNN:** To extract the ROI (region of interest) from the input image, R-CNN uses a selective approach. ROI is a rectangular shape that shows an object's edge in a picture. An image may have a lot of ROI. A neural network receives each ROI in order to generate output features. Several support vector machine classifiers are employed to identify the kind of item that is present in ROI.

- **Fast-RCNN:** Unlike R-CNN, which passes ROIs to a neural network to build output features, Fast-RCNN passes the entire image. Next, every ROI is separated from the network's output tensor using a brand-new technique known as ROI pooling. The Fast R-CNN generates its area proposals by selective search, just like the R-CNN.
- **Faster R-CNN:** This algorithm is different from Fast-RCNN in that it doesn't employ a selective search technique. Faster: RCN incorporates ROI generation directly into the neural network architecture.
- **Mask R-CNN:** While R-CNN is primarily concerned with object identification, mask R-CNN further incorporates segmentation capabilities. Additionally, it substitutes the new ROI align approach for the ROI pooling method. A fraction of a pixel can be represented by an ROI alignment.
- **R-FCN:** Compared to Fast R-CNN or Faster R-CNN, R-FCN is faster. This is due to the fact that Fast R-CNN or Faster R-CNN repeatedly apply an expensive per-region subnetwork.
- **HOG:** A feature descriptor is the Histogram of Oriented Gradients (HOG). The HOG method measures how often gradient orientation appears in specific areas of an image. The image is separated into tiny, interconnected areas called cells in this case, and a gradient-direction histogram is generated for each pixel inside a cell.
- **SPP:** Fixed-size input images are necessary for the convolutional neural network. Nonetheless, regardless of the size of the input image, the novel network structure known as spatial pyramid pooling can produce a fixed-length representation. For this reason, SPP (scalable parallel processor)-net algorithms are faster and have similar accuracy to R-CNN algorithms.
- **Single Shot Detector:** SSD has an extremely accurate object detection algorithm and operates at a much faster pace. This object detection system is based on regression. SSD detects several things in a picture in a single shot.
- **Yolo:** Convolutional neural networks (CNNs) are used to detect objects in real time; you only need to glance once. Because the method only examines the image once, making predictions only requires one forward propagation through the neural network.
- **BlitzNet:** This technique completes semantic segmentation and object detection in a single forward pass. Real-time calculations are made. The task of semantic segmentation involves parsing an image and assigning a class label to every pixel.

3. Experimental Methods and Materials

3.1 Maintenance Cost Reduction of IIoT Devices through Predictive Techniques

In different sectors, predictive maintenance guards against impending failures of equipment under 4.0 case studies before any systemic eventuation of catastrophic failures happen, preventing non-functionality. In fact, this leads to increased availability of equipment for lower maintenance costs.

Thus, sensors perform condition monitoring to create meaningful figures in assessing the health of devices or equipment by obtaining data to watch some of the critical parameters' readings, such as vibration, sound anomalies, air flow, and current. Advanced analytics and AI forecasting for machine failures or suggested maintenance before they become serious is one way of using predictive maintenance. Everything may be monitored, from individual parts to fully intelligent systems. An example in application is hospital or industrial environments, with HVAC units being even more critical since they are often in immediate square footage.

Thus, we present a use case wherein available data from sensors was used to provide a predictive maintenance solution:

A company has a set of digital devices that provide sensor data daily. The organization will like to build a predictive maintenance solution that will identify in advance when maintenance is due. This would save cost when measures are carried out on a reactive basis over preventive maintenance performed routinely or at events.

You have been charged with developing an ML prediction model that will predict the likelihood of a device failure. The expectation in this process is that great effort is put in place to reduce the number of false positives and false negatives. Failures represent the target, so it is 1 if it failed, and 0 if it did not.

3.2 Wearable IIoT Device for the Visually Impaired

With that being said, this project aims to reduce the challenges faced by the visually handicapped, thereby making this world a little more equal. The project is aimed at designing a smart wearable device that speaks, where DL and ML are used to analyze walking barriers, making it possible for blind people to walk around easily. It also offers other recognition capabilities in the form of text and face recognition [15]. When paired with ultrasonic sensors, the device warns the user before an obstacle, allowing him or her to move around freely, but with caution. The speech aid will also work in conjunction with an application companion [4], development of high empowerment of the already feature-rich device, with an incorporation of a number of new capabilities. There are several additional features in the mobile app that will make life easier for users. The voice assistant keeps updating the latest developments for the user. Most helpful in the SOS functionality, it will essentially send a message of distress to his or her preferred emergency contacts. Earphone supplements are then utilized for notifying the blind about the most recent activities on this device.

The wear-on wrist device has been developed meticulously to make the user experience a memorable one. One active hand of the user will be used to trace the engraved symbols on the buttons. The design of this particular gadget would make very fundamental activities of daily living easy to manage by individuals. A Raspberry Pi 4 embedded system module was added, which is responsible for managing computational tasks. The ultrasonic sound sensor utilizes ultrasonic sound waves and helps the device estimate the separation of an obstructing object to about 40cm, alerting users ahead of time to avoid any obstacles. The camera present in the Raspberry Pi is used along with the complete computerized computer vision faculty for measuring the local radius of the sensors. Rather, the appointment includes five switches in it. This also holds one rechargeable LiPo battery source. Accessories included in it are, but are not limited to, a Raspberry Pi camera to measure exactly how far away an object is in terms of sensor radius and deal greatly with computer vision. Besides this, it is composed of five switches, each with different functions, and a vibration motor to provide vibro-tactile feedback. The operational characteristics of the buttons are the following:

The Activate button toggles the voice assistant in an on- and-off state.

The Volume Up and Power buttons adjust the voice assistant's volume, respectively.

The Volume Down button decreases the volume of the voice assistant.

MobileNet-based mobile object detection has brought some projects into being. It is mainly intended for embedded and high-speed applications. Despite having been fast with a high frame rate in low-computation devices, it seems very criticizable because of the decrease of accuracy. To overcome such compromises, the MobileNetSSD architecture is designed. It is an object detection model responsible for extracting the category and location of an object within an input image. MobileNetSSD operates under the Single Shot Detector methodology, which is novel with its capability of requiring only one image for the identification of numerous objects in an image.

Following the feature extraction with the MobileNet backbone in the SSD , the process involves the generation of bounding boxes at each resolution. Extra feature layers are employed to reduce the resolution effectively. In the MobileNetSSD model, the results from six levels of resolution are concatenated to produce 3000 bounding boxes. Subsequently, a filtering process is applied to eliminate redundant bounding boxes using a technique known as non-maximum suppression (NMS).

A model of determining the maturity stage of the fruit has been developed, which relies on MobileNet SSD algorithm, and categorizes the fruits into three categories from the three different types (i.e., ripe, raw, and rotten) for each different fruit type (banana, apple, orange, mango, pomegranate, and papaya). The integrated mobile application was developed into voice assistant having additional functionalities that go beyond object detection. The key things include:

- **Web Scraping:** Using Beautiful Soup: Using Beautiful Soup, a Python package for HTML and XML parsing, data is scraped from the web, for instance, live cricket scores.

- **Weather Information Retrieval with the OpenWeatherMap API:** Employing the OpenWeatherMap API to retrieve weather information based on the desired location and the user's geolocation from their phone, providing detailed weather conditions.
- **Wikipedia Search Word Python Library:** A python library that makes it simple for you to get and parse Wikipedia information. This means getting Wikipedia information quicker by voice-based keywords. It evaluates the Wikipedia articles from the keywords through voice input.
- **Retrieving News Headlines with the News API:** This will help the Google Assistant to give the most recent headlines from the News API, which helps in the effective retrieval of News Headlines in recent days. Also, it uses the REST API with JSON search results obtained for diverse news articles currently available from a lot of sources along with covering historical news.

It has been suggested that one can produce more than one category of results with object recognition and information retrieval in today's voice assistant. Configuration-wise, it relies on Tesseract, which embeds Long Short-Term Memory (LSTM) [7]. In the Tesseract wrapper class, the Pytesseract [16] serves as an interface for the Optical Character Recognition (OCR) Engine to integrate the Tesseract Engine into the code [14]. LSTM is beneficial for its sequential processing ability in identifying patterns and contextual dependencies in text data, with improvements in the character recognition accuracy aspect.

We have accomplished this by browsing a site with the aid of opening web browser selenium and then getting ourselves a dataset. The Selenium open-source tool auto webdriver will search for all the URLs in the fetched HTML from a URL, save them in a set, and return this set. Now let us create a function that will accept the image URL and save the image in a particular folder. That is how images were made for all three classes of each fruit, i.e., rotten, ripe, raw. The fruits' total number selected for the job came down to six: apple, banana, mango, orange, papaya, and pomegranate. Thus, 1987 images from 18 categories were to be used as training images for our model.

Possessing some fundamental characteristics of mainly the firm ripeness detection model, data is pre-processed in which the necessary steps require first dividing material into image data first because it is necessary to utilize the object of Keras, the ImageDataGenerator class. Keras, an open-source library named after berries, will further help with creating access input for use into deep learning. The transformation applied was that the images got rescaled for normalization, which in actuality reduces the image to essentially a 224-shaped input. This is an assignment where we split up the dataset and image batches into label batches. The dataset shuffling happened, and as a result, validation sets have already been deployed with an approximately 20% fraction. Transfer learning means using a model of development as adapted for solving a problem in the new problem structure. So we will start developing our model by importing a pre-trained MobileNetv2 model hosted on Tensorflow Hub. This machine will work better because of its uses, for instance, in our embedded devices such as the Raspberry Pi. The two-phase training is performed when only a part of the data is trained. OC is a process with multiple subprocesses that must be accurately completed. In the first subprocess, we preprocessed the image OpenCV includes image pre-processing tools, while it reads the image grayscale since this is needed for the next two algorithms. The applied algorithm is the bilateral filter [10] which, besides being a picture smoothing filter to keep edges, removes all noise pixels by creating the average intensity from pixels into the weighted mean to minimize the transition of pixel intensities between any two neighbors. This allows one to have sharp edges. Next,we use the processing of the thresholding algorithm, where any black color is rendered ultra-black and white into snow-white thresholding. The threshold value is calculated as a mean of each pixel in the image.

The voice assistant's settings can be changed on the fly with the setting configurator screen in the mobile application. This feature offers a variety of customization options that users can use to alter their experience with the assistant. Forgetting is a natural process in which the brain replaces old information with newer, more valuable, and relevant information. Regretfully, you might need to recall certain unimportant details from time to time, like where you put your wallet, phone, or keys. If this happens, you'll have to spend time trying to find some of your regular items. Therefore, one of the main reasons people misplace or even lose personal items is simple forgetfulness.

This is done by using the FindMyDevice function. As a result, it really changes very much to knowing the location of the phone that is not tied with the individual. There is ease in identifying where the device is located because this functionality has been incorporated by clicking a single button on the mobile application's home page. When the device is said to produce sound for a specific amount of time, the device responds.

One way we suggest integrating this is a mobile application [2, 11] that connects modules directly with the voice assistant application in order to supply them with a whole new dimension. First of all, text messages could be produced simply without the phone having to be interacted with completely, along with tight cooperation with the assistant to accomplish the reading of the text message feature. In addition, it would enable the easier composition of text messages without signal interaction with the phone. There will also be a possible way to locate the voice assistant device if the user should go and misplace it or cannot locate it easily.

Activating the SOS feature-functionality [12] would provide help for the user in case of the below: one is that the user has been faced with trauma or the failure of one of the essential parts of the system. The device detects the circumstances and future distress signal is then stored behind an integrated mobile consequently instructing the application to send an SMS while informing five pre-determined contacts within the KN to the W for Pennsylvania information from this user's current location. While configuring the device in the mobile application, the Emergency SMS recipient contacts could also be chosen. Use the features of mobile app alerts or by sending timely messages to your consumers to keep alive a nonintrusive communication with them.

To go around the limitation stated above, we connected the Voice Assistant to the Mobile Application for the out-linked phone's messaging system. The user asks the Assistant for a message readout or types a text message. Upon voice assistant request by the user, the camera will capture the desired subject, convert it into a vector, send that vector to the Central Processing Unit for recognition, and return the audio file back to the user. An SSD algorithm is being utilized for the operation of the proposed object detection model [18]. The model employs a MobileNet back-end which supports mobile devices-such as Raspberry Pi-for the model. The model has been trained and can identify about those 1,000 major objects. Provides a list of such objects that are detected. A confidence threshold of 60 percent or more in detection is acceptable.

It is also considered taking care of an event in which an effect occurs on the user stumbling or falling over a minute obstruction, such as a raised platform. When such events at times result in death, practically speaking, all the nearest relationships should be duly informed about the happening soonest so that the person may easily await help. This really permits a system for the faulting or, rather, the fallen person. This device does mark the act on an event of fall and will alert the contacts preconfigured on the user falling in need of help with an example of detecting a fall function of the Fall Detection Device Module. Hardware, including a fall circuit, as shown in the image below, is needed to realize this module. Accelerometers and some radio wave sensors providing low power consumption will create a significant portion of this module on the circuits that can monitor the actions of the user. The most unique system can actually have an improvement in fall detection accuracy by integrating gyroscopes, infrared sensors, acoustic sensors and so on. It is, however, expected to relate data with the partner device mobile app regarding any fall, when an incident happens. The data will include the location of the user, activating alerts, over where the fall happens to the contacts saved in the mobile application. For instance, this trend must prevent such behavior in which the user has lately become more assured following the use of post-event damage management.

4. Results and Discussion

4.1 Predictive Maintenance Use Case of IIoT

Looking at the data, we clearly understood the fact of the data being that of 1,169 devices for which it had all collection periods from 1 January 2015 to 2 November 2015, and 9, as sensor readings, as metric readings. By time, the different fault pattern senses were recorded under the framework of metrics on each device. For one time in our case, time movements are captured. Actually, the data is seriously imbalanced. It amounts to 124,494 instances, with 124,388 cases where there was no failure and only 106 where there was the failure. This makes 8.5*10-4 as the imbalance ratio. This particular case can be solved with anomaly detection or outlier detection. Another approach to the problem may be transforming the metric values; however, we cannot talk about it as we are unaware of the domain. Big data has not been processed for DL models. Structures data can be subjected to the types of ML modeling.

In order to address the preceding issue, we imported the following libraries, namely, numpy, pandas, matplotlib, seaborn, statsmodels, sklearn, imblearn, etc., and then went ahead to import metrics, preprocessing, model_selection, and linear_model modules in Sklearn. From Imblearn, we used its ensemble and over-sampling modules.

Next the date is imported, and we get the brief description of DF. A column was added by which dates will be captured against discrete failures. We set it up in this way. So we turned it into a datetime object using dt.date. This is just pulling the day from date and naming it day. After analyzing the statistical properties of data, it has been deduced that there are two exactly the same sensor reading metrics; this is metric 7 and metric 8, which are not at all required metrics, so we need to remove them.

Three hundred and seven totals only rely upon different dates. Similarly, not all sensors are designed to offer values for different instances. Metric 1 is dated 12,387 times, while Metric 2 has only 560 data. Metric 3 has only 48 scores; metric 4 is for 115 cases; metric 5, there are just 60; but metric 6 has some 44,838 data. Metric 7 and 8: The track might go up to a maximum of 28 observations but approach for metric 9 could approach as high as 67 observations.

We conducted our EDA (exploratory data analysis), and later in BoxPlot, devices began to break mostly on Mondays. Devices such as S1F and W1F addressed most failures more compared to Z1F. From the graph given below, most complaints were recorded on Monday and least on Sunday.

Out of all the metric 1 value, the number of values are between 0–244140500, the range of metric 2 falls within 0–64968, metric 3 values lie between 0–24929, number of values metric 4 ranges within 0–1666, number of values metric 5 ranges from 1–90, metric 6 ranges from 8–689161, while metric 7 and 8 are from 0 to 832 and metric 9 lies between 0–70,000. It could be due to data loss when the sensor results in a value of zero, without having to have data losses due to the absence of no domain knowledge, even if the recorded data is questionable.

There are no null values in the data; no values are missing. One duplicate row has been removed.

Based on measuring the outlying values, metric numbers 2, 3, 4, 6, 7, and 9 are left skewed and they are being transformed. For meaningful outliers, z score can be used. For instance, the SD (standard deviation) method can be applied for the removal of outliers. Because too many outliers are presented and can make the DataFrame empty first, outliers can reduce the DataFrame.

On noting this, the data distribution by plotting a histogram was found to have a skewed distribution in metric 7 and 8. As the data is cyclical, the histogram doesn't suggest any relation between date and failure.

By plotting this FacetGrid to visualize the distribution of one attribute with another and the relationship of the two outcome classes in the chosen metric, we find that metric 1 is not representative of failing. Low metric 2, 3, 4, and 7 values tend to cause failure due to their high variability. As per the two subclasses, metric 4,9 seem to develop similar behavior. They do not discriminate between a failure and a non-failure class.

Then, we create a barplot to compare metrics between two subgroups of the outcome variables. We observe that metric 1, 5, 6 do not stand to be discriminative, while 2, 3, 4, and 7 account for potential prediction.

For collinearity, and multi collinearity, metrics 7 and 8 are considerably correlated.

On the scatter plots, we can infer that for metric 3 differing values could suggest failure, that for metric 7 there are circumstances that preclude any differentiating failure, and metric 9 could suggest failure. No linear relationship is white and visible between the metrics and the RP file.

After all these processes, the model is fitted to the processed data. Logistic regression is an interpretable probabilistic binary classifier; scaling is performed before hauling the first building of this model at place. Stratification is carried out since the set is imbalanced. Similarly, the RFE wrapper with Chi2 is the filter-based feature-selection method that accounts for the same number of features as relevant.

Using imbalanced data for simple logistic regression, we had to go with RoCAuC as our evaluation metric since we could not train our model on the minority classes. So we had to make balanced logistic regression training by computing an average AuCRoC on our train data at 0.768 and on our test data at 0.748. The chosen model is now set as the base classifier to get the integrated random forest use on top of it; DT. After checking feature_importance, we see that metric 2, 6, and 7 provides almost no information for failure prediction. Therefore, we go for a balanced logistic regression classifier with Cost-Sensitive Learning using RoCAuC value of.505 on train and 0.536 on test data. I designed my model even more by hyperparameter tuning through Randomized SearchCV as regards the outliers and set it to three folds. I could now obtain 0.93 for RoCAuC in the training set and 0.50 for the test set. This means that the model appeared to be overfitting. It is due to the fact that the samples in these situations are taken from different distributions for test and training.

In this work, we have tried oversampling, random under sampling, and the classifier called Easy Ensemble. Another alternative could be Isolation Forest, which looks into anomalies in this data at a later stage. Last but certainly not least, we forecasted probabilities of the samples as a way to predict which device would fail on which date and with what probability. Throughout the report, every effort was made to reduce the number of false negatives and positives, by a constant criterion.

4.2 Wearable Device for the Visually Impaired

By embedding Raspberry Pi technology, the visually impaired can see their experiences as direct users or even actuator operators, where they will use simple hand gestures thereby commanding their personal voice assistant, recognizing easily what plausible things might be surrounding him. With wearable Raspberry Pi technology, it is entirely possible to turn the lights on by just one simple hand movement-very useful with the visually impaired. It is therefore an advancement in the technology of designing assisting devices for individuals who have a low level of visual impairment. Perceptual usability becomes the chief concern here—investigating visually impaired individuals' needs and how useful the devices are in a particular field come third. The pitch: Whenever we stand at crossroads, we often face the challenge to make a decision. This can extensively impact the development of new products as well as their respective development process. Comprehension in Raspberry Tina K.P.' s mentioned device: This Raspberry Pi has been constructed with safety features such as unplugging function if the hearing protection is removed from the ears, and reset functions if the device becomes unresponsive.

And to be concerned about that word, we note here the application modules such as object detection [5], OCR, emergency response, messaging, fruit ripeness detection, web search, weather information, sport score, news information and system shutdown, respectively; the performance metrics like accuracy, response time, latency, precision, recall, and the F-1 score.

5. Conclusion and Future Scope

The chapter concludes with the powerful message of combining ML with IoT technologies for the bigger things out there in making another device wearable on Raspberry Pi through voice assistant for the blind. This is highly technological progress, having a device for object identification, in the navigation with advanced features such as emergency response, smartphone apps as well. A memorable part of Chapter 9 that should not be diminished calls for work to be done in the future for various improvements such as object recognition and surrounding obstacles but also voice assistant capacities that put it as a tie-up point so that whatever advances the visually impaired enjoy is also imitated as a greater benefit to society. The much hyped marriage of the impact on the lives of visually impaired people as also on society can profoundly be promoted from endeavours accomplished in compliance with the law and directed towards developments for improving the field of the IoT and Machines.

The plan quite literally pays dividends, providing lives an incredibly efficient, humane solution for people suffering from almost any level of blindness. Many aspects have indeed been contemplated thus far, but there still remain a lot of things which require attention regardless of being ML implementation; among them object recognition and OCR can comparatively be developed and differentiated much more through the use of machine learning with improved processes for accuracy. One can further expect the characteristics to be trained on numerous data for greater accuracy. The current system, being able to through giving answers in a haptic, acoustical, and auditory manner, helps in avoiding obstruction; however, new technologies, for example, depth sensing by using arrays of LiDARs (light detection and ranging) or through either stereo cameras, will be in a position to recognize impediments with far greater precision than the earlier; the AI, therefore, can acquire the above performance by plugging into other clever things for the installation of services like depth sensing, which can be through LiDAR or stereo cameras, that offer more precise recognition of barriers. The voice assistant could even use the advantages of being integrated with other smart devices and services, such as those smart home devices, calendar applications, and an intelligent virtual assistant. It quickly becomes clear that the user could control his smart home devices even more easily from anywhere else in the room, make appointments even without additional devices.

References

[1] Alberto Rizzoli, (June 10, 2021), *Ultimate Guide to Object Detection*. Retrieved from blog. https://www.v7labs.com/blog/object-detection-guide#types-and-modes-of-object-detection.

[2] Awad, M., Haddad, J.E., Khneisser, E., Mahmoud, T., Yaacoub E. and Mal li, M. (2018). Intelligent eye: A mobile application for assisting blind people, In: *2018 IEEE Middle East and North Africa Communications Conference* (*MENACOMM*) (pp. 1–6). doi: 10.1109/MENACOMM.2018.8371005.

[3] Banafa, A. (2023). Introduction to Internet of Things (IoT). In: *Introduction to Internet of Things* (IoT). River Publishers, pp. i–xii.

[4] Bhowmick, A., Prakash, S., Bhagat, R., Prasad, V. and Hazarika, S. (2014). IntelliNav: Navigation for Blind based on Kinect and Machine Learning. In: *Multi-disciplinary Trends in Artificial Intelligence* (*MIWAI '14*),. *8875,* 172–183. 10.1007/978-3-319-13365-2_16.

[5] Bressoud, T. and White, D. (2020). *Web Scraping*. doi: 10.1007/978-3-030-54371-6_22.

[6] Derwing, T.M., Rossiter, M.J. and Munro, M.J. (2002). Teaching native speakers to listen to foreign-accented speech. *Journal of Multilingual and Multicultural Development*, *23*(4), 245–259.

[7] Hochreiter, S. and Sch midhuber, J. (1997). Long Short-term Memory. *Neural Computation, 9*. 1735–1780. 10.1162/neco.1997.9.8.1735.

[8] Howard, A., Zhu, M., Chen, B., Kalenichenko, D., Wang, W., Weyand, T. Andreetto, M. and Adam, H. (2017). *MobileNets: Efficient Convolutional Neural Networks for Mobile Vision Applications*. arXiv.170404861.

[9] Javed, A., Awais, M., Shoaib, M., Khurshid, K.S. and Othman, M. (2023, FEB). Machine learning and deep learning approaches in IoT. *PeerJ. Comput. Sci.,* 6, (9) e1204. doi: 10.7717/peerj-cs.1204. PMID: 37346567; PMCID: PMC10280223.

[10] Kornprobst, P., Tumblin, J. and Durand, F. (2009). Bilateral Filtering: Theory and Applications. *Foundations and Trends in Computer Graphics and Vision,. 4,* 1–74. 10.1561/0600000020.

[11] Mambu, J. Y., Anderson, E., Wahyudi, A., Keyeh G. and Dajoh, B. (2019). Blind Reader: An Object Identification Mobile-based Application for the Blind using Augmented Reality Detection. In: *2019 1st International Conference on Cybernetics and Intelligent System (ICORIS)* (pp. 138–141). doi: 10.1109/ICORIS.2019.8874906.\

[12] Mohapatra, S., Rout, S., Tripathi, V., Saxena T. and Karuna, Y. (2018). Smart Walking Stick for Blind Integrated with SOS Navigation System, In: *2018 2nd International Conference on Trends in Electronics and Informatics (ICOEI)* (pp. 441–447). doi: 10.1109/ICOEI.2018.8553935.

[13] Moin, A., Challenger, M., Badii, A. and Gunemann, S. (2022). A model-driven approach to machine learning and software modelling for the IoT. *Software and Systems Modelling.* SpringerLink, Jan.

[14] Patel, C., Patel, A. and Patel, D. (2012). Optical Character Recognition by Open source OCR Tool Tesseract: *A Case Study. International Journal of Computer Applications,. 55,* 50–56. doi:10.5120/8794-2784.

[15] Pise, A. and Ruikar, S.D. (2014) Text detection and recognition in natural scene images. In: *International Conference on Communication and Signal Processing*, pp. 1068–1072, doi: 10.1109/ICCSP.2014.6950011.

[16] Saoji, S., Eqbal, A. and Vidyapeeth, B. (2021). Text recognition and detection from text images using PyTesseract. *Journal of Interdisciplinary Cycle Research,. XII,* 1674–1679.

[17] Shuai Q. and Wu, X. (2020). Object detection system based on SSD algorithm, In: *2020 International Conference on Culture-oriented Science & Technology (ICCST)* (pp. 141–144). doi: 10.1109/ICCST50977.2020.00033.

[18] Thomas, H.K. (2004). *Training Strategies for Improving Listeners' Comprehension of Foreign-accented Speech* (Doctoral dissertation). University of Colorado, Boulder.

3 Ethical Considerations in IIoT Development

Pratibha Sajwan[1*] and Swati Mude[2]

The technology will collect, transmit, and analyse a vast amount of data, including sensitive information regarding processes, equipment, and personnel. Therefore, making data privacy an important issue in developing the Industrial Internet of Things. The threat of privacy breach increases with IIoT combined with cloud computing and big data analytics because current protection mechanisms for data are inadequate. This includes, in addition, such issues as informed consent, anonymization, and ownership of data; these emphasize the importance of knowing how data are being collected and agreeing to such collection. Best practices toward resolving the issue would mean the implementation of transparent data privacy policies, encryption as well as other security methods, and regularly auditing data-handling procedures to ensure that such procedures comply with all applicable privacy laws. The literature indicates that risks are mitigated in protecting both individual and organizational data by invoking privacy-by-design principles and strong anonymization techniques.

1. Introduction

The Industrial Internet of Things (IIoT) is key to optimizing operations in automotive manufacturing through technologies like predictive maintenance and surveillance of workers. Predictive maintenance uses advanced sensor networks that monitor the performance of machinery and equipment day and night, helping manufacturers predict and avert potential failures before they happen. This minimizes expensive downtime while extending the life of critical assets such as robotic arms and conveyor systems so that production is as smooth as possible. In parallel, IIoT-enabled worker surveillance systems improve safety and productivity on the shop floor. Using wearable devices and AI-(artificial intelligence) driven analytics, manufacturers can monitor worker health, detect hazardous conditions, and enforce the right safety protocols. Although these innovations improve operational efficiency and workplace safety, they create the challenge of worker

[1-2] Department of Artificial Intelligence and Data Science, Thakur College of Engg. & Technology, Mumbai, India.
Email: swati.mude@tcetmumbai.in[2]
* Corresponding author: pratibhasajwanl8@gmail.com

DOI: 10.1201/9781003587903-3

privacy and data security that requires careful thought so that it is adopted in an ethical and responsible manner. Together, predictive maintenance and worker surveillance empower automotive manufacturers to increasingly embrace a more proactive and data-driven approach toward efficiency, sustainability, and safety in their production environment. Year-wise progress development of ethical IIoT development is provided in Table 1.

Table 1 Ethical IIoT Development Year-Wise Progress.

Year	*Key Development*
2015	Initial discussions around ethical concerns in IoT, focusing on data privacy and security.
2016	Introduction of guidelines for ethical data handling in IoT by industry groups.
2017	Early adoption of privacy-by-design principles in IoT devices.
2018	Emergence of standards and regulations for data protection (e.g., GDPR) affecting IoT implementations.
2019	Increased emphasis on transparency and consent in IIoT data usage.
2020	Growth in the use of AI ethics frameworks within IIoT applications.
2021	Development of industry-specific ethical standards for IIoT (e.g., manufacturing, healthcare).
2022	Enhanced focus on sustainability and ethical sourcing of materials for IIoT devices.
2023	Integration of ethical auditing tools in IIoT systems to ensure compliance and accountability.
2024	Future trends: Expected advancements in ethical AI for IIoT, with emphasis on fair and unbiased data usage.

2. Data Privacy

2.1 Importance of Data Privacy

Data privacy is one of the major concerns in IIoT since large amounts of data are being collected, transmitted, and analyzed. Much of this data will be sensitive in nature regarding processes, equipment, and employees. The ethical challenges for IIoT development include privacy issues. The systems developed in IIoT generate massive amounts of data that are sometimes sensitive in nature related to processes, products, and even individuals. Xu et al. (2018) consider that the merging of IIoT with cloud computing and big data analytics would increase the prospects of privacy breach due to ineffective data protection mechanisms. The literature calls for sound data anonymization techniques and adoption of privacy-by-design principles that would protect both individual and organizational data.

2.2 Ethical Concerns

- **Informed Consent:** Ensuring that all stakeholders, including employees and partners, are aware of data collection practices and consent to them.
- **Anonymization:** Implementing measures to anonymize data to protect individual identities.
- **Data Ownership:** Clarifying who owns the data generated by IIoT devices and how it can be used.

2.2.1 Best practices

- Develop transparent data privacy policies.
- Use encryption and other security measures to protect data.
- Regularly audit data practices to ensure compliance with privacy regulations.

2.3 Security

2.3.1 Security in IIoT

IIoT systems are vulnerable to cyber-attacks as they have many entry points, making security a prime ethical concern. Security is a crucial ethical aspect because of the ease through which a malicious hacker can attack these interconnected devices. According to Borgia (2014), the current technologies and techniques are becoming more susceptible to hacking, ransomware, and data tampering attacks on critical infrastructure. Literature indicates that good cybersecurity practices such as encryption, regular security audits, and secure communication protocols are put in place. In addition, Gubbi et al. (2013) assert the use of a multi-layered security approach comprising device authentication, network security, and secure data storage in countering the threat.

Industrial systems are generally networked and inherently contain key infrastructures like manufacturing, electricity, transportation, and utilities, hence security forms a core element of the Industrial Internet of Things. The IIoT extends the OT (operational technology) networks handling physical processes beyond the space of the traditional IT system; it, therefore, encompasses an enormous spectrum of new security challenges. Hereunder are the core security challenges and obstacles of IIoT (Figure 1).

1. *Weaknesses in Older Systems*: Most industrial systems were designed without security in mind while it is still today. Legacy OT systems are therefore at risk of attack from ransomware, malware, and unwanted access since they may not have built-in security capabilities.
2. *Device Access Control and Authentication*: All devices within the IIoT network need to be authorized and roles implemented in order to restrict access to data and control. Unprivileged access can be avoided with strong authentication methods, such as hardware-based security tokens or certificates.

3. *Encryption of Data*: Since large amounts of data are generated by IIoT devices, this generated data needs to be encrypted both while it is in transit as well as while at rest in order to safeguard it against tampering and interception. It is critical to employ secure communication protocols for example, TLS: Transport Layer Security.
4. *Security on Endpoints:* Every [5] IIoT device is an endpoint, and if not secured, could easily be used to commit some mischief. Thus, stopping known vulnerabilities from being exploited requires securing the firmware and utilizing secure boot techniques in addition to installing frequent updates.
5. *Segmentation of Networks*: In case a threat would begin to spread, IIoT networks need to be kept away from the enterprise and secured behind firewalls. Industrial systems should be kept isolated from other corporate systems and the Internet as much as possible.
6. *Anomaly Identification and Tracking*: In an IIoT environment, such threats could be detected before their actual incidence due to constant monitoring of any abnormal or suspicious activity. The tools and systems tracking the behavior of the devices as well as network traffic are: Intrusion detection systems.
7. *Security of the Supply Chain*: There are also several other IIoT devices and components offered by third-party vendors. In this regard, vulnerabilities could be embedded into the IIoT products in the production or distribution process; therefore, such products must have supply chain assurance in terms of security.

3. The Architecture of IIoT and Security Issues

3.1 Architecture of IIoT

Actuators: It comprises actuators, embedded systems, and industrial sensors, such as Ethernet, 5G, wireless, and many more in the form of networks and protocols. Communication layers are basically formed by networks and protocols. Analytics and data processing include real-time analysis through cloud and edge computing. The application layer forms AI-based systems, dashboards, and user interfaces (Figure 2).

- Security Issues at Every Level.
- Layer of Device and Sensor.
- Physical security: Protection against sabotage, theft, and tampering.
- Software vulnerabilities: Potential risks arising from out-of-date or unpatched software.
- Device identity management: Ensuring that each device has an authenticator.
- Communication Layer
- Data Interception: Unlawful interception of confidential data.

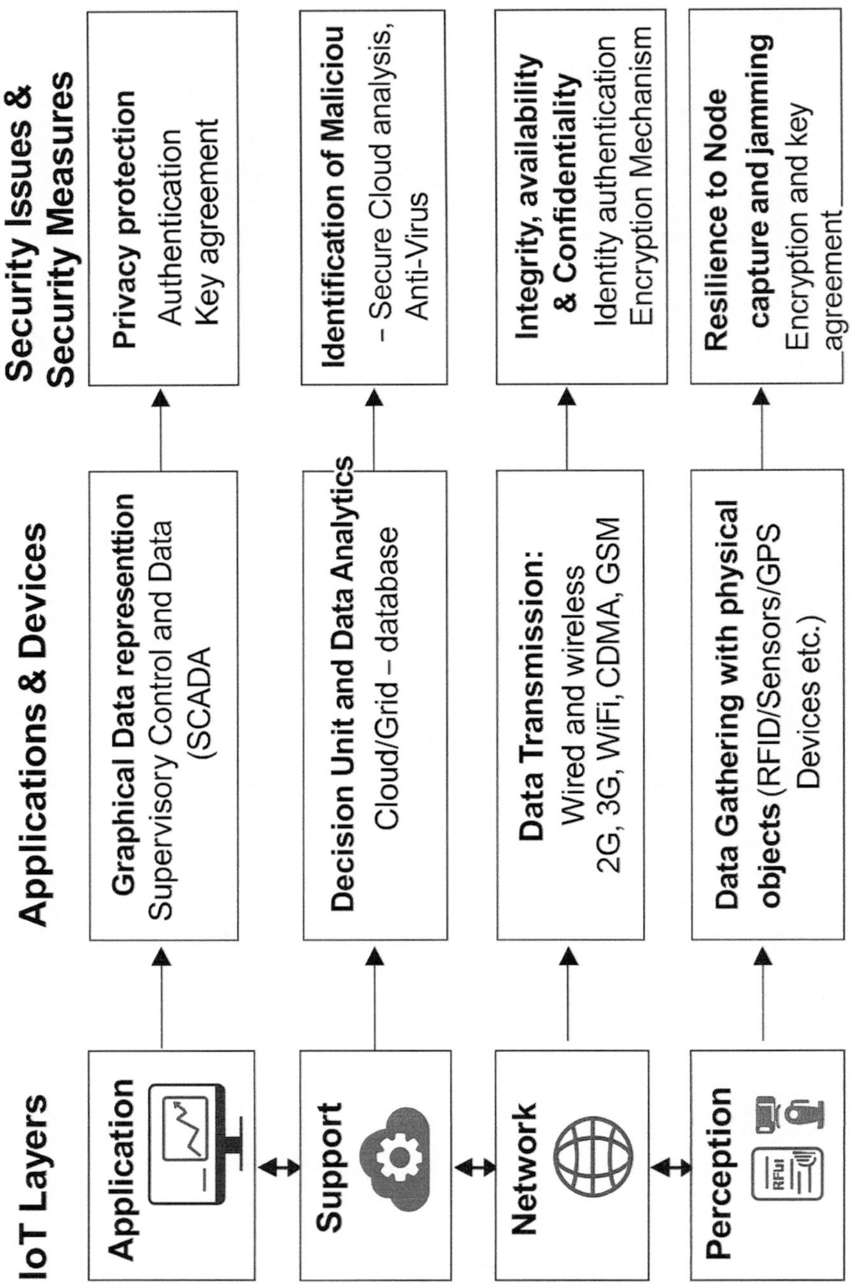

Fig. 1 Security Challenges with different layers.

- Denial-of-service (DoS) attacks: Attempts to disrupt the functionality of the network.
- Protocol weaknesses: It is susceptible to exploitation due to weak communication protocols.
- Data Processing Layer.
- Cloud security: The protection of cloud-based storage both in motion and at rest.
- Edge computing risks: Susceptibility of insecure edge devices.
- Malware and ransomware are malware and ransomware that can interrupt data analytics and automation.
- Application Layer.
- Access Control: Ensuring that only authorized persons access critical systems.

Data integrity refers to the process of preventing unauthorized alteration of data used for decision-making.

Contractors or employees who have evil intentions are known as insider threats.

4. Frequent Cyberattacks on Industrial Systems Posing a Threat to IIoT Security

Malware and ransomware are malicious software programs that lock down important systems until a ransom is paid. DDoS (distributed denial of service) attacks overwhelm the IIoT networks, causing them to be unusable. MITM (man-in-the-middle) attacks include data manipulation and device communication interceptions. Spoofing and replay attacks are those in which one pretends to be a trustworthy device or plays back previous messages to get unwanted access.

4.1 Physical Attacks

Manipulation of the hardware or device to make the operation fail. Thefts or unauthorized physical access to electronics.

Insider Threats: Employees using their access privileges to confidential information or systems [7].

4.2 Security Standards and Procedures

The most basic security standards established by Information Security Management ISO/IEC 27001 include management systems to protect information assets.

IEC 62443: Industrial automation and control system security.

NIST Cybersecurity Framework: It controls to reduce the risk of cybersecurity threats of IIoT.

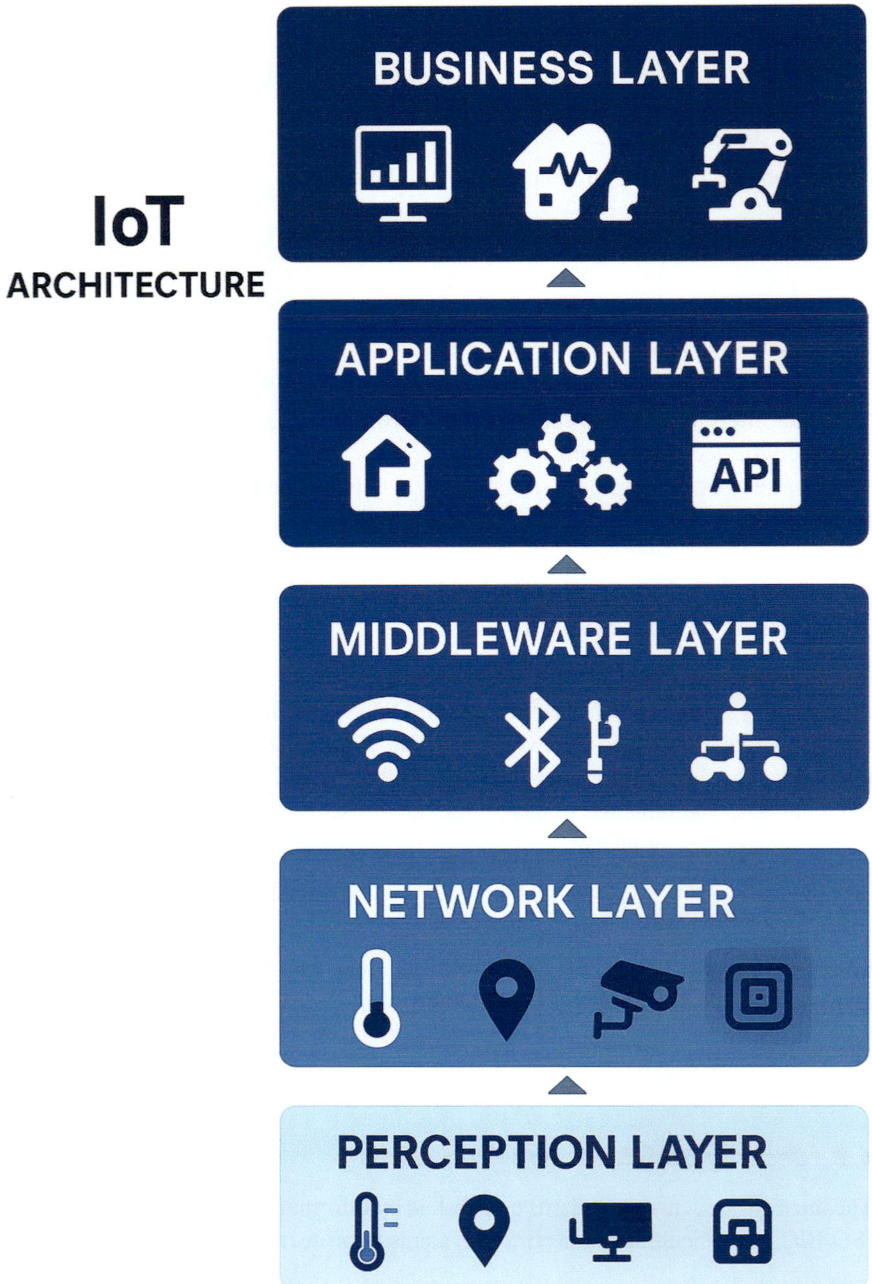

Fig. 2 The IIoT structures with different layers.

IIoT Security Practices for Communication TLS/SSL: Encryption of the messages between hardware and cloud servers.Transporting safe data over low-bandwidth environments using MQTT with TLS.

IPSec: This encrypts the data as it moves through networks. The encryption standard applied for encrypting data in motion as well as at rest is Advanced Encryption Standard or AES.

Infrastructure: It uses PKI to authenticate devices.

4.3 Ethical Concerns

Issues of data ownership and consent assume a highly important role in the IIoT landscape. Often, IIoT systems involve multiple stakeholders from device manufacturers to service providers and end-users and creates complicated questions on who owns the data and how it may be used. Wan et al.'s study (2016) puts across one of the crucial findings in IIoT on the grounds that clear legal frameworks about data ownership do not exist, bringing various disputes and misuse of data. Ethical IIoT development is also characterized by clear policies about data ownership, user consent, and transparent usage practices regarding the data.

- **Vulnerability Management:** Addressing vulnerabilities in IIoT devices to prevent unauthorized access.
- **Impact of Breaches:** Considering the potential impact of security breaches on operations, safety, and stakeholders.
- **Responsibility:** Determining who is responsible for ensuring the security of IIoT systems.

It has to do with the automation of systems, the analytics of real-time data, and the improvement in operational efficiency brought about by linked systems, sensors, and industrial devices. As many benefits the IIoT possesses, numerous questions of labor, privacy, security, and environmental sustainability stand at the surface of this approach. The aforesaid issues need to be addressed, and only through the responsible application of IIoT technologies shall one hope for its just implementation [3].

4.3.1 Data Ethics and Privacy

- *Data Collection and Use:* IIoT systems collect data from the workers, the equipment, and the operational procedure. The sharing of data, in many cases with third-party vendors, has become a source of concern as to who should be granted access to it and what is being done with the collected data.
- *Lack of Consent:* There is a problem that seems to regard consent and ownership of data. Most workers in industrial settings lack knowledge of the scope of personal data being collected and processed.

- *Exposure to Sensitive Data:* In the case of data breaches, a critical ethical risk is present since the IIoT devices in crucial infrastructure or health care often handle very sensitive information. If this said information falls into unauthorized access, it could have devastating impacts on public safety and personal privacy.

4.3.2 Safety and Responsibility

Online Safety Deficiencies: Cyber-attacks can even disrupt industrial processes besides resulting in huge financial losses, and in some extreme cases, loss of human lives when these IIoT systems come into play. There's an ethical issue involved in knowing who is in control of managing such systems and what is being done to prevent potential risks about the users and the organizations.

It is often ambiguous as to whom the responsibility[2] lies for a breach in security-whether it's the end user, system integrator, or the device vendor. This raises an ethical issue wherein offences lead to injuries to either human life or the environment.

Privacy vs. Security Trade-offs: Sometimes, the privacy of individuals may clash with the necessity for more robust security of IIoT. For example, in real-time monitoring of IIoT devices as a method to detect vulnerabilities, this could be an issue concerning the monitoring of employees or customers

4.3.3 Implications for Labor and Workforce

- *Loss of Jobs:* Most of the IIoT systems result in automation, which decreases the amount of human labor in industries like manufacturing, transportation, and energy. Automation, although increasing productivity, raises some ethical considerations involving job loss and what will happen to humans whose jobs are no longer needed because of automation.
- *Education and Re-skilling:* It is ethically the responsibility of the governments and industries to support re-skilling programs such that employees can move into new positions with a chance to use IIoT technology. If this does not happen, then unemployment will increase, and income distribution may worsen.
- *Workplace Monitoring:* IIoT systems are mainly used for maximum productivity and safety through the real-time monitoring of worker performance. This raises a number of ethical issues with regards to the scope of monitoring, possible privacy violations, and implications for employee autonomy and well-being.
- *Informed Consent:* Workers don't really know the depth of monitoring within an IIoT system. It creates a very negative work environment, and there are questions raised ethically when people are monitored without knowledge or consent.

5. Impact on the Environment: Resource Consumption

IIoT enhances productivity and waste reduction. Yet, its production, consumption, and disposal processes are harmful to the environment. The exploitation of raw materials and energy, besides e-waste, or electronic garbage, is unethical.

- **Sustainability:** In the light of IIoT systems deployed by businesses, it is morally the responsibility of these businesses to take into account the environmental consequences of these technologies and to reach for sustainability in deploying eco-friendly measures such as recycling and energy-efficient equipment.
- **Greenwashing:** "Greenwashing" is the term that refers to the practice of some businesses using IIoT to portray themselves as more environmentally responsible than they actually are. In doing so, it brings about moral questions related to misrepresenting the stakeholders and customers regarding environmental benefits from the deployment of IIoT.

5.1 Best Practices

Best Practices for IIoT Security Security by Design: To provide strong protection through all aspects, security has to be included right from the initial stages of design for IIoT devices and systems. Frequent updates and patching: Firmware and software updates should be installed on time in order to resolve known vulnerabilities of IIoT systems. Zero Trust Architecture, or ZTA, is a concept that stipulates no system, person, or device should be trusted by default. It requires constant verification before access is granted.

Vulnerability management and risk assessment: Assessments have to be carried out periodically to find vulnerabilities and get an idea about the threats that may occur. Real-time monitoring and incident response: Though a cyber attack's recovery is possible within no time with the incident response plan, real-time monitoring helps in recognizing anomalies or breach [1].

5.2 Societal Impact and Equity

The mainstream effects of IIoT on society must be taken into account. The literature suggests that the consequences of increased inequality are possible for those in developing regions who lack ready access to technologies. In addition, environmental concerns about increased device manufacturing as well as increased energy consumption in association with IIoT are an emerging ethical concern. According to the research, IIoT is developed by implementing green approaches; this involves equipment that uses low energy, which also enhances waste minimization by supporting the concept of the circular economy.

5.3 The Social Effect

Automation has societal implications that extend beyond the workplace and impact entire communities. Ethical considerations here include:

Impact on Local Economies: Manufacturing, warehousing, and logistics are three sources of employment that, in some areas, provide a large number of jobs, and these types of occupations are often hit disproportionately by automation. Automation-related job displacement can also put a community in financial trouble if it is primarily dependent on a single industry. This could mean higher unemployment and lower local economic activity. This is especially worrisome in places where there might not be many alternatives for alternative employment.

Top on this list is the community support system, which may be strained by job losses brought on by automation—local community support systems, including healthcare, social services, and welfare programs [6].

To provide support networks to the displaced workers, there has to be a collaboration by firms, governments, and local organizations toward an ethical approach to automation. For example, such public-private agency partnerships can be helpful in the development of training centers or investment in infrastructure for more business opportunities to be established in the local area.

Well-being and Mental Health: The loss of a job or threat of unemployment due to automation causes major psychological effects on employees and their families. This may lead to stress, despair, and anxiety from the loss of identity, purpose, and financial security. Along with retraining, ethical automation will also provide mental health resources and therapy for the affected individual.

6. Social and Economic Equity

Sustainability in IIoT has social and economic dimensions that include equity and access. The adoption of IIoT can increase inequalities, especially in regions with no or very limited access to advanced technologies or where the workforce lacks the skills to adapt to new industrial processes. This may lead to a digital divide where some groups are left behind in the transition to smart industrial systems.

Ethically, IIoT development needs to be carried out with due regard for inclusivity, ensuring access to the gains of technology by all. This will require investment in education and training programs to upskill the workforce—especially in disadvantaged communities—and opportunities for equal participation in the digital economy. There needs to be collaboration between policymakers and industry leaders who ensure that the regulation framework of IIoT, along with relevant business standards regarding the realization of rights from various stakeholders, has to be duly made.

6.1 Ethical Issue

Resource use is based on the IIoT manufacturing and deployment of ecological footprint.

E-Waste: The electronics part of the IIoT devices which have malfunctioned or become obsolete to ensure the development through the sustainable aspects along with the targets.

7. Compliance Regulatory

Necessity to be compliant

The basic compliance with the regulations primarily guarantees that IIoT systems are operated ethically and legally.

7.1 Regulatory Adherence

- **Avoid Legal Penalties:** IIoT systems often work with sensitive data, interact with critical infrastructure, and have potential safety implications. "Compliance with GDPR, HIPAA, or industry specific standards, such as ISO/IEC 27001, shall help to avoid legal penalties, operational disruptions."
- **Ensuring Safety and Security:** Most of the regulations mandate provisions to ensure safety and security of IIoT systems, protection against cyber threats, and safe operation of devices, especially in manufacturing industries, healthcare, energy, and other general sectors [5].

7.2 Data Privacy and Protection

- **Protection of Sensitive Information:** The IIoT systems collect and process massive amounts of data, that are related to persons and operations. Compliance with the laws for protection of this data Ensures that data collection, storage, and processing are secure.
- **Gaining Trust:** Privacy laws support the building of trust with customers, partners, and stakeholders in that it ensures respect for data.

7.3 Continuity and Reliability of Operations

- **Counter-Measures for Risk Mitigation:** Compliance to operational standards helps identify and mitigate risks caused by possible system failure, downtime, or some data breaches for reliable and continuous operation of the IIoT systems.
- **Standards and Interoperability:** Compliance to industrial standards ensures that the IIoT systems will be integrated and function properly with

other systems and technologies, thus increasing overall efficiency and reducing compatibility problems.

- **Responsible and Ethical [3] Usage**
  - *Ethical Operation:* Proper use of IIoT systems so that technology is not used to harm anyone, any business, or the environment; principles of transparency, accountability, and fairness should be applied in machine decision-making.
  - *Environmental Regulations:* Generally, IIoT systems are part of monitoring and controlling environmental effects. With regard to environmental law, these systems can be agents of sustainable development.

7.4 Market Access and Competitiveness

- **Compliance with Market Requirements:** There are some markets or industries, where compliance acts as a requirement to enter that market, mainly because most of the customers and partners demand compliance toward certain standards under their procurement requirements.
- Companies' demonstration of compliance is yet another potential difference, which presents the commitment to quality, safety, and security.

7.5 Avoiding Liability

- **Decrease Legal Liability:** Compliance decreases legal liability for damages or harm brought about by the operation of IIoT systems since adherence to standards can provide a degree of legal protection or defense in case of an incident.

7.6 Ensuring Customer Confidence and Satisfaction

- **Enhancing Customer Assurance:** By complying with relevant standards and regulations, organizations reassure customers that their products and services meet required safety, security, and quality benchmarks.

8. Ethical Concerns

- **Adherence to Standards:** Ensuring that IIoT systems comply with industry standards and regulations.
- **Accountability:** Accountability for the organization's compliance with regulations
- **Transparency:** Transparency in terms of compliance practice and any breaches that may happen

8.1 Best Practices

- Keep track of relevant regulations and industry standards.
- Develop and implement compliance programs and policies.
- Regularly audit and review compliance practices to identify and address gaps.

8.2 Predictive Maintenance and Worker Surveillance: Automotive Manufacturing

- **Predictive Maintenance:** IIoT uses sensor data of the machinery to predict and avoid failures of the equipment. Critical assets such as robotic arms, conveyor systems, and CNC machines are installed with sensors in automotive manufacturing to monitor temperature, vibration, and operational efficiency.
- It gives live data, that speaks to a core cloud platform for predictive analysis regarding the potential malfunctions of assets, through artificial intelligence. With predictive maintenance, maintenance costs will decrease and unscheduled downtime reduction. Thus, this optimizes machinery lifetime supporting productive continuity. Predictive maintenance aids in transforming from scheduled to condition-based maintenance that can better optimize the inventory for spare parts followed by the potential waste reduction [4].
- **Worker Surveillance:** It is about assuring the well-being and productive output of a worker in manufacturing. IIoT-enabled wearable devices, which include smart helmets and vests, monitor the status of workers—vital signs and movement—plus environmental conditions. These devices may monitor fatigue levels, posture deviations, and adverse health effects due to unsafe behaviors. Real-time alerts are therefore provided to the supervisors and staff members regarding the proper implementation of safety procedures. Monitoring cameras with AI can identify hazardous practices. However, it should be designed in such a way that a balance is maintained between the enhancement of safety and privacy since complete monitoring raises questions of privacy and workers' autonomy.

9. Summary and Conclusion

The development and implementation of IIoT technologies also raise important ethical considerations that are to be thoroughly thought out so that the appropriate development and sustainability are ensured. The first among these is data privacy. While IIoT systems collect tons of data on machines and people, it's imperative to ensure sensitive information's security. The manufacturers should ensure strict application of cybersecurity. Data breach and unauthorized access prevention measures; they must also establish explicit policies regarding ownership and usage of data. The other very crucial concern is job displacement. With automation and predictive technologies, it may reduce the workforce in specific industries due to increased efficiency. A firm needs to balance technological advancement with the human business needs to balance the pace of technological development with the human touch and retraining programs for helping employees find their way into new, tech driven jobs. The firms can mitigate all the negativistic effects on employment through a culture of continuous learning. Finally, the environmental impact of deploying IIoT is something that should not be ignored. While predictive maintenance and automation can reduce waste and energy consumption, the production and disposal of IoT devices and sensors give rise to their own ecological problems. Companies must make sure that the practices involved in IIoT are sustainable; for example, the use of recyclable materials and energy-efficient devices. Another critical aspect is regulatory compliance, as governments and industry bodies continue to come up with standards to regulate the use of IIoT. Manufacturers must be up to date with these regulations to ensure that their systems are in compliance with changing data protection, worker safety, and environmental laws. Finally, the ethical issues just identified can only be worked out through sustained ethical reflection and the adoption of best practices. In such a dialogue with stakeholders, it should not just be industry experts or merely workers, policymakers, and society at large. Doing so will bring IIoT its full realization of good, as well as its potent drivers for innovation while focusing on people and the environment.

References

[1] Binns, R. (2018). Fairness in Machine Learning: Lessons from Political Philosophy. In: *Proceedings of the 2018 Conference on Fairness, Accountability, and Transparency*, pp. 149–159.

[2] Floridi, L. (2013). *The Ethics of Information*. Oxford University Press.

[3] IEEE. (2020). *Ethically Aligned Design: A Vision for Prioritizing Human Well-being with Autonomous and Intelligent Systems*. IEEE Standards Association.

[4] Borgia, E. (2014). The Internet of Things vision: Key features, applications, and open issues. Computer Communications, *54*, 1–31. https://doi.org/10.1016/j.comcom.2014.09.008.

[5] Gubbi, J., Buyya, R., Marusic, S. and Palaniswami, M. (2013). Internet of Things (IoT): A vision, architectural elements, and future directions. *Future Generation Computer Systems*, *29*(7), 1645–1660. https://doi.org/10.1016/j.future.2013.01.010.

[6] McKinsey Global Institute. (2017). *A Future that Works: Automation, Employment, and Productivity*. McKinsey & Company. https://www.mckinsey.com/featured-insights/future-of-work/a-future-that-works-automation-employment-and-productivity.

[7] Wan, J., Li, D. and Zou, C. (2016). Industrial IoT: A survey on the enabling technologies, applications, and challenges. *IEEE Communications Surveys & Tutorials*, *18*(3), 1628–1651. https://doi.org/10.1109/COMST.2016.2565681.

[8] Xu, L.D., He, W. and Li, S. (2018). Internet of Things in industries: A survey. *IEEE Transactions on Industrial Informatics*, *10*(4), 2233–2243. https://doi.org/10.1109/TII.2014.230075.

4 Sustainable Design Principles for IoT Systems

Minal Bhoir[1*], Kavita Maskar[2] and Achala Khandelwal[3]

The numerous noteworthy improvements in human productivity, convenience, and safety have resulted from the incorporation of Internet of Things (IoT) technologies into contemporary life. However, there are sustainability issues associated with the quick spread of IoT devices. In order to address these issues, "Sustainable Design Principles for IoT Systems" offers a framework for creating IoT systems that are socially beneficial, energy-efficient, and ecologically conscious. Key concepts including reducing energy use, maximising resource utilisation, guaranteeing long-term scalability, and advancing data security and privacy are outlined in this article. By utilising Internet of Things (IoT) technology, the system combines smart connection with gas sensors, allowing users to receive real-time alerts and notifications on connected devices in the event of a gas leak. A very sensitive gas sensor, such the MOS Gas Sensor, which is adept at identifying a broad range of flammable and combustible gases, forms the basis of the system. A microcontroller board, like Arduino or ESP8266, is interfaced with the gas sensor and serves as the main processing unit for gathering and analysing data. A Wi-Fi module is built into the microcontroller board to provide Internet of Things communication. Through this integration, a smooth data exchange network may be formed between the gas leak detection and cloud-based servers or MQTT brokers. Through specialised mobile applications or web-based dashboards, users may remotely check the status of the gas leak detection, the levels of gas concentration, and receive real-time alerts on their PCs, tablets, or smartphones. The SMART gas leak detector is always looking for indications of gas leaks in the surroundings The system immediately sounds an alarm and notifies the linked devices over the Internet when the gas concentration rises above a certain threshold. This proactive strategy guarantees quick reactions, allowing users to take prompt action to reduce possible risks.

[1-3] Assistant Professor, Dept. of Engineering Science and Humanities, Thakur College of Engineering and Technology, Mumbai, India.
Email: kavitamhaskar12@gmail.com[2]; achalakhandelwal@gmail.com[3]
* Corresponding author: minalbhoir1524@gmail.com[1]

DOI: 10.1201/9781003587903-4

1. Introduction

The Internet of Things (IoTs) have revolutionized every thing, from smart cities and homes to automation in the industries to the technological advancements in health care. Although the advantages of IoT are numerous, the growing deployment of IoT devices creates major sustainability challenges. These challenges range from an increase in energy usage, electronic waste, and possible risks to data confidentiality and protection. In order to tackle these challenges, a paradigm shift is essential: the shift to sustainable design practices where environmental and social aspects are more important than emerging technologies.

"Sustainable Design Principles for IoT Systems" discusses the growing necessity for sustainable development in the IoT domain. In this paper, we describe a comprehensive set of principles for the design of efficient and environmentally sustainable IoT systems. Important principles of it include energy efficiency, resource optimis ation, scalability, and strong data protection strategies. Implementing these principles in designing and deploying the IoT systems can help developers find a balance between the negative impact on the environment, while also extending the longevity of their solutions. [10, 11]

The "Sustainable Design Principles for IoT Systems" can be applied towards developing novel methods of leakage detection in industries and households, as they leverage the potential of IoT devices. Gas leaking has some serious risks such as an explosion, fire rage among others. Hence, having an automated gas leaking detection system is fundamental for saving property and life. This modern ingenious system provides the functionalities of gas leak detection by combining modern microcontroller boards like Arduino or ESP8266 together with very sensitive gas sensors. The gas sensors used in this system have the capability of detecting huge varieties of combustible and flammable gases and hence provide accurate detection mechanisms with much wider coverage. To enable the device to provide real-time notifications and updates over the internet, the Wi-Fi module is embedded into the microcontroller which makes it possible for the gas leakage detector to connect to the MQTT broker or even cloud servers. Such a connected network allows people to get real-time notifications and alerts about the concentration levels of gases through their tablets, computers, or smartphones using suitable applications or web-based dashboards. The "Smart Gas Leak Detector with IoT Connectivity" aims at establishing uninterrupted and real-time tracking. The overarching goal of a "Smart Gas Leak Detector with IoT Connectivity" spans around the automated and uninterrupted management of any gas leakage cases. This system makes sure that users take appropriate measures to mitigate any dangers arising out of gas leakage by instantly informing users of possible gas leaks and giving them real-time warnings. Alongside the near immediate notifications, this allows for users to view historical analytics such as gas concentration over a long period of time, assisting them with devising strategies. At its core, this project seeks to develop a robust and intelligent solution for identifying gas leaks by leveraging the power of the internet and IoT technology. Since there has been an increase in the interest towards smart home automation, the "Smart Gas Leak Detector with IoT Connectivity" adds to the safety of the residents and the devices using the technology that can be deemed as an innovative step towards IoT development [1, 2].

2. Proposed Scheme

"Sustainable Design Principles for IoT Systems" is an advanced design that seeks to offer solutions to the major problem of detecting gas leaks in residential, industrial, or commercial buildings. Hiccups result whenever the gasses leak, which in turn is able to have someone altered in people who are inside a building, thus this system will enable designs that are able to detect gasses in a quicker manner. This subsequent solution works towards changing how gasses are again shifted by the use of the latest IoT techniques with a combination of high-defined gas sensors. With the application of the IOT, the system allows for on-time surveillance, external oversight, and timely notifications that are needed to prevent any gasses from leaking. Modern gas sensors form the system's centre, with design capable of detecting various flammable and combustible gases at a high level of sensitivity and accuracy. These sensors are connected to a microcontroller like Arduino or ESP8266, which acts as the brain of the system. With a built-in Wi-Fi module, the microcontrollers can easily transfer data securely to cloud-based servers or MQTT brokers. All that data can let users check the concentration of gas and receive notifications on tablets, smartphones, and computers anywhere and anytime they are. This implementation represents a step forward in early detection of any potential gas leaks and avoids accidents. It has an on-site and constant measure of the levels of gas concentration in the general environment, and systems of alarms and notifications that activate when gas leaks are detected. By responding in seconds like this, the user can take immediate action like shutting off gas or evacuating the occupants which significantly pulls down the risk of gas explosions or intoxication [3, 4,]. Besides that, the system provides a historical data and analytics options, allowing users to access information on gas concentration trends and patterns. This knowledge helps making smart choices, taking steps ahead of time and using gas more efficiently in industrial applications. The device, "Smart Gas Leak Detector with IoT Connectivity", is expected to provide a smart home automation system, industrial safety system, and commercial safety system for a safe environment of users and workers. With its combination of IoT technologies and gas sensor embedding, it is a smart gas detector, which is a breakthrough solution in gas leak detection, designed to enhance safety standards and protect lives and properties against possible gas-influenced threats [5, 9] (Figure 1). The gas leakage detector with IoT connectivity is shown in the block diagram.

3. Requirements and Description for the Components

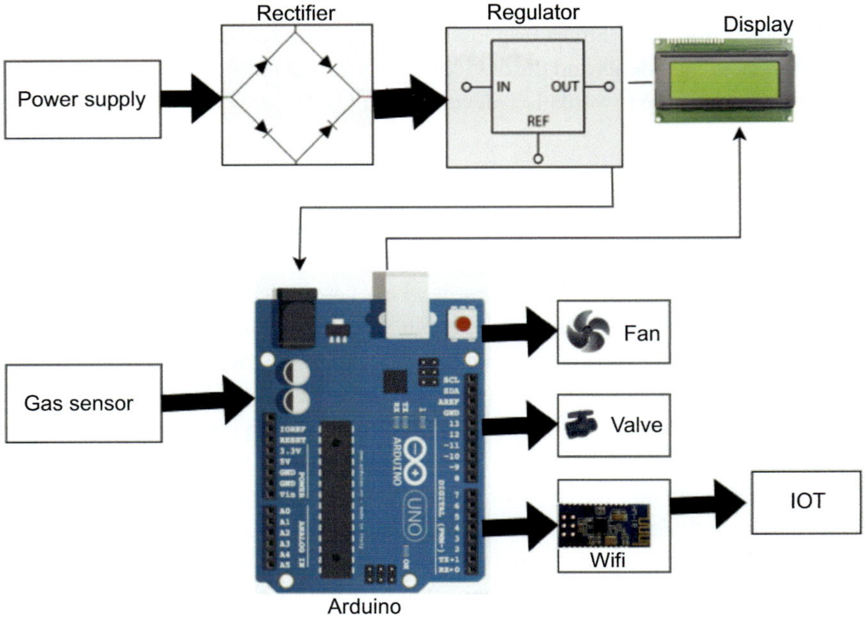

Fig. 1 Block diagram of an IoT-connected gas leak detector.

3.1 Gas Sensor

3.1.1 Requirement

The detection of flammable and combustible gases by metal oxide semiconductor gas sensors (e.g., MOS Gas Sensor).

3.1.2 Explanation

The core of the system is the gas sensor, which provides real-time data on gas concentrations and accounts for the detection of gas leaks. Select a sensor that satisfies the particular gas detection needs of your use case.

3.2 Microcontroller

Requirement

Here we can can used different type of Arduino board and advance microcontroller with connection of gas sensor.

Explanation

As the system's brain, the microcontroller and Arduino board handles communication, analysis, and data processing. In order to connect to the Internet of Things, it exchanges data with the Wi-Fi module and reads data from the gas sensor.

3.3 Wi-Fi Model

Requirement

Wi-Fi modules or Arduino boards with integrated Wi-Fi.

Explanation

The gas leak detector can connect to the internet and speak with cloud servers or MQTT brokers thanks to the Wi-Fi module. It makes data exchange and real-time remote monitoring easier.

3.4 Power Supply

Requirement

An appropriate power source, such as a USB cable and adapter, is required to power the gas sensor and microcontroller.

Explanation

Ensure a stable and reliable power source to operate the gas leak detector continuously. USB power is commonly used for Arduino boards.

3.5 Breadboard and Jumper Wire

Requirement

Jumper wires and a breadboard are used to connect components.

Explanation

The microcontroller, Wi-Fi module, and gas sensor can all be connected and prototyped easily using the breadboard.

3.6 Enclosure

Requirement

A protective enclosure for the microcontroller, gas sensor, and other electronic parts.

Explanation

The components are shielded from physical harm, moisture, and dust by an enclosure. Select a material and size that are appropriate for the intended setting.

3.7 Internet Connection

Requirement

IoT communication is made possible by a steady internet connection.

Explanation

To enable real-time data transmission to the cloud, make sure the gas leak detector has access to a dependable internet connection (such as Ethernet or Wi-Fi).

3.8 MQTT Broker

Requirement

Communication between the gas leak detector and linked devices is facilitated by an MQTT broker.

Explanation

To manage data exchange and provide remote access to gas concentration data and alerts, set up an MQTT broker on an IoT platform (such as Google Cloud IoT Core and AWS IoT).

3.9 Mobile Application

Requirement

Use a web-based dashboard or mobile application to remotely monitor gas concentration and get alerts about gas leaks.

Explanation

Connect the broker to a mobile app or web dashboard that can show real-time gas concentration data and alert notifications.

All these components put together make a Smart Gas Leak Detector with IoT Connectivity This setup allows the user to monitor the gas leaks in real time and over remote access for a safer life and peace. [6, 7, 8]

4. Conclusion

We understand the need to develop IoT solution that are sustainable, which is why if you are interested to know new innovative trends do flag up the "Sustainable Design Principles for IoT Systems" research paper that emphasizes on sustainable IoT solution from the onset itself. Advanced gas sensors, microcontroller technology, and Internet of Things (IoT) connectivity for safety as exemplified by the proposed smart gas leak detector. From the use of Real-Time Monitoring of the system itself to providing IMMEDIATE alerts right away, this ensures proactive responses to gas leakages, preventing potential hazards and saving individuals and properties. the "Smart Gas Leakage Detector with IoT Connectivity" is an example of the collaboration of modern technologies in solving major safety issues. It is a useful tool in the field of home automation and safety systems because of its real-time monitoring, remote accessibility, and data analytics capabilities. In addition to improving personal safety, this project supports the larger goal of building intelligent, networked communities. Such intelligent systems play an ever-more-important role in guaranteeing the safety and well-being of our living areas as technology advances.

References

[1] Kumar Keshamoni and Sabbani Hemanth (2017). Smart Gas Level Monitoring, Booking & Gas Leakage Detector over IoT. In: *2017 IEEE 7th International Advance Computing Conference (IACC)*.

[2] P. Spachos, Liang Song and Dimitrios Hatzinakos. (2014). Gas Leak Detection and Localization System Through Wireless Sensor Networks. In: *2014 IEEE 11th Consumer Communications and Networking Conference (CCNC)*.

[3] Babu Prasanth, V. (2014, Nov.). Cloud Connected Smart Gas Leakage Detection And Safety Precaution System. *International Journal of MC Square Scientific Research, 6* (1).

[4] Asmita Varma, Prabhakar, S. and Kayalvizhi, Jayavel. (2017). Gas Leakage Detection and Smart Alerting and Prediction Using IoT. In: *Internet of Things and Applications (IOTA), International Conference*. IEEE.

[5] Mohammad Reza Akhondi, Alex Talevski, Simon Carlsen and Stig Petersen (2010). Applications of Wireless Sensor Networks in the Oil, Gas and Resources Industries. In: 2010 *International Conference on Advanced Information Networking and Applications*. IEEE.

[6] Ashish Shrivastava, Ratnesh Prabhaker, Rajeev Kumar and Rahul Verma. Gsm Based Gas Leakage Detection System. *International Journal of Technical Research and Applications, 1*(2), 42. EISSN: 2320–8163.

[7] Tyler Kersnovski, Felipe Gonzalez, and Kye Morton. (2017). *A UAV System For Autonomous Target Detection and Gas Sensing*. Yellowstone Conference Center, Big Sky, Montana. IEEE.

[8] Vana Jelicic, Michele Magno, Davide Brunelli, Giacomo Paci, Luca Benini and Fellow (2013, Jan.). Context Adaptive Multimodal Wireless Sensor Network for Energy-Efficient Gas Monitoring. *IEEE Sensors*. doI: 10.1109/JSEN.2012.2215733, IEEE 2013.

[9] Chet Sandber, Jim Holmes, Ken Mccoy and Heinrich Koppitsch (1998, Sept.-Oct.). The Application of a Continuous Leak Detection System to Pipelines and Associated Equipment. *IEEE Transactions on Industry Applications*, *25* (5).

[10] Smith, J. and Anderson, P. Energy-Efficient and Scalable IoT Framework for Smart Gas Leak Detection Systems (2022). *Journal of Sustainable Computing: Informatics and Systems*, *28*, 100451.

[11] Johnson, M. and Lee, H. (2023). A Comprehensive Approach to Sustainable IoT: Enhancing Gas Leak Detection with Data Analytics and Resource Optimization. *International Journal of Environmental Informatics*.

5 Case Studies
Responsible IIoT Implementations Smart Bin Waste Prediction Using Machine Learning

Aditeya Prajapati[1], Sushan Uchil[2], Vishal Patwa[3], Neha Kapadia[4*] and R.G. Mehta[5]

The amount of waste that a community or municipality will produce can be predicted using a technique called smart bin waste prediction. The system is capable of accurately predicting future trash levels by analyzing patterns in waste generation through the use of machine learning algorithms. Waste management companies can then utilize this information to streamline their disposal and collection schedules, cutting expenses and increasing productivity. It has been demonstrated that utilizing machine learning for smart bin waste prediction is a dependable and efficient way to manage garbage, and it has the potential to completely change the waste management sector.

1. Introduction

Due to the fact that garbage creation is increasing along with the global population and changing consumption habits, waste management is becoming an increasingly important issue for communities and governments [1]. Accurate forecasts of future trash levels are essential for efficient waste management and reduction of waste's detrimental effects. Smart bin waste prediction is a technique that makes use of machine learning algorithms [3] to predict how much waste a community or municipality will produce.

Responsible Industrial Internet of Things (IIoT) implementations focus on ensuring that IIoT technologies are used in ways that are ethical, secure, and beneficial to all stakeholders. Time series prediction and Responsible IIoT are interconnected in several important ways, particularly in the context of how IIoT systems utilize predictive analytics [2] to enhance industrial processes. An outline of their relationship is presented herewith.

[1-4] Dept. of Information Technology, Thakur College of Eng. & Tech, Mumbai, India.
[5] Dept of Computer Science & Engineering, SVNIT, Surat, India.
Email: aditeyakprajapati@gmail.com[1]; uchilsushan36@gmai.com[2]; vishal10031patwa@gmail.com[3]; rgm@coed.svnit.ac.in[5]
* Corresponding author: E-mail: nehap03@gmail.com

DOI: 10.1201/9781003587903-5

1.1 Predictive Maintenance

Time Series Prediction: Time series forecasting techniques are used to predict when equipment or machinery might fail based on historical sensor data, which is collected in real time. In a responsible IIoT implementation, predictive maintenance helps avoid unexpected downtimes, reduce costs, and extend the lifespan of equipment, all while ensuring that maintenance schedules are optimized to minimize disruptions.

1.2 Operational Efficiency

Time Series Prediction: Predictive models analyze patterns in operational data to forecast future conditions, such as demand, production rates, and inventory levels.

IIoT Context: Responsible IIoT systems use these predictions to optimize processes, reduce waste, and improve resource management.

1.3 Anomaly Detection

Time Series Prediction: Models can detect anomalies or deviations from expected patterns in real time, which might indicate potential issues or inefficiencies.

IIoT Context: In responsible IIoT implementations, anomaly detection helps quickly identify and address problems, enhancing system reliability and safety. It also supports security by identifying potential breaches or unusual activities.

1.4 Data Privacy and Security

Time Series Prediction: Predictive analytics rely on large volumes of data, which can include sensitive information. Proper handling and protection of this data are crucial.

IIoT Context: Responsible IIoT practices involve implementing strong data privacy and security measures to protect the data used in time series predictions. This includes data encryption, access controls, and compliance with data protection regulations.

1.5 Transparency and Ethics

Time Series Prediction: The accuracy and fairness of predictive models depend on the quality and representativeness of the data used.

IIoT Context: A responsible approach ensures that time series predictions are based on unbiased, high-quality data and that the methods used are transparent. This helps maintain trust and fairness in decision-making processes.

1.6 Scalability and Integration

Time Series Prediction: Predictive models need to scale with increasing volumes of data and be integrated with existing systems for actionable insights.

IIoT Context: Responsible IIoT systems ensure that time series prediction tools are scalable and interoperable with other industrial systems, allowing seamless integration and enhancing overall system efficiency.

1.7 Human Factors

Time Series Prediction: Predictive analytics can generate complex insights that need to be interpreted and acted upon by human operators.

IIoT Context: Responsible IIoT implementations focus on user-friendly interfaces and adequate training to ensure that users can effectively interpret predictive insights and make informed decisions.

The smart bin waste prediction system works by analyzing patterns in waste generation and making accurate predictions of future waste levels [10]. This information can then be used by waste management organizations to optimize their collection and disposal schedules, reduce costs, and improve efficiency. For example, by knowing how much waste will be generated [8], waste management organizations can adjust their collection routes to ensure that they are not collecting partially full bins, which can be a waste of resources and lead to higher costs.

Machine learning algorithms used in smart bin waste prediction can include decision trees, random forests, and neural networks, among others. These algorithms can analyze large amounts of data and identify complex relationships between different factors, such as weather conditions, population density, and waste generation patterns. This allows for more accurate predictions of future waste levels and helps to minimize the risk of waste being left uncollected.

In conclusion, smart bin waste prediction using machine learning has the potential to revolutionize the waste management industry. By providing accurate predictions of future waste levels, waste management organizations can optimize their collection and disposal schedules, reducing costs and improving efficiency. Additionally, by minimizing the risk of waste being left uncollected, smart bin waste prediction can help to improve public health and reduce environmental impacts.

2. Overview of Smart City IOT Applications for Waste Management

Smart city IoT applications are playing an increasingly important role in improving waste management in cities [5] [6]. By integrating IoT technology with existing waste management systems, cities can collect and analyze data in real-time, making more informed decisions about waste collection and disposal.

Smart bins are one of the main ways that IoT is being used in trash management [7]. Sensors built inside smart bins can determine the bin's fill level and send that information to a central management system. This allows waste management organizations to optimize their collection routes and minimize the number of partially full bins that are collected, reduce costs and improve efficiency. In addition, smart bins can also be equipped with other sensors, such as air quality sensors, that can provide additional information about the environment around the bin [9] [11].

Another important application of IoT in waste management is route optimization. By collecting data on waste generation patterns and traffic conditions, waste management organizations can optimize their collection routes to minimize the distance traveled and reduce the carbon footprint of waste collection. This can also help to improve public health by reducing the amount of time that waste is left uncollected.

IoT technology can improve recycling process optimization in addition to collection route optimization. For example, smart recycling bins can be equipped with sensors that identify the type of material being deposited and sort them accordingly. This can reduce contamination of recyclable materials and increase the effectiveness of the recycling process.

Monitoring the effects of waste management on the environment is another application of IoT technology. For example, sensors can be used to monitor air and water quality near landfills and waste management facilities, allowing authorities to take action to mitigate any negative impacts. In addition, IoT technology can be used to monitor the emissions from waste management facilities, helping to reduce their carbon footprint and improve sustainability.

Finally, IoT technology can also be used to educate the public about waste management. For example, smart bin systems can provide feedback to users on the amount and type of waste they are generating, encouraging them to recycle and reduce waste. This can also help to raise awareness about the importance of waste management and encourage more sustainable behavior.

In conclusion, smart city IoT applications have the potential to transform the way waste is managed in cities. By collecting and analyzing data in real-time, waste management organizations can make more informed decisions about waste collection and disposal, improving efficiency and reducing the environmental impact of waste management.

3. Materials and Models

3.1 An Intelligent Waste Management System

It combines machine learning and the IoT in certain ways [4]. Waste collection and real-time tracking are two applications for the Internet of factors. Smart garbage bins equipped with sensors can be positioned in various locations to gather rubbish [12]. Dustbins that are Arduino and microcontroller based are capable of collecting rubbish in a limited space and are not able to classify it according to its features [13]. Hence, an IoT-based waste management system could be chosen to continuously collect and monitor the waste. However, the existing systems no longer offer information on hazardous trash that emits a foul odor and poisonous fumes. Thus, it may be suggested to use an intelligent waste management system to keep an eye on both hazardous and non-hazardous trash. Sensor-based dustbins, such as those with ultrasonic sensors, can be positioned in various locations with sensors [14].

However, current methods did not focus on toxic waste, which generates foul odors and dangerous gases. Thus, it is possible to suggest intelligent-based waste management systems to keep an eye on both hazardous and non-hazardous trash. Sensor-based dustbins, such as those with ultrasonic sensors, can be positioned in various locations with sensors.

3.2 *Pursuit of Forecasting the Trash Levels of the Smart Bins*

The research began by acquiring the dataset from the Australian government portal. The dataset contained the following parameters, namely, coordinates, latest fullness, reason, Serial Number, Description, Age threshold, fullness, Timestamp. Out of all these parameters only three of the parameters were useful, namely, Latest Fullness, Fullness Threshold, and timestamp other parameters like coordinates, serial number, etc. were omitted as it did not add value to the forecasting of the latest fullness and showed constant as well as redundant data points causing to reduce the accuracy of the model. So, the trash level values in the latest fullness and threshold fullness values were represented in the integer form ranging from 1–10 (1 being the lowest and 10 being the highest value). Based on the data visualization, the data showed a considerable amount of missing values, outliers, and inconsistencies. The data was cleaned by using a performing series of pre-processing techniques.

Linear Regression-Linear: A popular statistical technique known as "linear regression" aims to provide light on the relationship between a dependent variable and one or more independent variables by converting the calculated statistics points into a linear equation. It functions by reducing the differences between the expected and actual values. Linear regression makes it easier to estimate the relationship between the variables by identifying the slope and intercept of the best line.

By doing so, it not only sheds light on the direction and intensity of the relationship, but also enables the forecast of the dependent variable's value based on the independent variables. In essence, linear regression serves as a valuable tool for identifying patterns and trends in data, making projections, and exposing links between distinct variables, ultimately aiding in a more comprehensive understanding of the data.

3.3 LSTM Model

Processing that uses LSTM network and recurrent neural networks (RNNs) with long short-term memory (LSTM) do the opposite of taking input in an advanced layer. They are frequently employed for sequence prediction and time series analysis tasks, such as smart bin waste prediction. The issue of vanishing gradients in typical RNNs is addressed by LSTMs, which enables them to retain data from lengthy sequences and identify long-term dependencies.

In the context of smart bin waste prediction, LSTMs can be used to analyze time series data collected from smart bins and predict the fill level of the bins at future points in time. The LSTM community may be skilled on historical data of bin-fill levels and use that information to make predictions about future fill levels.

One of the key advantages of LSTMs for smart bin waste prediction is their ability to handle the unpredictability of waste generation patterns. By analyzing multiple historical sequences, the LSTM network can learn the underlying patterns of waste generation and make predictions accordingly. In addition, LSTMs can handle non linear relationships between features, making them well-suited to the complex and non linear relationships that often exist between bin-fill levels and waste generation patterns.

However, it's important to note that LSTMs can be computationally expensive, particularly for large and complex data sets. Furthermore, the process of training LSTM models can be difficult and necessitates careful selection of Hyperparameters, such as the variety of hidden layers and neurons in each layer.

To sum up, long short-term relationships in waste formation patterns may be captured and complex time series data can be analyzed using LSTMs, making them an effective tool for smart bin waste prediction. To guarantee precise predictions and avoid overfitting, attention must be taken in the selection and training of LSTMs. In order to overcome the issues with teaching long-term dependencies caused by gradient problems in deep neural or recurrent networks, Hochreiter and Schmidhuber introduced the LSTM architecture. The memory block's design aims to manage the data stream via nonlinear gating units while maintaining its state over time. The form of an LSTM block [19] which includes the doors, yield $h(t)$, entrance flag $x(t)$, enactment capacity, and peephole relationships. The output block is frequently connected to the entrance block and each gate. centered on categorizing and forecasting data according to its properties. Binary forms are used to indicate the structured variable in this case. It can be divided into ordinal, multinomial, and binomial categories. The two types of established variables—for instance, win and loss, skip and fail, etc., —can occur in binomial regression.

The established variable in multinomial regression should have three or more valid values in an unordered fashion. The established variable in an ordinal regression should have three or more possible values in an ordered fashion. In logistics regression, the following presumptions are taken into account. The target values are represented by these binomial regressions in terms of binary values. There are no longer any couple of collinearities, meaning that each variable is independent of the others. These variables ought to be appropriately defined. It is applicable to large models. This case, a linear function is the input.

An LSTM community can only do forward passes on consecutive data due to its design, which eventually suggests the unidirectional form of statistics processing. One natural way to alleviate this restriction is to employ a cloned copy of the LSTM association, but in the opposite sequence as suggested by Alex Grave [20]. Thus, creating a backward layer by combining the LSTM front and backward opposite order. The inputs from the input layer are used to follow the forward layer, while the output layer receives the results produced by the backward layer.

3.4 Exponential Smoothing Model

Exponential smoothing aids time series forecasting by giving more weight to recent data points, allowing for more accurate predictions of future values. When the time series data shows no discernible trends or seasonality patterns, this strategy is quite helpful. It captures the underlying patterns and changes in the data over time by giving previous observations exponentially diminishing weights [15]. This method has the benefit of not requiring a lot of computation, which makes it simple to use and perfect for short-term forecasting. It helps in minimizing the impact of random noise and fluctuations, allowing analysts to focus on the underlying trend within the data. The simplicity and adaptability of exponential smoothing make it a valuable technique for forecasting applications in various fields, providing reliable estimates and insights into future trends.

3.5 SARIMAX Model

An expansion of the well-known ARIMA model, SARIMAX (Seasonal Autoregressive Integrated Moving Average with Exogenous Regressors) adds characteristics to account for seasonality and outside influences [16] [17]. SARIMAX aids in identifying intricate seasonal trends and outside regressors that might affect the time series data. To effectively model and forecast the data, it takes into account a number of factors, including the seasonal period, differencing, and autoregressive/moving average terms. SARIMAX is well-suited for managing a variety of real-world applications because it provides a thorough knowledge of the underlying patterns and trends in the time series data by taking into account both the seasonal and non-seasonal components [18]. It offers a strong foundation for time series data modeling and prediction, allowing analysts to take into account a variety of influencing elements and increase the precision of their projections.

4. Results and Discussion

The outcomes and a summary of all the machine learning techniques and algorithms utilized to produce precise trash prediction for smart bins are shown in this part. Since the time series data used to estimate the smart bins was available, it was sourced from the Australian government's website about the smart bins that are deployed in their cities.

Initially, regression-based time series forecasting techniques were employed to predict waste levels. Multiple algorithms, including linear regression and decision tree regression, were explored. However, these algorithms proved to be extremely imprecise. Subsequently, traditional time series algorithms such as ARIMA and SARIMA were employed; however, these algorithms produced an RMSE (root mean square error) of approximately 4.25, significantly exceeding the anticipated result. Graphical analysis was conducted to acquire an understanding of the data's characteristics.

Based on the analysis, an LSTM approach was decided to be followed as it can capture long-term dependencies in the data, which is crucial for many time series patterns. LSTMs have memory cells that can learn and remember patterns over extended time intervals. The overall fitting of the data was good but the predictions performed by LSTM were inaccurate; the LSTM model was not able to capture the uncertain dips and spikes of the latest Fullness. Based on the past performance of the classification algorithms as compared to regression algorithms is more reliable performance. So we made the data points of the sensor data concise by assigning class labels to certain fixed trash levels. For example, trash levels from 0–3 will be assigned a 'low' class label, 4–7 will be assigned a 'medium' class label, and 8–10 will be assigned a 'high' class label accordingly. After assigning the class labels to the respective trash levels, analysis visualization was performed that showed that the dataset was imbalanced having the presence of the minority data points due to which the model was not able to capture the uncertain changes as the presence of these data points were very less as compared to the other regular data points. So to counter this SMOTE (synthetic minority oversampling technique) analysis was performed and the Random Forest algorithm was performed to check the effectiveness of implementing the class labels and SMOTE by giving a test data to be predicted by the random forest algorithm giving an accuracy of around 70% (precision and recall were also considered and the various class labels were predicted effectively) being the highest accuracy every achieved out of the implemented algorithms. Based on the seasonality inference achieved form the regression algorithm, seasonal component was added to the data and then effectiveness check was performed using random forest algorithm giving an accuracy of 94%, and then forecasting were performed using SARIMAX (before forecasting ADA fuller, PACF, PDQ tests were performed for optimal order) giving RMSE between 1.0–1.62 (varies at each compilation). This variation was due to the synthetic instances created by SMOTE to counter that ADASYN was implemented ensuring more consistent results for imbalanced data yielding an RMSE of 1.4 was achieved.

Another approach of Exponential smoothing +SARIMAX (E-SARIMAX) was implemented as is especially beneficial for gathering and projecting data that heavily relies on seasonality and short-term variations. It offers a straightforward and understandable method for extracting underlying patterns from noisy time series data and SARIMAX is a flexible and effective model for processing time series data with many elements, such as seasonality, trends, and exogenous variables. In this approach, the train data was fitted using exponential smoothing, then residuals were calculated by original_train_data -model_fitted_train_data. These residuals were then used by SARIMAX for forecasting. Basically, the exponential smoothing is used to capture the short-term seasonality and the SARIMAX is used to capture the long-term seasonality which exponential smoothing was not able to capture in the form of residual forecasting. The forecasting of both the algorithms were combined and yielded an RMSE of 0.71.

Table 1 Comparison of various ML models used for Time Series Prediction.

Model	*When to Use*	*Disadvantages*
Regression model	When the time series data exhibit a clear trend (upward or downward with respect to time).	Strong assumptions need to be made. Sensitive to outliers.
Exponential smoothing	Simple and moderate trends and seasonality components.	Sensitive to outliers.
ARIMA	When the data exhibit stationary behavior (the mean and variance remain constant over time).	Requires more data. Strong restrictions and assumptions.
SARIMA	Can predict the seasonality components.	Needs the stationary series.
ANN	Suitable for any nonlinear relationship. High predictive power.	Lack of interpretability Requires a lot of parameters.
CNN	Able to extract features automatically. Suitable for image data.	Need parameter tuning. Need a lot of sample data.
RNN/LSTM	Effective for capturing long-term dependencies and complex patterns in time series data.	Need a large amount of calculation.

5. Conclusion

In the context of the Waste level Forecasting system, the utilization of classification algorithms showcased a substantial performance advantage over regression algorithms. This significant disparity in results is attributed to the ability of classification methods to compact the data effectively. By employing various techniques, such as balancing imbalanced data, seasonality, the classification approach streamlined the prediction process and demonstrated superior accuracy. This outcome underscores the importance of tailored algorithm selection and preprocessing techniques to ensure optimal predictive outcomes in waste management applications.

5.1 Social Impact and Practical Impact

- With overflowing trash bins, there will be a decrease in litter on the streets. This improvement benefits both residents and visitors enhancing the quality of life.
- It can lead to a cost-effective waste management practice. By predicting when trash bins will reach their capacity, collection schedules can be optimized. This means trips for garbage trucks, reduced fuel consumption and lower expenses, for waste management companies. It is a situation where everyone gains, helping both the environment and saving money.

References

[1] Uganya, G., Radhika and Vijayaraj, N. (2021). A survey on internet of things: Applications, recent issues, attacks, and security mechanisms. *Journal of Circuits, Systems, and Computers*, *30* (5), Article ID 2130006.

[2] Yelamarthi, K., Aman, M.S. and Abdelgawad, A. (2017). An application driven modular IoT architecture. *Wireless Communications and Mobile Computing*, *2017*, Article ID 1350929, 16 pp.

[3] Anh Khoa, T., Phuc, C.H., Lam, P.D. et al. (2020). Waste management system using IoT-based machine learning in university. *Wireless Communications and Mobile Computing*, *2020*, Article ID 6138637, 13 pp.

[4] Washburn, D., Sindhu, U., Balaouras, S., Dines, R.A., Hayes, N. and Nelson, L.E. (2009). Helping CIOs understand "smart city" initiatives. *Growth*, *17* (2), 1–17.

[5] Giffinger, R. and Gudrun, H. (2010). Smart cities ranking: An effective instrument for the positioning of the cities? *ACE - Architecture, City, and Environment*, 4 (12), 7–26.

[6] Sosunova, I. and Porras, J. (2022). IoT-Enabled Smart Waste Management Systems for Smart Cities: A Systematic Review. *IEEE Access*, *10*, 73326–73363. [CrossRef.]

[7] Baby, C., Singh, H., Srivastava, A., Dhawan, R. and Mahalakshmi, P. (2017). Smart bin: An intelligent waste alert and prediction system using machine learning approach. In: *2017 International Conference on Wireless Communications, Signal Processing and Networking* (*WiSPNET*), 771–774. https://doi.org/10.1109/WISPNET.2017.8299865.

[8] Hussain, A., Draz, U., Ali, T., Tariq, S., Irfan, M., Głowacz, A., Daviu, J., Yasin, S. and Rahman, S. (2020). Waste Management and Prediction of Air Pollutants Using IoT and Machine Learning Approach. *Energies*. https://doi.org/10.3390/en13153930.

[9] Pereira, W., Parulekar, S., Phaltankar, S. and Kamble, V. (2019). Smart Bin (Waste Segregation and Optimisation). In: *2019 Amity International Conference on Artificial Intelligence* (*AICAI*), 274–279. https://doi.org/10.1109/AICAI.2019.8701350.

[10] Ferrer, J. and Alba, E. (2018). BIN-CT: Urban Waste Collection based on Predicting the Container Fill Level. *Bio Systems*. https://doi.org/10.1016/j.biosystems.2019.04.006.

[11] Wijaya, A., Zainuddin, Z., and Niswar, M. (2017). Design a smart waste bin for smart waste management. In: *2017 5th International Conference on Instrumentation, Control, and Automation* (*ICA*), 62–66. https://doi.org/10.1109/ICA.2017.8068414.

[12] Catania, V. and Ventura, D. (2014). An approach for monitoring and smart planning of urban solid waste management using smart-M3 platform. In: *Proceedings of 15th Conference of Open Innovations Association FRUCT*, 24–31. https://doi.org/10.1109/FRUCT.2014.6872422.

[13] Haribabu, P., Kassa, S., Nagaraju, J., Karthik, R., Shirisha, N. and Anila, M. (2017). Implementation of a smart waste management system using IoT. In: *2017 International Conference on Intelligent Sustainable Systems* (*ICISS*), 1155–1156. https://doi.org/10.1109/ISS1.2017.8389367.

[14] Haribabu, P., Kassa, S., Nagaraju, J., Karthik, R., Shirisha, N. and Anila, M. (2017). Implementation of a smart waste management system using IoT. In: *2017 International Conference on Intelligent Sustainable Systems* (*ICISS*), 1155–1156. https://doi.org/10.1109/ISS1.2017.8389367.

[15] Akinadé, O. and Oyedele, L. (2019). Integrating construction supply chains within a circular economy: An ANFIS-based waste analytics system (A-WAS). *Journal of Cleaner Production*. https://doi.org/10.1016/J.JCLEPRO.2019.04.232.

[16] Hannan, M., Arebey, M., Begum, R. and Basri, H. (2012). An automated solid waste bin level detection system using a gray level aura matrix. *Waste Management*, *32* (12), 2229–2238. https://doi.org/10.1016/j.wasman.2012.06.002.

[17] Arebey, M., Hannan, M., Begum, R. and Basri, H. (2012). Solid waste bin level detection using gray level co-occurrence matrix feature extraction approach. *Journal of Environmental Management*, *104*, 9–18. https://doi.org/10.1016/j.jenvman.2012.03.035.

[18] Aziz, F., Arof, H., Mokhtar, N., Shah, N., Khairuddin, A., Hanafi, E. and Talip, M. (2018). Waste level detection and HMM based collection scheduling of multiple bins. *PLoS ONE*, *13*. https://doi.org/10.1371/journal.pone.0202092.

[19] Ralf C. Staudemeyer and Eric Rothstein Morris. (2019). Understanding LSTM: A tutorial into Long Short-Term Memory Recurrent Neural Networks. https://doi.org/10.48550/arXiv.1909.09586.

[20] Alex Grave (2005). Framewise Phoneme Classification with Bidirectional LSTM Networks. In: *Proceedings of International Joint Conference on Neural Networks*, Montreal, Canada, July 31–August 4, 2005.

6 Regulatory Landscape for Responsible IIoT Practices

Gauravi Suraj Naik[1*], Janhvi Vinod Kushwaha[2], Pallavi Manoj Jha[3], and Ansh Santosh Mishra[4]

The Industrial Internet of Things (IIoT) is transforming industrial sectors to integrate advanced technologies. Connectivity, analytics, and automation in machines and systems are promoting unprecedented levels, operational efficiency, and innovation. This change presents opportunities and challenges. This is especially true in the area of data privacy, cyber security and compliance. It is IIoT technology. It is widespread because the industry faces maritime tariffs, which have a complex regulatory situation to warrant implementation. This article explores several regulatory challenges related to IIoT, including Compliance with Data Protection Laws, Cyber Security Guidelines Safety, and Environmental Guidelines Regulations to determine existing directions and emerging trends. This article is intended to provide basic information. A clear view of the best practices and governance structures required to operate ethically and effectively are also discussed. Using IIoT technology, this study is illustrated through a literature review and industry practices. This is because responsible IIoT practices can increase organizational efficiency, promote sustainability, and can be driven by innovation, while reducing risk and ensuring compliance with regulatory changes.

1. Introduction

The Industrial Internet of Things (IIoT) represents a major advancement in industrial technology. It combines cutting-edge analytics, connectivity, and automation to transform sectors such as manufacturing and logistics. Interconnected devices can integrate sensors, actuators, and communications devices into industrial systems to create and exchange IIoT data in real time, an innovation that creates expansive networks and provides unprecedented performance opportunities: Predictive Maintenance and Better Decision-Making. However, the rapid expansion of IIoT technologies comes with a complex regulatory situation. Due to many sectors as

[1-4] Internet of Things Thakur College of Engineering and Technology, Mumbai, India
Email: janhvikushwaha94940@gmail.com[2]; pallavijha703@gmail.com[3]; anshmishraaa0017@gmail.com[4]
* Corresponding author: naikgauravi84@gmail.com

DOI: 10.1201/9781003587903-6

more people adopt IIoT solutions, they face regulatory challenges. Cybersecurity and environmental impact models including data privacy. Each of these issues needs careful consideration to balance the benefits of technological innovation with the need to protect public interests. The enormous amounts processed by IIoT systems often contain sensitive information about operational processes. It might include the devices and sensitive personal information.

Regulators are responding with guidance and best practices, such as the NIST Cyber Security Box and ISO/IEC 27001, to implement security controls and incident response processes. To help protect IIoT systems from breaches, you need regulations from agencies such as the European Union Cybersecurity Infrastructure Assurance Agency (CISA) and the European Union Information Network. Information such as the EU's General Data Protection Regulation (GDPR), and the California Consumer Privacy Act (CCPA) in the United States, and to ensure transparency and accountability of dice. The increasing attack surface of IIoT systems also highlights important cybersecurity concerns. Emerging challenges are an ongoing challenge [1] [2].

Beyond data privacy and cyber security, integrating IIoT technologies must also address security issues. Compliance with sector-specific safety standards set by the International Electrotechnical Commission (IEC) and the American National Standards Institute (ANSI) consider it extremely important to ensure the safe operation of machinery and equipment to prevent accidents from occurring. Environmental considerations are also an important factor in the IIoT regulatory scenario as industries adopt IIoT solutions to promote efficiency and reduce waste. However, regulators are increasingly focused on the environmental impact of these technologies. Regulations such as the European Union's Waste Electrical and Electronic Equipment (REEE) Directive and the Restriction of Hazardous Substances (RoHS) Directive addresses issues such as energy use. E-waste disposal and resource use. The goal is to reduce the environment in which two IIoT devices work together. This is because IIoT practices involve understanding and meeting these complex challenges [5]. It is still being developed and integrated into various industrial sectors. Therefore, it is imperative that decision-makers provide industrial policy. Leaders and technologists join forces to develop and enforce progressive regulations. These structures evolve for privacy of two dice: Cybersecurity, Safety, and Environmental Impact. To ensure that IIoT innovations are deployed responsibly and sustainably, the technique in this article explores these regulatory challenges in depth. Consider the current direction trends that occur such as Best Practices for Responsible Use of IIoT Technology. IIoT refers to the use of internet-connected devices, sensors, and actuators in industrial environments to improve and increase operational efficiency, automate, and enable predictive maintenance and real-time analytics.

These devices collect large quantities of dice, which are processed and used to increase production line efficiency, and manage cadeia auxiliary. Improving the overall performance of the Embora system to IIoT provides many benefits and also recommends various advantages [6].

1.1 Challenges Related to Cyber Security, Data Privacy, and System Collaboration

The Industrial Internet of Things (IIoT) represents a major shift in the way industries operate, manage and optimize operations perfectly and integrate physical machines with networked smart devices. This connection allows for real-time dice trading and advanced analytics to achieve unprecedented levels of automation efficiency and decision accuracy. However, with the rapid advent of IIoT, the need to act responsibly has become even more important. This ensures that the associated benefits are not overshadowed by risks such as cybersecurity vulnerabilities, data privacy concerns, and regulatory compliance challenges responsible IIoT practices aim to address those concerns. To include strict security measures, they follow a strict regulatory framework, and promote transparency in the use of information and governance. A unique scenario that IIoT offers is the ability to create total automation. Predictive analytics and machine learning work alongside operational processes; however, to avoid preventive issues such as equipment failure or insufficient energy efficiency, it is necessary to utilize this potential in a manner that ensures both operational excellence and excellence in operations. However, moral behavior and ecological sustainability are also matters of concern.

Scalability is supported by IIoT's passive assistance, supporting companies to expand and increase performance without sacrificing interoperability or security, and create new avenues for innovation. It provides a safe space for technological experimentation in sectors such as transformation industries. Healthcare and Energy Responsible adoption of IIoT has the potential to revolutionize the way we produce. Supply chain and resource management at the same time complies with global sustainability and safety standards. The transformation industry relies heavily on IIoT for smart factories and automation. Standards such as ISO 50001 for energy management and IEC 61508 for functional safety provide guidelines for safe and efficient use of IIoT in industrial environments. For taking care of our health, integrating IIoT devices such as connected medical devices and remote monitoring systems, strict regulations are required. The Food and Drug Administration (FDA) has issued guidelines for medical devices, including the Health Insurance Portability and Accountability Act (HIPAA), which guarantees the availability of data collected from two patients.

Revolutionizing or defining the enthusiasm and enabling of smart grids?

For predictive maintenance and emergency energy monitoring in the energy sector, regulations from agencies such as FERC in the United States and the European

Network of Transmission System Operators (ENTSO-E) in the European Union have established protocols for maintaining safety and protection.

1.2 Cyber Operations of IIoT-enabled Infrastructure

Meanwhile, many industries are moving towards smart factories and ecosystems. Therefore, responsible IIoT practices are becoming increasingly important. This is to ensure that advances in automation and efficiency do not lead to increased risk or uncontrolled environmental impacts. By combining a weighted approach with regulatory compliance and risk management and ethical considerations, IIoT can serve as a powerful tool to drive industrial progress and social responsibility. This balanced approach combines new technology with a regulatory approach. This paves the way for IIoT, not just as a production mechanism. It is also the cornerstone of sustainable and responsible industry growth.

2. Literature Survey

The governance framework for IIoT includes international and local formats. International standards such as ISO/IEC 27001 focus on information security management, ISO/IEC 30141 focuses on the IIoT reference architecture, and IEC 62443 covers information security. industrial cyber. This includes the General Data Protection Regulation (GDPR) in Europe and the California Consumer Privacy Act (CCPA) in the United States. Different industries have various sectors.

Organisations like the National Institute of Standards and Technology (NIST) and the International Automation Society (ISA) create best models for manufacturing.

Regulations from agencies like the International Energy Agency (IEA) and particular standards for cybersecurity and energy management are crucial in the energy sector. Data theft is one of the ethical issues with IIoT, not to mention adherence to moral data processing. Reducing the amount of data and protecting individual rights, creating accountability procedures for data breaches and surveys including ensuring openness in IIoT operations is important. A key part of the IIoT architecture is risk management and security. This includes following established cybersecurity best practices such as the NIST Cyber Security Fund and the Cyber Security Act of 2015, as well as reviewing established risk assessment models. They can provide information on how to manage potential vulnerabilities for IIoT environments. Constraints also form the basis of operational strategies. Organizations must comply with relevant laws and regulations, and use strategies such as audit and certification processes. To ensure compliance with regulations, it is important for regulators to understand compliance and the legal consequences of non-compliance. As the IIoT situation continues to evolve, it is important to keep up with the latest trends in regulatory developments and technological advancements. New data governance regulations and new privacy concerns will shape the future of IIoT as new technologies require advanced approaches. Check this issue completely, we must provide academic literature. Recent historical reports and official documents from regulatory agencies focus on practices to adapt to technological advances and changes in countries of origin. In terms of predictive management, specific studies such as **Mobley (2020)**, go beyond the basic concept of error prediction and explore how IIoT systems can provide management under similar circumstances, specific usage patterns, and environments. This self-care strategy helps in saving money and reduces equipment downtime.

Notable contributions to IIoT security include the work of **Shroff et al. (2014)** discussing the integration of augmented reality (AR) systems with IIoT to provide real-time security information. Giving information to employees across devices can be used to improve facility safety. The main focus of important literature, such as that of **Jeschke et al. (2017)**, is the revolutionary role of data-driven decision-making to improve not only the efficiency of management but also the overall strategic direction of the company. They argue that IIoT's ability to use real-time big data analytics will help companies gain a competitive advantage or whether it help them to act quickly in response to changes in the market or our results? Meanwhile in cybersecurity, **Gao et al. (2020)** show that the combination of IIoT systems and edge computing creates new vulnerabilities and highlights the importance of developing cybersecurity infrastructure to protect against these new threats. From a positive energy perspective a unique perspective can be found in the literature of **Tao et al. (2019)**, where we explore how IIoT systems can efficiently manage energy, and help machines and groups of machines increase energy efficiency. This not only reduces waste but also helps companies. Sustainability goals can be better achieved without central control.

Similarly, researchers such as **Wazid et al. (2019)** discussed interoperability in a number of contexts examining the role of blockchain in ensuring secure communication between IIoT devices from multiple manufacturers. Blockchain accounting technology is seen as a facilitator of interoperability without compromising security. Adaptation studies such as **Madakam (2015)** offers a unique perspective on how IIoT systems can help companies. How to comply with the regulations? Effectively integrate compliance controls into operational processes. This improves the audit process and reduces the risk of non-compliance. This is true in corporate and regulatory environments. In terms of scalability, **Gobi et al. (2013)** have a specific discussion of IIoT systems that allow future proofing and ensure that it can be easily scaled to include new technologies such as 5G without having to change the system. Based on the concept of customer satisfaction, **Lu et al. (2020)** explores the role of IIoT in enabling mass customization, where IIoT systems facilitate the mass production of personalized products. This increases customer loyalty and satisfaction.

In terms of cost reduction, a separate study conducted by **Pfeiffer et al. (2017)** was more open about the general economics surrounding automation It focuses on how IIoT-based resource allocation algorithms help industries maximize the use of raw materials and reduce indirect costs. Asset utilization is another area where IIoT has made unique progress. Based on the research of **Palattella et al. (2016)**, they analyze digital twin physical assets in real time. To guarantee the greatest uptime of equipment activities, it can be used to optimize resource allocation and simulate various usage situations. **Raj et al. (2020)** also wrote a paper on Sustainability and Environmental Impact. Real-time analysis, monitoring and mitigation of furnace gas emissions using IIoT systems helps various stakeholders. To reduce carbon dioxide emissions and comply with environmental regulations. In terms of risk reduction, **Singh et al. (2021)** is a prominent example of integrating machine learning algorithms and IIoT systems to create predictive risk models that can predict or risk. Through action we can take and recommend real-time mitigation strategies, maximize safety and productivity. Finally, turning to the literature on market diversification, **Kelberg et al. (2020)** specifically argue that responsive IIoT initiatives not only create operational efficiencies, but it also serves as a differential to improve the company's reputation for innovation, safety, and security. This is especially relevant in industries where technical management is the main reason for attracting customers and investors. One of the most important regulations is the EU Anti-Cyber Act (CRA). This sets strict home security requirements.

Additionally, the CRA or Industrial Internet Consortium (IIC) [6] has issued guidelines highlighting best practices for data security in IIoT systems. These guidelines address issues such as integrity and confidentiality. Emphasizing the importance of protecting teeth throughout their lifecycle, the CII guidance is designed to complement existing structures such as the IoT Security Maturity Model and the Industrial Internet Security Structure to help organizations protect sensitive information.

Effective silence as the IIoT scenario expands, the need for strong compliance and governance structures becomes critical. Organizations must follow rules and regulations, ensure proper integration and security of IoT solutions, and meet demanding testing and certification requirements. Effective management techniques can reduce the risks associated with cyber vulnerabilities and help ensure successful and reliable IoT initiatives. The integration of IIoT technologies presents challenges and opportunities for compliance. The continuous monitoring and real-time data collection has been made possible by IIoT which can help organizations meet regulatory requirements more efficiently. However, the increasing complexity of IIoT systems also increases their vulnerability to cyber threats. This requires a proactive approach to ensuring and managing compliance. Collaboration between sector stakeholders is essential to improving IIoT security and compliance. Structures such as IEC 62443 provide valuable guidance on securing IIoT systems, emphasizing it is necessary for a secure development lifecycle and integration of different technologies. Collaboration across the industry allows for the adoption of best practices that benefit everyone. This guarantees that security measures are in place.

3. Benefits of IIoT Practices

Responsible IIoT practices offer numerous benefits.

Predictive maintenance an advantage. This is because IIoT systems use real-time data collection and analysis to detect potential equipment failures. This reduces unplanned downtime. These systems also improve safety by monitoring conditions and alerting operators of potential hazards. This helps reduce accidents in the workplace.

Data-driven decision-making is made easier. As IIoT devices generate vast amounts of real-time data to improve operational planning and strategic planning, regulatory frameworks also enhance cybersecurity. They protect critical infrastructure from potential attacks that disrupt operations or information that is sensitive or can be compromised

IIoT promotes energy efficiency by tracking and optimizing energy use, reducing waste, reducing costs, and supporting sustainability goals. Standards also ensure compatibility between devices from different manufacturers. It makes it easy to integrate compliance with cyber security laws. Privacy and security can be made easier and help businesses avoid legal penalties and improve their reputation.

IIoT systems are simple and flexible by helping companies expand their operations without compromising safety or efficiency. These systems improve product quality through real-time monitoring, strengthen quality control, and also reduce waste.

IIoT in supply chain management helps in monitoring assets in real-time, and improves inventory management and operational efficiency. An improved IIoT environment promotes innovation and provides a safe environment for testing new business models and technologies.

IIoT also improves resource utilization by reducing downtime and increasing equipment efficiency through real-time monitoring. Helping companies, we can operate at higher frequencies causing distortion or abnormalities.

IIoT systems reduce risk by using predictive analytics and monitoring to predict device failures and security breaches. They are also involved in sustainability projects. They increase the efficiency of resource use and reduce environmental impacts which are consistent with the objectives and responsibilities of the company. In the end, will adding IIoT functionality differentiate it or market it? It positions the company as a leader in innovation, safety, and efficiency that will attract customers and investors.

Another notable benefit is the **improvement in product quality** along with real-time monitoring of the production process. They help companies with prompt assistance and bug fixes, guarantees consistent product style, and reduces waste.

Customer satisfaction is another benefit as IIoT helps companies to respond to needs more efficiently, improve product quality, and responsiveness.

Reducing risk through predictive analytics is increasingly urgent which helps companies anticipate possible failures or security breaches and take proactive steps to avoid them.

Standards and regulations help ensure that IIoT devices from different manufacturers communicate well and work together and promote interoperability and flexibility between systems.

All in all, the benefits of responsible IIoT operations will span across operational productivity, innovation, cost savings and sustainability, making them an integral part of new industrial processes [72].

4. Challenges of Administering IIoT Practices

The IIoT brings noteworthy progressions in operational productivity, real-time observing, and data-driven decision-making. In any case, the administrative scene administering dependable IIoT practices presents complex challenges. One of the essential concerns is information protection and proprietorship. IIoT gadgets create and collect enormous sums of delicate mechanical information, raising basic questions around who possesses that information and how it is utilized. Administrative systems just like the GDPR within the European Union force strict necessities for information taking care of, commanding unequivocal assent from clients and clarity on information proprietorship [1] [2] [4]. In mechanical situations, where information can be shared between numerous stakeholders—such as producers, providers, and conclusive customers—determining possession and utilization rights gets to be lawfully challenging.

Another challenge worth considering is cyber security and the lack of interest in data. This is because the IIoT structure is highly interconnected. This creates a sufficient attack surface for cyber threats. Compromised infrastructure can result in disappointments such as idle time in the infrastructure. This leads to financial and security risks.

Regulators are beginning to emphasize the need for strict cybersecurity conventions; however, tracking dangers in advance can be a significant challenge. Many controls available, particularly in departments such as manufacturing, vitality, and health, are outdated or sector-specific to deal with the real-time cyber threats faced by IIoT devices. Interoperability and standardization pose a challenge to capable IIoT appropriation as IIoT gadgets regularly work on diverse conventions, equipment designs, and stages. The need of standardization leads to troubles in coordination of gadgets from diverse merchants, and the nonattendance of administrative guidelines complicates worldwide arrangements. Administrative agencies have struggled to keep up with IIoT mechanical advances, resulting in regulatory fragmentation that moves from one region to another. This creates complexity for companies trying to ensure regulatory compliance in their multinational operations.

Finally, regulations and safety obligations for IIoT-based structures are still being developed. Integrating IIoT into infrastructure such as controlled factories, production lines, and transportation requires safety agreements to predict failures caused by equipment failures or disturbances. However, deciding the obligation in the event of such a failure is questionable. For example, in the event of a breakdown of IIoT-controlled machinery, questions arise. Is the manufacturer program provider or is the administrator responsible? However, the acting administrative system has not fully addressed these concerns. This eliminates legal liability loopholes.

One of the main issues with the IIoT is concerns about privacy and data ownership. IIoT devices continuously collect, transmit, and analyze large amounts of data. This usually includes confidential information, such as proprietary manufacturing processes or customer perceptions. This situation raises an important question: What do you give with two dice? Is it the organization that generates the data, or the processing suppliers or platforms that manage and utilize it? Additionally, how can organizations ensure that data is used responsibly? This is especially true in areas with strict data protection regulations, such as the GDPR in Europe. The GDPR requires companies to obtain the consent or explicit consent of the consumer. Two people consume and use this form before playing dice and grant individuals the right to access, correct, or exclude their information. In an industrial context where multiple stakeholders, such as suppliers, operators and customers, have access to data, achieving compliance can be a significant challenge. Managing these complexities can result in operational delays or increase operating costs for companies looking to adopt IIoT systems.

Cybersecurity and data integrity also raise serious concerns. This is because IIoT devices are connected to a wide network. The potential for cyber threats has increased significantly. Cyber attacks can disrupt entire production operations affecting energy supply or even make important infrastructure such as wastewater treatment plants at risk; this is unlike traditional IT systems that can easily receive patches and security updates. Protecting IIoT devices can be more complicated.

Many are integrated deeply into production systems, function in demanding environments, or depend on outdated software not designed with cybersecurity considerations. Organizations such as the NIST in the U.S. have formulated cybersecurity frameworks to assist in IIoT implementation; however, these standards often emerge reactively, following significant incidents. So there are countless companies still being revealed, especially those who fail to invest in advanced cybersecurity solutions. Ensuring cybersecurity compliance across regions or sectors can be expensive and delays, especially in multinational operations. Interoperability and mapping are another key challenge. Remarkably, the IIoT scenario is such. Segmented with devices from multiple manufacturers using various communication protocols. Hardware configuration and software platform Lack of registration complicates continuity. Integrating various IIoT systems for enterprises, for example, manufacturing plants can use sensors from one supplier's machinery from another and software from another facilitate communication between these systems can be problematic and lacks a universal format.

Exacerbating this problem, however, regulators have recognized the need for standardized protocols. It struggles to keep up with the rapid advancements in IIoT technology, which often results in inconsistencies. Regulations vary between countries or industries. This makes IIoT implementation a challenge for companies. Global Solutions, for example, a factory located in Europe may comply with a set of supervisory regulations. Installations similar to the United States should be consistent with those elsewhere. This leads to increased operational inefficiency and expenses. Security and liability are additional concerns. IIoT systems are becoming more integrated. Critical infrastructure such as energy distribution networks, automatic vehicles, and health system audits. Failure in any system can have serious and even fatal consequences.

In summary, the regulatory environment for IIoT is complex and fraught with challenges that must be faced to ensure its responsible and safe use [8]. Implementation and embracing IIoT has significant potential to revolutionize the industry. Various companies must comply with ever-changing regulations related to data privacy: cyber security working together with responsibility and ethics, policy makers, and companies.

We must work together to create clear, forward-looking regulations for the future that strike a balance between incentivizing innovation and protecting public safety and privacy. Without such measures, the full potential of IIoT may remain unrealized, with its associated risks overshadowing its advantages.

5. Survey Responses

5.1 Harmonization of Regulations

There is a dire need for international harmonization of IIoT regulations. Try going to a concert on cyber security, data protection, and security models across regions to promote interoperability and facilitate [3]. This is because many companies operate in many markets.

Dynamic control box

To keep up with technological advancements, the regulatory framework must be flexible and dynamic. Policymakers should further consider regulatory sandboxes that allow testing of new IIoT proposals. Technology controlled the environment before it was widely used. This approach creates an incentive for innovation, and guarantees safety and protection simultaneously.

5.2 Accelerate Safety Certification

Government agencies and industry must work together to develop innovative certification schemes that guarantee safety, privacy, and reliability of IIoT products and services. These certifications will have to be serious, and help ensure IIoT solutions meet the required formats for use in critical infrastructure.

5.3 Cooperation between the Public and Private Sectors

Joint efforts between government agencies and private companies are essential to develop effective regulations. Because cooperation between the public and private sectors can help bridge the gap between innovation and governance, involving industry stakeholders in the process of overseeing and approving regulations, they must be practical and effective.

Conclusion

The integration of the IIoT is a significant evolution in the industry. Technology provides better operational efficiency, predictive maintenance, and better decision-making capabilities. However, the rapid expansion of IIoT technologies requires a different approach. Compliance and ethical considerations ensure compliance with data privacy regulations, e.g., GDPR and CCPA, and implementing effective cybersecurity measures are critical to protecting data confidentiality, information, and maintaining operational integrity. It also meets sector-specific safety standards. Environmental laws and regulations help reduce the risks associated with new technologies. The literature and case studies reviewed show that the development towards IIoT offers considerable benefits, such as, increased productivity, energy efficiency, and being able to compete in the market; acting responsibly is essential. To face the challenges involved, it is important to adopt an open governance and ethics framework. As the industry harnesses the full potential of IIoT technology and drives sustainable growth and innovation, In the final analysis the success of IIoT implementation depends on a balanced approach. It combines technology development with regulatory oversight and risk management [7].

References

[1] Grassi, L. and Lanfranchi, D. (2022). RegTech in public and private sectors: The nexus between data, technology, and regulation. *Journal of Industrial Business and Economics, 49*, 441–479.

[2] Firmansyah, B. and Arman, A.A. (2022). A Systematic Literature Review of RegTech: Technologies, Characteristics, and Architectures. In: *Proceedings of the 2022 International Conference on Information Technology Systems and Innovation (ICITSI)*, Bandung, Indonesia, 8–9 November (pp. 310–315).

[3] Butler, T. (2017). Towards a standards-based technology architecture for RegTech. *Journal of Financial Transformation*, *45*, 49–59.
[4] Gasparri, G. (2019). Risks and Opportunities of RegTech and SupTech Developments. *Frontiers in Artificial Intelligence*, *2*, 14.
[5] Dziawgo, T. (2021). Supervisory Technology as a New Tool for Banking Sector Supervision. *Journal of Banking, Finance, and Economics*, *15*, 5–13.
[6] IIoT Consortium. (n.d.). *Global Standards Strategy White Paper*. Retrieved from IIC Global Standards Strategy.
[7] Digital Guardian. (n.d.). *Highlighting Data Protection Best Practices for IIoT Systems*. Retrieved from Digital Guardian.
[8] IIT Kanpur. (n.d.). *IoT Standards and Regulations: Compliance and Governance in IoT Solutions*. Retrieved from IIT Kanpur Knowledge Hub.

7 Human-Centric Design in IIoT Solutions

Sonal Dubal[1], Usha Gupta[2], Sony Jha[3], Priyanka Musale[4] and Pratibha Sajwan[5*]

In the changing digital industrial paradigm, companies increasingly focus on upgrading through put and innovating to gain competitiveness. The growing pressure to maximize efficiency, increase productivity, and save costs has further led to widespread adoption of IoT (internet of things) technologies in manufacturing. IoT is a powerful solution to those challenges by allowing connections, data transfer and the intelligent implications of diverse stages of production. This study explores the Transformative role of industrial IoT in operations and processes. It goes into the details of how IIoT utilizes sensors, processing power, and connectivity to provide real-time monitoring and predictive maintenance, asset tracking, and other aspects with improved utilization of machines. Cloud-based IoT applications are essential in accessing and analyzing data and then making informed decisions that increase agility and responsiveness of operation. Architecture and components of IoT, which include perception, network, processing, and applications layers, are discussed to provide a comprehensive understanding of IIoT frameworks. The benefits outlined in the research, such as enhanced machine uptime, predictive maintenance, and just-in-time manufacturing capabilities, highlight the significant impact of IIoT on cost savings, productivity gains, and overall operational efficiency. Moreover, the study underlines the importance of IIoT in knowledge sharing, standardization of processes, and efficient management of data silos in the manufacturing environment. With integration of intelligent assets, data infrastructure, analytics, and human expertise, IIoT is presented as a powerful enabler in the extraction of actionable insights toward continuous improvement of manufacturing operations. This research therefore provides important insight into strategic adoption and implementation of IIoT toward sustainable competitive advantage in the digital age.

[1-4] Department of Internet of Things, Thakur College of Engg. & Technology, Mumbai, India.

[5] Department of Artificial Intelligence and Data Science, Thakur College of Engg. & Technology, Mumbai, India.

Email: sonaldubal468@gmail.com[1]; usha.gupta@tcetmumbai.in[2]; sony.jha@tcetmumbai.in[3]; priyanka.pekhale@tcetmumbai.in[4]

* Corresponding author: pratibha.sajwan@tcetmumbai.in

DOI: 10.1201/9781003587903-7

1. Introduction

The Industrial Internet of Things (IIoT) has dramatically changed industries by creating interlinked systems of machines, people, and processes in ways that were not possible before. While technological progress has been at the forefront, it is now largely a matter of how well IIoT solutions match human needs and behaviors. Human-centric design becomes essential to ensure that these advanced systems are functional, user-friendly, safe, and adaptable to those who use them. In the recent digital industrial, smart, and competitive world, currently companies are increasing their throughput to enter and to create innovative markets which results in key for success. Due to coercion of technology, pressure to increase efficiency and productivity while reducing cost, the Internet is an adventure and the role of Internet of Things (IoT) in the manufacturing sector came in to play [1]. The Internet of Things (IoT) has developed as a potential solution to these challenges. The IoT describes devices with sensors, processing ability, software, and other technologies that connect and exchange the data with other devices and systems over the Internet or other communications networks [2]. Additionally, IoT can also support manufacturers to create revenue streams and business models, resulting in them to stay competitive. A Deloitte survey shows, "More than 86% of survey respondents believe smart factory initiatives will be the main driver of manufacturing competitiveness in the next five years, though only 51% are making/have made investments" [3].

Expertise in IoT contributes to enhanced business insights [4], spanning from raw material supply to inventory control, asset evaluation for maintenance planning, production quality improvement, and monitoring of goods delivery and fleet operations to ensure optimal customer service. Cloud-based IoT applications play a crucial role by providing real-time access to both external and internal data, facilitating faster decision-making processes for organizations [5]. Industrial IoT (IIoT) is a result of focusing on the industrial applications of IoT. IIoT refers to the interconnectivity of sensors, instruments, and devices combined with industrial computers for applications such as manufacturing and energy management [6, 7]. The criticality of sophistication in IIoT devices comes in, particularly in managing high-stakes machinery, requiring sensitive sensors to provide accurate data that can be automated, visible, and analyzed for the benefit of manufacturers.

2. Human-Centric Design in IIoT

Human-centric design places the needs, abilities, and limitations of end-users at the forefront of every stage of the design process. In contrast to engineering-based solutions that are most focused on their method, the insights of users, stakeholders, and the wider social context combine for intuitive, accessible, and efficient solutions.

2.1 Human-Centric Design Principles

- **Empathy:** Developing a deep understanding of the user's environment, tasks, and challenges.
- **Inclusion:** Designing with a focus on diverse user groups, considering varying skill levels and physical abilities.
- **Iterative Process:** Continuously testing and refining the design based on feedback from users.
- **Usability:** Ensuring the system is simple to learn and operate.
- **Safety and Comfort:** Prioritizing both the physical and psychological well-being of users.

2.2 Significance of Human-Centric Design in IIoT

In the context of IIoT, human-centric design addresses the complexity of industrial settings where human-machine interaction is crucial. It ensures that systems are not only functional but also enhance productivity, minimize errors, and boost overall user satisfaction.

2.3 The Value of Human-Centric Design in IIoT

With the trend of increasing IIoT adoption, it becomes more relevant to focus on human factors. In incorporating human-centric design into IIoT solutions, there are critical advantages as described below:

- **Improved User Experience:** A well-thought-out IIoT system should be very easy to navigate, such that users can get along with interacting with it with minimal effort, without much learning. Through UX, companies reduce the learning curve and raise the chances of adoption for new technologies.
- **Efficiency and Productivity:** Human-centered design focuses on the workflows of the users and optimizes interfaces and interactions to make processes more efficient. This leads to higher productivity because users can complete tasks more quickly and with greater accuracy.
- **Minimize Operational Risks:** Human factors design ensures that the possibilities of user mistakes are minimized as much as possible, and industrial users avoid cost downtime or accidents resulting from the equipment. Safety aspects and clear and intuitive interfaces are components of a human-centered IIoT solution.
- **Aiding Decision-Making:** Human-centric IIoT solutions provide the users with timely and relevant information for more informed decision-making. Visualization tools, dashboards, and alerts are built to present data in an understandable and actionable form.

2.4 Process of Human-Centric Design for IIoT Solutions

The process of integrating human-centric design into IIoT solutions requires the following steps:

- **User Research and Needs Assessment:** This stage begins with an understanding of users' needs, behaviors, and pain points. This can be done by interviewing people, observing their behaviors, and carrying out contextual studies in the industrial environment.
- **Prototyping and Iteration:** Developing prototypes allows designers to test ideas and gather user feedback. This iterative process is crucial for refining the design to ensure it meets user needs and expectations.
- **Usability Testing:** Usability testing is vital to uncover potential issues users may face when interacting with the system. Testing in real-world conditions ensures the solution is both robust and user-friendly.
- **Implementation and Training:** Even a well-designed system will fail if the users are not properly trained. Human-centric design creates robust training programs that enable users to utilize the IIoT solution to its fullest potential.
- **Ongoing Improvement:** Human-centric design is a process that never stops. As the users use the system, there will be emerging challenges and opportunities for improvement. The key to maintaining the relevance and effectiveness of the solution is continuous feedback loops and the updating of the system as required.

3. IIoT Architecture and Components

The classical architecture of an IIoT system is represented by the commonly adopted four-layer model accepted in industry. These layers include perception, network, processing, and application layers [8].

3.1 The Perception Layer

The perception layer encompasses sensors, actuators, devices, and physical objects utilized for data gathering in the Industrial Internet of Things (IIoT). This layer is alternatively known as the edge layer or device layer, a concept akin to the commercial IoT framework. Within IIoT, the sensors employed are termed smart sensors, and a distinctive attribute is their capacity to communicate with the network, often employing wireless signals in contrast to the typical wired industrial networks [9].

Smart sensors also possess the salient features of being self-identifiable and flexible [10]. IEEE standards are critical to setting identification norms, and through this, the smart sensors produced by different vendors can be interlinked without creating any issues with critical sensor information easily being available. Controllers within IIoT, categorized under the perception layer, are typically embedded and dedicated to specific functions. In contrast to robust Programmable Logic Controllers (PLCs) that oversee entire processes, many contemporary IIoT-controlled processes rely on decentralized embedded PLCs. Some view decentralized control systems as precursors to IIoT controllers.

3.2 Network Layer

The network layer of the IIoT is dedicated devices, which transmit composed data packets over the internet [11]. From smart sensors, up the data flow, the second device to encounter is an IIoT Gateway.

These are known as Edge Gateways; IIoT Gateways store and process locally collected data before forwarding it to the next level of the network. Let's get a fact straight, however-Edge Gateways are not process controllers; they do all this on data before forwarding it to the cloud. Despite this, they necessitate high processing power due to the substantial volumes of processed data. Certain gateways aim to bring older processes into modern IIoT compatibility [12]. These gateways might utilize traditional industrial network protocols (such as RS-485, Modbus, etc.) to extract information from existing control systems, subsequently converting the data into IIoT-ready packets using standard formats like FTP or MQTT.

After the Edge Gateway sends the locally processed data, it travels through many network devices before reaching the cloud. Selection of devices during this stage depends mainly on customer preferences and standards and is within the IT management sphere. This stage is also the implementation ground for security measures to safeguard downstream devices from external attacks.

3.3 Processing Layer

Transitioning into the cloud domain, the data has now arrived at a database or server within a data center [13]. This is the most critical phase because significant computing power is usually needed to run data analytics and machine learning algorithms, making it the central hub for critical data processing. In this layer, aside from algorithms, there are web services and other cloud-based technologies that help in the overall process. Web services are very important in describing the transfer of data from devices to other interfaces and programs. It is important to note that data storage at a remote server location is often more secure than at a local level. When data is stored locally, the on-site IT staff bears the sole responsibility for security. In contrast, outsourcing data management to a specialized company can raise the level of security measures by a significant degree.

3.4 Application Layer

In this sense, the application layer refers to the implementation of graphical user interfaces that help human beings understand information obtained from preceding steps. Generally speaking, this is a layer hosting user programs with tasks like generation of reports, analysis, and statistics. Current markets include multitudes of solution providers offering software applications.

4. Need of Implementation of IIOT in Manufacturing

1. IIoT is bringing a revolutionary period for manufacturers in which connected machines are capturing and transmitting real-time data with a level of accuracy and consistency that is unprecedented.
2. The introduction of IIoT allows organizations to break down the data silos, thus creating pervasive access to information at every level [14]. The output actionable data has wide-scale benefits.
3. Operators, supervisors, and engineers gain valuable insight into production processes.
4. Engineers use signals from process, operator, and machine data to drive continuous improvement initiatives to increase efficiency on the plant floor.
5. Management makes informed business decisions based on data, thereby reaping even more benefits from IIoT.
6. People at all levels of the organization are empowered to identify problems and inefficiencies sooner in the process, which means that they can proactively optimize operations.
7. The use of data-driven decision-making eliminates guesswork, which is a significant advancement in problem-solving methodologies within manufacturing processes.

4.1 Additional Benefits that IIoT can Bring to Manufacturers

(a) Improved usage of machines: The usage of Industrial IoT technology allows manufacturers the opportunity to put their machines in connection with the internet. This connectivity provides manufacturers with valuable insights into real-time machine health as well as key performance indicators (KPIs) that include metrics such as overall equipment effectiveness (OEE) and overall process effectiveness (OPE)[15]. Using this information, the manufacturer will be able to identify root causes for unplanned downtime even before they happen and can thus act proactively. Besides this, it helps improve machine utilization by determining whether preventive measures of equipment maintenance are necessary.

(b) Predictive maintenance: The use of real-time data from IIoT-connected systems proves instrumental in predicting defects in machinery. This predictive capability enables manufacturers to proactively implement preventative measures, mitigating potential issues before they manifest.

This has led to increased machine uptime and productivity. Equipment failures are prevented at the same time, thereby reducing process time, minimizing rework and scrap, and reducing unplanned downtime. All these improvements help in saving considerable amounts of money for manufacturers.

(c) Asset tracking: The manufacturer can track products through the supply chain and inform stakeholders about any damage or possible damage to goods.

(d) Facility management: IoT-based green sensors can be used to monitor conditions like vibration, temperature, humidity, etc. that might detect status affecting the operations or leading to overwear of equipment.

(e) Just-in-Time Manufacturing: The availability of real-time data reporting is a significant enabler for the implementation of Just in Time manufacturing [16]. This ability allows for an immediate adjustment to processes, allowing for the removal of waste and ensuring that production ends on time and in rhythm with materials in process and raw materials. Therefore, this method helps bring planned production closer to actual production and hence improves the overall efficiency of operations.

(f) Setting up remote assets: The connection between devices means that information on remote assets can be accessed centrally from the central point, remote assets are monitored and controlled, offering more control in operation.

(g) More intuitive user interfaces: The use of integrated software allows operators, engineers, and managers to view data via Human-Machine Interfaces (HMIs), which are carefully designed to be intuitive, even for personnel who are not sophisticated in IT. More importantly, the use of these interfaces centralizes data from various sources, allowing personnel to effectively use tools without requiring higher amounts of training or reliance on IT staff.

(h) Sharing knowledge across plants: Institutionalizing knowledge through a system ensures that important information stays within the workforce over time. Centralized knowledge not only helps in preserving institutional memory but also contributes to standardization of processes. This is very important for the organization in terms of continuous improvement. In addition, standardized and centralized knowledge enables experts to respond to problems quickly and effectively, irrespective of their location [17].

Data silos and tribal knowledge, which is experiential knowledge transmitted orally without standardization or documentation, pose significant inefficiencies for manufacturers.

(i) Process and behavior monitoring: Leveraging data gathered from IoT-enabled devices and software provides managers with valuable insights into employees' performance. This data enables the identification of bottlenecks and areas for improvement within processes [18]. For instance, managers may discover patterns where employees consistently make mistakes or encounter defects during specific steps. Armed with this information, process engineers can conduct root cause analysis to ascertain potential improvements and utilize the data as a benchmark to measure progress.

These advantages culminate in substantial business impacts, manifesting as cost savings, enhanced quality, and increased operational efficiency.

5. IIoT Statistics

Here are some statistics about industrial IoT or IIoT [19], which show the existing and expected trends.

- The global market for IIoT has surpassed $263 billion in value in 2021 and has since risen to over $323 billion as of 2022. Projections show that the market of the industrial IoT is likely to go beyond a trillion dollars in value by 2028.
- The use cases of IIoT, according to the McKinsey Global Institute, are supposed to contribute an economic value that could be nearly $4 trillion to $11 trillion by the year 2025.
- By 2030, the IIoT market is expected to contribute over $14 trillion to the global economy through various value-added benefits.
- Top industries like transportation and logistics, utilities, and the discrete manufacturing sector are anticipated to invest $40 billion each in IoT transformation initiatives.
- Industries are pursuing IoT applications as a strategic avenue to attain various enterprise benefits. About 54% of IIoT protocols and applications are motivated by cost savings as a major revenue generator. On the other hand, around 23% of Chinese companies have embraced the adoption of IoT to boost their competitiveness.
- Almost 58% of producers believe IoT to be a tool that must drive digital transformation within industrial operations.
- Simultaneously, around 80% of senior executives in all industries opine that IoT is the most important factor facilitating growth in various business sectors.
- According to estimates, investments in IoT by companies may reach $15 trillion by 2025. Currently, projections indicate that expenditures on IoT solutions may reach more than $1 trillion by 2023.

6. IIoT Components

The IIoT is the integration of intelligent assets, data communication infrastructure, software, analytics, and human involvement to extract actionable insights from manufacturing operations.

6.1 Intelligent Assets

Intelligent assets include connected devices such as sensors, controllers, and edge devices, application software, and security elements. This classification includes contemporary assets specifically designed with inherent local intelligence and communication capabilities. Moreover, these capabilities can be incorporated into existing legacy assets. Each of these assets is equipped with the ability for connectivity, embedded intelligence, and the provision of support for analytics. These assets generate data and facilitate the exchange of information throughout the entire value chain. Examples of intelligent assets in IIoT include:

- Plant instrumentation
- Machines
- Equipment
- Systems or other resources connected with sensors, processors, memory, and communications capability

6.2 Common Types of Equipment and Devices

Sensors contribute by providing fresh data sourced from existing assets or integrated into new or existing machines, designed for external accessibility through commonly used protocols and communication technologies.

Edge devices are hardware devices that control the flow of data at the boundary between two networks, or in other words, entry or exit points. Their functions include transmission, routing, processing, monitoring, filtering, translation, and storage of data as it is transmitted between networks.

Positioned closer to endpoints, edge devices efficiently collect, process, and store data, optimizing the utilization of network resources.

IoT gateways are the most common type of edge devices in manufacturing environments. Sensors and many other devices are also categorized as edge devices.

Embedded systems denote computing systems dedicated to a singular purpose, distinct from general-purpose computing. Embedded devices possess their own computing capabilities, including processors, memory, operating systems, and communication capabilities. Within IIoT, numerous embedded computing devices collaboratively operate with other devices within the system.

6.3 A Data Communications Infrastructure

Assets in an IIoT system must communicate using the Internet and other network technologies. Often, IIoT systems are implemented on cloud infrastructures, like Amazon Web Services. Cloud computing refers to the processing, management, and storage of data through a distributed network of remote servers, which is not traditional since it doesn't rely on a local server.

Software

IoT software is the one that checks all the data that is being collected from your equipment and devices. It also provides an interface through which users can interact with the IIoT system, thereby acting as a catalyst for individuals to make informed decisions and enhance their performance. Cloud-based software offers a good number of benefits over on-premise systems for manufacturers. These include lower total cost of ownership, greater reliability, speed, and flexibility.

People

A frequently underestimated yet crucial component of any IIoT system is the human factor. Individuals interact with the system by making decisions grounded in the data and analysis produced by the other elements of the IIoT framework. Improved data quality and the availability of more advanced analytics tools empower individuals to establish stronger connections with plant equipment, machines, systems, and fellow personnel. Consequently, this increased connectivity results in more informed and quantifiable decision-making.

7. Factors of IIOT in Manufacturing

1. **Predictive Maintenance:** The IoT enables producers to monitor the condition of machinery and equipment in real-time. Predictive maintenance algorithms can predict when equipment is likely to fail by collecting and analyzing data from sensors.
2. **Asset Tracking and Management:** RFID tags and sensors track where and when assets are at a production site [20-21]. This enhances the visibility of supplies, reduces potential losses of items, and facilitates the optimal utilization of such assets.
3. **Quality Control and Defect Detection:** Sensors and cameras connected to the IIoT can scan and inspect products constantly [22]. It can be seen from this perspective that an early identification of problems can result in higher product quality and less probability of defective products hitting the market.
4. **Supply Chain Optimization:** Internet of Things helps in supply chain visibility and efficiency [23]. Manufacturers can optimize the supply chain processes by minimizing lead times, and it increases overall efficiency through real-time data on inventory levels, production progress, and demand changes.
5. **Data Analytics for Process Optimization:** The huge amounts of data produced by IIoT devices can be analyz ed to detect patterns, optimize manufacturing processes, and make data-driven decisions. This results in enhanced efficiency, less waste, and better overall performance.

7.1 Predictive Maintenance and Asset Management

The IIoT plays a multifaceted role in the manufacturing sector, with a particular emphasis on the critical aspect of asset management. The common theft of assets, accounting for over 30% of companies' declaration of bankruptcy, remains a key challenge that manufacturers are bound to face. It thus represents a great risk to the manufacturing firms and jeopardizes the general sustainability of the business operation [24]. Fortunately, there is a promising way to monitor and supervise assets in IoT solutions – a proactive step to counterbalance this problem. In addition, IoT-enabled devices enable predictive maintenance by sensing warning signs, allowing the managers to develop timelines for scheduled maintenance and prepare equipment service even before potential problems occur.

7.2 Asset Management

This includes the tracking of information as well as physical assets of a business. The phrase "physical assets" can include machinery, IT equipment, tools, or even vehicles, depending on the type of business. Traditional manufacturers used legacy systems, like handwritten lists, notes, and excel sheets, to keep records of asset information. However, the methods were not very efficient and prone to mistakes, offering many opportunities for alteration of information. The implications of asset theft for manufacturers cover issues such as loss of personal data, hacking accounts, disruption in daily operations, and the eventual financial challenges presented by replacing such assets.

IIoT emerges as a valuable solution for effective asset management in manufacturing. IoT-enabled asset management software, representing a significant advancement over legacy systems, facilitates automated and intelligent asset tracking. Various sensors, including temperature, pressure, proximity, humidity, acceleration, accelerometer, and gyrometer, are strategically attached to assets in the manufacturing unit. These sensors collect information, which is then analyzed, transformed into valuable insights, and transmitted to the cloud for further evaluation and action.

Asses Management could be done in four different steps:

1. Data acquisition system through sensors
2. Data Processing
3. Data Analytics and Machine Learning
4. Dashboards and Reports

7.3 IoT-enabled Predictive Maintenance and Asset Management

1. **Location and Usage Tracking:** Utilizing IoT solutions enables the automatic tracking of location data and implementation of geofencing solutions. Asset managers get real-time insights into the usage hours of all machines, know their occupancy status, and whether they are free or surpassing threshold values. IoT devices help deploy a smart asset monitoring system, which provides immediate visibility into asset utilization, load distribution, and operational cycle times, thereby increasing productivity.
2. **Condition of Machine:** Manufacturers use IoT-enabled devices for monitoring the condition of machines and receive multiple real-time alerts

before potential machine breakdown [25]. This ensures that preventive measures are taken in time so that there will not be a disruption in operation. Manufacturers and line managers can immediately take actions and identify problem areas using predictive maintenance solutions.

3. **Warehouse Inventory and Material Flow Tracking:** The IoT-enabled device plays a pivotal role in real-time tracking of inventory in the warehouses, thus minimizing the threat of asset theft. Asset managers find it easy to monitor how many assets are available within the inventory. In addition, IoT devices enable the smooth monitoring of asset entries and exits from the factory inventory, which would thereby effectively control and enlighten the flow of materials for asset managers.

8. IoT Manufacturing : Applications

There is a huge scope for IIoT applications to be integrated into manufacturing processes in order to enhance sustainable practices [26]. Through IIoT technology, the industrial manufacturer improves their operational efficiency by reducing waste generation and pollution levels. These are some of the main IIoT applications promoting sustainable manufacturing.

1. Energy Management

(a) *Smart Grids and Monitoring*: The deployment of IIoT sensors in energy systems provides real-time monitoring of energy consumption and grid performance [27]. This empowers manufacturers to optimize energy usage efficiently.

(b) *Demand Response Systems*: IIoT facilitates the integration of demand response systems, allowing for dynamic adjustments to energy consumption based on grid conditions [28]. This aids in the reduction of peak loads and promotes more efficient energy utilization.

(c) Using these applications of IIoT, the manufacturers will improve their operations along with their concern for the manufacturing process, focusing on a sustainable and environmental aspect.

2. Predictive Maintenance

- *Condition Monitoring*: In condition monitoring, IIoT sensors attached to the machines constantly monitor the status of equipment conditions. This proactive approach enables the prediction of potential failures thus allowing timely maintenance interventions [29]. Thus, the application of IIoT in condition monitoring minimizes unplanned downtime, reduces resource wastage, and extends the overall lifespan of equipment.

3. Resource Optimization

- *Monitoring and Control*: The integration of IIoT enables the tracking of assets, equipment, and inventory in real-time, ensuring that there is no wastage of resources [30].

- *Supply Chain Visibility*: IIoT provides a better view of the supply chain [31]. Improved visibility leads to optimization of logistics, reduction in transportation emissions, and excess inventories, which results in more efficient and sustainable supply chain activities.

4. Waste Reduction

- *Smarter Manufacturing Operations*: The deployment of IIoT applications enables the production procedures to be fine-tuned such as waste being limited [32]. By using raw materials efficiently and minimizing defects, IIoT enhances manufacturing efficiency.
- *Environmental Monitoring*: IIoT sensors are used to monitor environmental conditions, ensuring that the manufacturing process is in line with sustainability regulations.

Moreover, these sensors highlight areas of improvement in environmental practices, thus making the manufacturing process environmentally friendly [33].

5. Water Management

Incorporation of IIoT devices into the manufacturing system enables the ability to monitor and optimize water usage [34]. This benefit assists in the conservation of water and lets there be less negative environmental impacts from the manufacturing system.

6. Quality Control

- *Real-time Quality Monitoring*: Utilization of IIoT applications can enable the rteal-time monitoring of product quality [35]. This feature reduces the defects and also reduces the rework and finally leads to a reduction in total waste in the manufacturing process.

7. Emissions Monitoring

- *Air Quality Monitoring*: The use of IIoT sensors in the manufacturing plant enables the continuous monitoring of air quality [36]. This ensures that environmental standards are maintained and helps to create a healthier work environment.

8. Collaborative Robotics (Cobots)

- *Secure and Productive Automation* [37]: The implementation of IIoT-connected collaborative robots in manufacturing would ensure greater efficiency at the same time ensuring the safety of the workers. Therefore, these cobots are flexible and adaptable towards changing production demands and thus reduce the need for manual interventions and optimize overall workflows.

9. Life Cycle Assessment

- *Product Traceability*: The integration of IIoT allows full traceability of the product throughout its whole lifecycle. This ability enables manufacturers to evaluate and improve the sustainability of their products–from the production phase to end-of-life [38].

With these IIoT applications [39], the manufacturer can effectively participate in a sustainable practice while reducing its adverse effects on the environment and constructing resilient operations.

10. Cyber-physical Manufacturing Systems

Cyber-physical manufacturing systems represent the convergence of digital technologies with the physical processes within the manufacturing environment [40]. These systems integrate computers, networks, and physical parts into intelligent, connected manufacturing ecosystems. Notable characteristics include the following: real-time processing of data, automation, integration of sensors, and flexibility. These new manufacturing systems benefit from the full capabilities of recent innovations like IoT and AI (artificial intelligence). These systems try to increase efficiency, reduce mistakes, and make the best use of resources. The overall aim is to build intelligent manufacturing environments that can change dynamically in accordance with the base principles of Industry 4.0.

11. Energy Efficiency Management

IoT is used for efficient energy management in the manufacturing process [41], which proved the usage of IoT for energy consumption optimization in additive manufacturing. They designed an IoT framework to track and analyze energy consumption in the selective laser sintering process. They minimized energy consumption throughout the whole process by optimizing each build through a control system. The IoT also allows for real-time monitoring of energy efficiency on manufacturing shop floors. Energy data is wirelessly collected and transmitted from shop floors for analysis. This method will help identify abnormal energy consumption patterns that can be given timely feedback as soon as it occurs.

Caputo et al. (2016) [42] discussed the study based on the best practices in energy management that could be applied to everyday operations. The enabling technologies for the green IoT application, including RFID (Radio Frequency IDentification), sensor networks, and the internet, were explored along with their impacts on energy consumption and environmental concerns. The research further classified IoT applications into environmental aspects. In addition, [43] incorporated the IoT technology in ESER evaluations. They designed an IoT-enabled system for life cycle assessments in ESER based on the IoT sensing capabilities in real-time data collection and management.

This system enables all the energy consumption data and environmental influence information to be gathered throughout the entire life cycle of the manufacturing plant, thus ensuring uninterrupted data interfacing between an ESER assessment system and established enterprise information systems.

12. Manufacturing Operations Management (MOM)

The success in production processes in the context of global manufacturing depends much on the MES that are also referred to as MOM systems [44]. In production plants, the integration of comprehensive software solutions helps them achieve real-time expected performance levels and quality and flexibility. The debates surrounding the terms about MES and MOM highlight what sort of software is currently being applied to the operational productions [45]. We will define our view in an attempt to clarify this issue. The MES is computerized systems used in manufacturing to guide, track, and document the transformation of raw materials into finished products at specific workstations (process lines/units), including data exchange with the automation layer (such as SCADA, DCS, PLC, CNC). On the contrary, MOM systems are an extension to the MES and include all the functionalities of manufacturing processes within a defined region. These cover FCS, digitizing manual workflow, CMMS, EAM for maintenance, QMS, and LIMS, and WMS and more. We anticipate that MOM will continue to play a vital role in the manufacturing enterprise's IT landscape, positioned at the critical juncture where revenue-generating products come into existence. MOM already manages rapidly flowing streams of disparate unstructured and structured data, transforming it into useful, targeted information in near real-time. It is envisioned that CPS and CPPS data that will come up in the future will call for even greater capability from MOM.

9. IIOT Use Cases

The article first talks about the basics of the IIoT, its importance, and recent market trends as the basis of further exploring practical use cases across different industries. For those interested in understanding the industrial Internet of Things from a hands-on viewpoint, common questions include: "What are examples of industrial IoT. Below are some prominent instances illustrating the practical applications of the IIoT [46].

1. Tom Farms

The agricultural sector is the most significant adopter of IoT for industrial applications, and farming is at the forefront. A case study in this space is Bytes and Bushels, which discusses the application of IoT in large-scale farming in real life [46]. More precisely, the case study focuses on Tom Farms, which is a family-owned agricultural business in the United States. Tom Farms was an innovative leader by using IoT for efficient distribution of their agricultural products. In response to the farm crisis of the 1980s, Tom Farms took strategic steps such as increasing their lands and research and development for assimilation of new technologies in the production process.

Notably, Tom Farms embraced cutting-edge solutions such as self-driving tractors, GPS data utilization, and sensors, accompanied by the addition of iPhone apps for precision irrigation, which has been a rather forward-looking move in the agricultural world.

2. Airbus

Another interesting example in the industrial IoT protocols category is Airbus. This aerospace company utilizes IoT smartly to create a future-facing factory that increases the productive capacity while reducing complexities involved in these operations. Airbus uses integrated sensors installed in several tools and machines so that it can track the process of assembly very minutely, thus not letting any wrong piece go ahead.

Additionally, the company has made use of wearable technology, such as industrial smart glasses for its workers. This is to improve safety and productivity in the workplace.

3. Tesla IoT Car

Tesla Motors foresees the development of driverless automobiles [46] and this dream is solely dependent on the confluence of AI and IoT . The innovative automobile manufacturer employs IoT application-based solutions in making automobiles individually customized according to the specifications of its customers. Internet of Things is the technological backbone of all these critical features of Tesla's self-driving cars, in the context of autopilot mode, connectivity features, and a mobile application used to control the vehicle.

4. General Electric

One of the best-known cases in IoT case studies in the industrial sector is the case of General Electric, or GE for short. All the opportunities by which General Electric would leverage its operations through the combined power of IoT and industrial analytics are an area of great interest to derive exclusive value benefits. For example, in industrial machinery, GE doesn't stop at control and monitoring with IoT. Strategic use of IoT applications makes it explore the arena of predictive maintenance. This method allows GE [46] to foresee and anticipate possible failures, thus allowing for a more proactive and efficient maintenance plan. General Electric, or GE, has adopted the Internet of Things, or IoT, to enhance the productivity of its jet engine production processes. Through the installation of sensors on jet engines, GE systematically gathers vital information about the performance of the engines and their maintenance requirements. This precious dataset will be used to develop and fine-tune the design of future engines. It shows how IoT technologies help improve the manufacturing process.

Proficy Smart Factory is an all-inclusive MES software suite that offers both cloud-based and on-premises solutions. It is an opportunity to revolutionize manufacturing operations, tapping into data integration, IIoT, machine learning, and predictive analytics. Proficy Smart Factory MES software is created to provide performance management in every aspect of modern connected enterprise while integrating the seamless digital and physical realms of manufacturing.

9.1 *Impact of Proficy Smart Factory*

- **Cost Reduction:** Through the strategic alignment of resources, energy, and efficiency management, your manufacturing operations can effectively reduce costs. This involves optimizing the level of inventory, identifying and eliminating costly production gaps, reducing energy and utility waste, and gaining useful insights into throughput and critical processes.
- **Superior Quality:** Implement an effective quality management system that can capture detailed quality data and traceability information. At the same time, critically analyze the process-oriented data as well as the product-oriented data. The materials should be accurately classified as conforming or non-conforming, out-of-spece products, or those associated with recalls.
- **Accelerated Production:** Optimize on-time delivery performance and improve responsiveness to late-stage changes. Attain real-time visibility of all work-in-progress orders, thereby increasing the chances of successful new product introductions (NPIs) and significantly reducing the design-to-delivery cycles to enhance productivity.

1. Caterpillar

Industrial IoT applications research provides some important examples, including Caterpillar, a well-known heavy-equipment manufacturer worldwide for its pioneering introduction to the concept of the IIoT [47]. Caterpillar, or CAT for short, has successfully traversed the IoT adoption landscape in different industrial environments and has thus enjoyed significant benefits from this technological integration. In a bid to realize a more significant level of operational efficiency, Caterpillar is fast launching IoT and augmented reality applications. This deployment is supposed to provide real-time information for machine operators to know their levels of fuel, among other crucial information about their machine health status. Apart from land-based equipment, Caterpillar's marine asset intelligence business segment has utilized sensor-driven analytics in the protection of assets on vessels and boats in shipping. The diversified application of IoT technology reveals Caterpillar's commitment to industrial innovation and efficiency.

2. Hitachi

Hitachi is among the key contributors in the area of IIoT applications and contributes to the list of IoT case studies about industrial automation. A unique feature in Hitachi's IIoT use cases is the availability of a dedicated manpower of 16,000 IoT technology professionals [46]. One of the differentiation factors of Hitachi is an advanced IoT platform called Lumada, along with a wide portfolio of IoT-based products.

Apart from the overall IoT packages, Hitachi had been implementing development for an innovative IoT-based model of production that focused particularly on streamlining the lead times during production, that is, in the area of infrastructure production tasks. The combined efforts of Hitachi in IoT technology highlight its dedication towards promoting innovation in industrial automation.

3. PTC Thingworx

One of the great examples in the IIoT case study landscape is Amazon, a retail giant. The company is innovating in IoT to transform logistics and warehousing operations. An example of this is with connected robots in its fulfillment centers identifying which shelves have what products and bringing those to human workers for picking.

4. Intel

Intel is now becoming part of the Industrial Internet of Things to revolutionize the management of electronic medical records. Working with Intel, NEXCOM proposed an IoT-based medical informatics system, which is believed to eliminate barriers in communication between IoT networks and medical devices. This system allows for the easy documentation of data from medical devices into electronic medical records and ensures safe transfer to the private cloud infrastructure of healthcare providers.

5. John Deere

One of the most famous tractor manufacturing companies is John Deere, which is a good example in the IIoT. Through these IoT technologies, the company launches its future tractors into complete autonomy. This is another part of John Deere's project, and their ability to integrate GPS into a system gives massive support in self-driving tractor technology. John Deere also implements telematics to optimize maintenance activities and predict them for general efficiency and effectiveness in agricultural machines.

6. Amazon

One of the great examples in the IIoT case study landscape is Amazon, a retail giant. The company is innovating in IoT to transform logistics and warehousing operations. An example of this is with connected robots in its fulfillment centers identifying which shelves have what products and bringing those to human workers for picking. Integration of IoT-connected robots would not only enhance operational efficiency but also reduce the costs involved in the logistics and warehousing processes for Amazon.

Table shows a representative list of companies that implement IoT solutions in industrial case studies.

BMW incorporates IoT in its manufacturing facilities to monitor vehicles throughout the production process. This enables BMW to enhance inventory management and prevent production delays by precisely tracking the progress of each car.

Philips uses IoT during manufacturing to help improve product quality. Since the statistic is continuously checked during production, Philips will

Table 1 IIoT case studies.

Company	*Details*
Vale Fertilizantes	Vale implemented the GE Predix platform very well with respect to optimizing maintenance strategies, which improved the asset reliability and produced a notable outcome by getting rid of 25 lost production days during one year. The savings had also reached an impressive amount, about $1.4 million. The shift towards preventive maintenance was particularly impactful, with corrective maintenance reduced to zero between 2014 – 2015. As a result, the flow of weak acid surpassed 13 cubic meters per hour, showcasing the positive outcomes of Vale's utilization of the GE Predix platform [46].
BMW	BMW employs Amazon AWS to power its Car-as-a-Sensor (CARASSO) system, which gathers sensor data to provide drivers with dynamically updated map information. Leveraging AWS allows CARASSO to seamlessly adapt to swiftly changing load requirements. The ambitious goal for 2018 involves processing data gathered by a fleet of 100,000 vehicles covering a distance exceeding eight billion kilometers. This showcases BMW's strategic use of Amazon AWS for handling the evolving demands of its innovative CARASSO initiative [47-49].
Sandvik Coromant	Sandvik pioneers the creation of advanced predictive analytics through the utilization of the Microsoft Azure platform. In this innovative approach, it seamlessly integrates with internal shop floor control tools to gather crucial machine and tool data. The collected data is then transmitted to Azure, where real-time analysis takes place using cutting-edge machine learning algorithms. This process not only enables real-time optimization of operations but also facilitates the establishment of predictive maintenance schedules and alarms, showcasing Sandvik's commitment to leveraging technology for enhanced manufacturing efficiency [50].
Toyota Tsusho	By utilizing the Amazon AWS, Toyota Tsusho has launched the innovative traffic information broadcasting system called TSquare. What distinguishes TSquare from others is that it is providing real-time traffic data in Bangkok and the six suburban provinces. AWS has been very significant in processing immense traffic data with efficiency so that scalability and reliability are delivered in delivering the information to end-users accurately and on time. This initiative marks a commitment towards harnessing innovative cloud computing techniques for improving road management and real-time information offered to the masses [51].
Samsung	Leveraging Microsoft Azure, the company pioneered the creation of S-NET Cloud, a cloud-based platform facilitating remote energy management for air conditioners. This innovative system contributes to energy conservation by optimizing cooling and heating operations through the utilization of the air-conditioner sensor, operational data, and indoor environmental information. Beyond energy efficiency, the S-NET system excels in detecting equipment malfunctions and conducts integrated remote maintenance and management using real-time data analytics. This exemplifies the company's commitment to utilizing cutting-edge cloud technology for enhancing energy efficiency and ensuring seamless operations of air conditioning systems [52].

Contd.

Table 1 *Contd.*

Company	*Details*
Cummins Power Generation	Cummins innovatively designed the Cummins Power Command Cloud—a remote monitoring solution for generators and power systems—on Microsoft Azure. This cloud-based system showcases the capability to efficiently monitor millions of power systems and generators across the globe. The implementation of such a solution not only enhances service quality but also plays a crucial role in saving lives. Moreover, this forward-thinking approach fosters the development of more innovative products with the ultimate aim of improving the overall quality of life. Cummins' commitment to leveraging cloud technology for remote monitoring signifies a dedication to excellence, global impact, and continuous improvement [53].
Echelon	Echelon pioneered the creation of an adaptive streetlight control system utilizing the IBM Watson IoT platform. This innovative system aims to enhance the energy and operational efficiency of high-efficiency lighting systems by incorporating adaptive lighting control. City managers can benefit significantly from this intelligent solution, as it allows for dynamic adjustments in street lighting based on real-time weather data, activity levels, and the time of day. The implementation of such a system reflects Echelon's commitment to smart technologies that contribute to sustainability, energy savings, and improved urban management [54].
Marathon Petroleum	Marathon employs data collection for analysis through the GE Predix platform, establishing a collaborative strategy to optimize asset performance management and optimization. The integration of IoT technology provides Marathon with essential services, support, and the flexibility to design programs that align with its evolving requirements. This utilization of IoT showcases Marathon's commitment to leveraging advanced technologies for enhancing asset performance, ensuring operational efficiency, and meeting long-term strategic objectives [55].
Daimler	Daimler has developed the Detroit Connect system on the Microsoft Azure platform to gather performance data from vehicles in operation and store it securely. Through the Detroit Connect portal, fleet managers gain access to comprehensive fault-event details, allowing swift response to occurrences. This capability enhances flexibility, reduces costs, and fosters enduring relationships with customers, showcasing Daimler's commitment to leveraging technology for efficient fleet management and customer satisfaction [56].
INNOVYT	The company engineered IoT solutions utilizing the Microsoft Azure and Amazon AWS platforms to enable real-time fleet tracking, alerts, and advanced analytics. These solutions provide valuable insights into driving behavior, empowering fleet managers to make informed decisions and implement strategies for enhancing overall fleet performance [57].
LightInTheBox	The company leverages Amazon AWS to develop a resilient website for its customers, ensuring high availability and cost savings in operational expenses. With the integration of IoT technology, the company achieves the flexibility to handle transactions from various locations, optimizing computing resources to effectively minimize costs [58].

Contd.

Table 1 *Contd.*

Company	*Details*
TraceLink	The company created the Life Science Cloud platform, utilizing Amazon Web Services (AWS) to ensure compliance across the worldwide life science network and global pharmaceutical supply chain AWS plays a crucial role in enabling the company to fully meet the requirements of numerous pharmaceutical companies and their collaborative partners [59].

immediately notice if it has any form of potential problems in the output, thus performing interventions early in the production phase before products hit the shelves at retailers. In the realm of medium-sized businesses leveraging IoT in manufacturing, Cimpress, which specializes in customized print apparel, uses IoT-based software to customize print products for its customers [60–63]. Owens Corning, a manufacturer of roofing, insulation, and composite materials, uses IoT for safety data tracking, which has improved employee safety.

Conclusion

The advent of the IoT has revolutionized the industrial landscape, offering solutions to the pressing challenges faced by manufacturers in today's competitive environment. Increasing number of companies are therefore looking at the IoT technologies for streamlining operations and getting competitive advantage by boosting efficiency, productivity, and cost-effectiveness. IoT technology integrates the various devices, sensors, processing capabilities, and software to facilitate the communication and exchange of data between systems and devices in such a manner that manufacturers can enhance processes, control quality, and improve overall productivity.

Furthermore, the adoption of cloud-based IoT applications enables organizations to access real-time data, which enables quicker decision-making and response times.

Additionally, IIoT has further enhanced these capabilities, providing advanced solutions for high-stakes industries. The four-layer model includes perception, network, processing, and application layers, which is a framework for designing and implementing IIoT systems that enable automation, visibility, and advanced analytics. In general, the IIoT is transforming manufacturing by connecting intelligent assets, data infrastructure, software, analytics, and human expertise, all on their path to actionable insights and optimized manufacturing operations. As the smart factory initiative continues to gain momentum globally, with companies making significant investments into it, engaging the power of IoT and IIoT technologies will be the only way to stay competitive in future growth for the industrial sector.

References

[1] Fleisch, E. (2010). What is the internet of things? An economic perspective. *Economics Management and Financial Markets*, *2*, 125–157.

[2] Nagy, J., Oláh, J., Erdei, E., Máté, D. and Popp, J. (2018). The role and impact of Industry 4.0 and the internet of things on the business strategy of the value chain–the case of Hungary. *Sustainability*, *10*, 3491.

[3] Zuehlke D. (2010). Smart Factory: Towards a factory-of-things. *Annual Reviews in Control*, *34*(1), 129–138.

[4] Ding, Y., Zeng, L. and Zhou, S. (2006). Phase I analysis for monitoring nonlinear profiles in manufacturing processes. *Journal of Quality Technology, 38* (3), 199–216.

[5] Quiescenti, M., Bruccoleri, M., La Commare, U., Diega, N.L. and Perrone, G. (2006). Business process-oriented design of Enterprise Resource Planning (ERP) systems for small and medium enterprises, *International Journal of Production Research, 44* (18–19), 3797–3811.

[6] Tao, F., Zuo, Y., Da Xu, L. and Zhang, L. (2014). IoT-Based intelligent perception and access of manufacturing resource toward cloud manufacturing . *IEEE Trans. Ind. Inform.*, *10*, 1547–1557.

[7] Max DePree (2019). Introduction to IoT in manufacturing. *White Paper in Connected Industry Growth Partners.*

[8] Antonio Armenta. (2022, Aug. 22). Components of the Industrial Internet of Things. Technical Article.

[9] Shi, J. (2006). *Stream of Variation Modeling and Analysis for Multistage Manufacturing Processes.* CRC Press, Taylor & Francis Group.

[10] Pantelopoulos A. and Bourbakis, N.G. (2010). A Survey on Wearable Sensor-Based Systems for Health Monitoring and Prognosis. "*IEEE Transactions on Systems, Man, and Cybernetics, Part C (Applications and Reviews),* 40 (1), 1–12.

[11] Zhong, L., Wang, L. and Xu, X. (2017). An IoT-enabled Real-time Machine Status Monitoring Approach for Cloud Manufacturing. *Procedia CIRP, 63,* 709–714.

[12] Wan, C., Zheng, H., Guo, L., Xu, X., Zhong, R.Y. and Yan, F. (2020). Cloud manufacturing in China: A review. *Int. J. Comput. Integr. Manufacturing*, *33*, 229–251.

[13] Georgakopoulos, D., Jayaraman, P.P., Fazia, M., Villari M. and Ranjan, R. (2016). Internet of Things and Edge Cloud Computing Roadmap for Manufacturing. *IEEE Cloud Computing, 3* (4), 66–73, 2016.

[14] Caputo, G., Marzi, M. and Pellegrini. (2016). The Internet of Things in Manufacturing Innovation Processes: Development and Application of A Conceptual Framework. Forthcoming in *Business Process Management Journal, 2.*

[15] Tahera Kalsoom, et.al. (2021). Impact of IoT on Manufacturing Industry 4.0: A New Triangular Systematic Review. *MDPI Journal Sustainability*, *13.*

[16] Brozzi, R., Forti, D., Rauch, E. and Matt, D.T. (2020). The Advantages of Industry 4.0 Applications for Sustainability. Results from a Sample of Manufacturing Companies. *Sustainability*, *12*, 3647.

[17] Castelo-Branco, I., Cruz-Jesus, F. and Oliveira, T. (2019). Assessing Industry 4.0 readiness in manufacturing: Evidence for the European Union. Comput. *Ind.*, *107*, 22–32.

[18] Blackburn, M. and Denno, P. (2014). Virtual Design and Verification of Cyber-Physical Systems: Industrial Process Plant Design. *Procedia Computer Science*, *28*, 883–890.

[19] www.statista.com/statistics/611004/global-industrial-internet-of-things-market-size.

[20] Safa Ahmed, et.al. (2018, Jan.). Overview for Internet of Things: Basics, Components, and Applications. *Journal of University of Anbar for Pure Science*, *12*, 47–58,

[21] Ashton, K. (2009). That 'internet of things' thing . *RFiD Journal*, *22*(7), 97–114.

[22] Anderl, R. (2015). Industrie 4.0–Technological Approaches, Use Cases, and Implementation. *Automatisierungstechnik*, 63(10), 753–765.

[23] Zhou, L., Chong, A.Y.L. and Ngai, E.W.T. (2015). Supply chain management in the era of the internet of things. *International Journal of Production Economics, 159*, 1–3.

[24] Xu, L.D., He, W. and Li, S. (2014). Internet of Things in Industries: A Survey. "*IEEE Transactions on Industrial Informatics", 10* (4), 2233–2243.

[25] Bakar, L.J.A. and Ahmad, H. (2010). Assessing the relationship between firm resources and product innovation performance: A resource-based view. *Business Process Management Journal*, *16*(3), 420–435.

[26] Ng, I.C., Wakenshaw, S.Y. (2017). The Internet-of-Things: Review and Research directions. *Int. J. Res. Mark.*, *34*, 3–21.

[27] Tan, Y.S., Ng, T. and Low, J.S.C. (2017). Internet-of-Things Enabled Real-time Monitoring of Energy Efficiency on Manufacturing Shop Floors. *Procedia CIRP, 61*, 376–381.

[28] Lin, D., Lee, C., Lau, H. and Yang, Y. (2018). Strategic response to Industry 4.0: An empirical investigation on the Chinese automotive industry. *Ind. Manag. Data Syst.*, *118*, 589–605.

[29] Sanders, A., Elangeswaran, C. and Wulfsberg, J. (2016). Industry 4.0 implies lean manufacturing: Research activities in industry 4.0 function as enablers for lean manufacturing. *J. Ind. Eng. Manag.*, 9, 811–833.

[30] Holmström, J. (1997). Product range management: A case study of supply chain operations in the European grocery industry. *Supply Chain Management: An International Journal*, *2*(3), 107–115.

[31] Rymaszewska, A., Helo, P. and Gunasekaran, A. (2017). IoT powered servitization of manufacturing: An exploratory case study, *International Journal of Production Economics,* in press.

[32] Denkena, B., Henning, H. and Lorenzen, L-E. (2010). Genetics and Intelligence: New Approaches in Production Engineering. *Production Engineering–Research and Development*, *1*(4), 65–73.

[33] Damanpour, F. (2010). An integration of research findings of effects of firm size and market competition on product and process innovations. *British Journal of Management*, *21*(4), 996–1010.

[34] Tian, G.Y., Yin, G. and Taylor, D. (2002). Internet-based manufacturing: A review and a new infrastructure for distributed intelligent manufacturing. *Journal of Intelligent Manufacturing*, *13*(5), 323–338.

[35] Denkena, B., Lenz, A.T. and Lorenzen, L-E. (2009). Agile Planning for Gentelligent Production. In: *Proc. of the 3rd Int. Conf. on Changeable, Agile, Reconfigurable, and Virtual Production CARV-09*, Munich, Germany, 79–88.

[36] Surana, A., Kumara, S., Greaves, M. and Raghavan. (2005). U.N. Supply-Chain Networks: A Complex Adaptive Systems Perspective. *International Journal of Production Research, 43*(20), 4235–4265.

[37] Yang, H., Bukkapatnam, S.T.S. and Barajas, L.G. (2013). Continuous flow modelling of multistage assembly line system dynamics. *International Journal of Computer Integrated Manufacturing, 26* (5), 401–411.

[38] Barthelmey, A., Storkle, D., Kuhlenkotter, B. and Deuse, J. (2014). Cyber Physical Systems for Life Cycle Continuous Technical Documentation of Manufacturing Facilities. *Procedia CIRP*, *17*, 207–211.

[39] R. Sanchez-Iborra, R. and M. Cano, M. (2016). State of the Art in LP-WAN Solutions for Industrial IoT Services. *Sensors, 16* (5), 708–721.

[40] Bauernhansl, T. (2013). Industry 4.0: Challenges and Opportunities for the Automation Industry. In: *7th EFAC Assembly Technology Conference 2013*, Davos, Switzerland, January 18–19. Presentation.

[41] Podgórski, D., Majchrzycka, K., Dąbrowska, A., Gralewicz, G. and Okrasa, M. (2017). Towards a conceptual framework of OSH risk management in smart working environments based on smart PPE, ambient intelligence and the Internet of Things technologies. *International Journal of Occupational Safety and Ergonomics,* vol. 23, no. 1, pp. 1–20, 2017.

[42] Monostori, L. et al. (2016). Cyber-physical systems in manufacturing. *Elsevier, CIRP Annals–Manufacturing Technology*, CIRP–1555.

[43] Qin, J., Liu, Y. and Grosvenor, R. (2017). A Framework of Energy Consumption Modelling for Additive Manufacturing Using Internet of Things. *Procedia CIRP, 63,* 307–312.

[44] Shaikh, F.K., Zeadally, S. and Exposito, E. (2017). Enabling Technologies for Green Internet of Things. *IEEE Systems Journal, 11* (2), 983–994.

[45] Tao, F., Zuo, Y., Xu, L.D., Lv, L. and Zhang, L. (2014). Internet of Things and BOM-Based Life Cycle Assessment of Energy-Saving and Emission-Reduction of Products, *IEEE Transactions on Industrial Informatics, 10* (2), 1252–1261.

[46] https://www.ge.com/digital/stories/vale-fertilizantes-saves-million-production-losses-asset-performance-management.

[47] Soujanya Mantravadi and Charles Mqller (2019). An Overview of Next-generation Manufacturing Execution Systems: How important is MES for Industry 4.0. *Elsevier*, *13*, 588–595.

[48] www.101blockchains.com/top-iiot-case-studies.

[49] https://aws.amazon.com/solutions/case-studies/bmw.

[50] https://customers.microsoft.com/en-us/story/sandvik-coromant-process-manufacturing-sweden.

[51] https://aws.amazon.com/solutions/case-studies/toyota-tsusho/.

[52] https://enterprise.microsoft.com/en-ca/articles/industries/manufacturing-and-resources/remote-energy-management-solution-based-microsoft-azure-iot/

[53] https://customers.microsoft.com/en-us/story/keeping-the-power-on-when-you-need-it-most

[54] http://news.echelon.com/press-release/corporate/echelon-enables-outdoor-lighting-enhance-public-safety-through-ibm-watson

[55] https://customers.microsoft.com/en-us/story/daimlertrucks

[56] https://www.ge.com/digital/stories/marathon-petroleum-develops-collaborative-strategy-optimizing-apm

[57] http://innovyt.com/azure-big-data-solution

[58] https://aws.amazon.com/solutions/case-studies/LightInTheBox/

[59] https://www.tracelink.com/insights/the-tracelink-life-sciences-cloud-community

[60] Becheikh N., Landry R. and Amara N. (2006). Lessons from innovation empirical studies in the manufacturing sector: A systematic review of the literature from 1993–2003. *Technovation*, *26*(5), 644–664.

[61] Sony, M. and Naik, S. (2019). Key ingredients for evaluating Industry 4.0 readiness for organisations: A literature review. *Benchmarking*, *2019*.

[62] Kruth, J.P., Leu, M.C. and Nakagawa, T. (1998). Progress in additive manufacturing and rapid prototyping. *CIRP Annals-Manufacturing Technology*, *47* (2), 525–540.

[63] Hui Yang, S., Kumaral, S., Bukkapatnam and F., Tsung. (2019, Jan.). The Internet of Things for Smart Manufacturing: A Review. Article in *IISE Transactions, 51* (11).

8 Data Governance and Responsible Analytics in IIoT

Shaurya Chaturvedi[1], Sujata Alegavi[2*], Sukruti Kaulgud[3], Archana Belge[4], Sunil Khatri[5] and Sumit Kumar[6]

The goal of the Industrial Data Space (IDS) is to establish a network of interconnected industrial systems, devices, and services via the safe and trustworthy exchange of data. IDS is a critical enabler of the developments that characterize Industry 4.0, the fourth industrial revolution, including the convergence of cyber-physical systems, the Internet of Things (IoT), and cloud computing. The primary objective of IDS is to provide a standardized, vendor-neutral, and interoperable environment in which computers, devices, sensors, and people may freely share and receive information from one another. The foundations of this ecosystem are the values of autonomy, openness, and safety. To be data sovereign, an organization must be free to choose who gets access to its data and for what purposes. In a fully interoperable environment, information may flow freely across all participating platforms. Safeguarding information implies preventing tampering, misuse, and leakage. Blockchain, edge computing, and AI are just some of the technologies used by IDS to make data sharing safe and efficient. It also provides a collection of common data models, protocols, and interfaces for use in connecting disparate services. Many sectors may benefit from IDS, including manufacturing, logistics, energy, healthcare, and more. IDS may allow for real-time monitoring of production processes, predictive machine maintenance, and quality control in manufacturing, for instance. Using real-time data, IDS may improve logistics by tracing items and trucks. Integrating renewable energy sources and optimizing energy use are two

[1] *MS* - Astronomy and Astrophysics, The Australian National University, Bruce, Australian Capital Territory, Australia.

[2] *Associate Professor* - IoT, Thakur College of Engineering and Technology, Mumbai, Maharashtra, India.

[3] *Associate Professor* - E&TC, Thakur College of Engineering and Technology, Mumbai, Maharashtra, India.

[4] *Associate Professor* - MME, Thakur College of Engineering and Technology, Mumbai, Maharashtra, India.

[5] *Assistant Professor* - IoT, Thakur College of Engineering and Technology, Mumbai, Maharashtra, India.

[6] *Assistant Professor* - E&CS, Thakur College of Engineering and Technology, Mumbai, Maharashtra, India.

Email: 221shaurya@gmail.com[1]; sukruti.kaulgud@tcetmumbai.in[3]; archana.belge@tcetmumbai.in[4]; sunil.khatri@tcetmumbai.in[5]; sumit.kumar@tcetmumbai.in[6]

* Corresponding author: sujata.alegavi@gmail.com

DOI: 10.1201/9781003587903-8

energy-related goals that IDS may help achieve. In conclusion, IDS is a crucial catalyst for the digitalization of sectors and has the potential to radically alter the ways in which we create, use, and interact with products and services.

1. Introduction

Often referred to as the "Fourth Industrial Revolution", Industry 4.0 is the next step in the increasingly digitalization of manufacturing. Factors including the development of digital infrastructure, analytics, human-machine interfaces, and robotics. The Cyber-Physical Systems (CPS) and the Industrial Internet of Things (IIoT) are the cornerstones of Industry 4.0. CPS are autonomous, intelligent systems that use computer-based algorithms to monitor and operate physical things like equipment, robots, and vehicles. Warehouses and transportation hubs must also become 'smart' to meet the needs of the fourth industrial revolution. But logistics is only one part of Industry 4.0. ERP's ability to communicate with other back-end systems grants unprecedented insight into and management of a company's inner workings. To succeed in the modern, digital economy, Industry 4.0 is essential [1, 2]. In Figure 1, we see the nine technological cornerstones that support Industry 4.0. These developments bridge the gap between the analog and digital worlds, making it possible to design intelligent, self-sufficient systems. Businesses and supply chains can make better use of the state-of-the-art equipment they have at their disposal by adopting the principles of Industry 4.0.

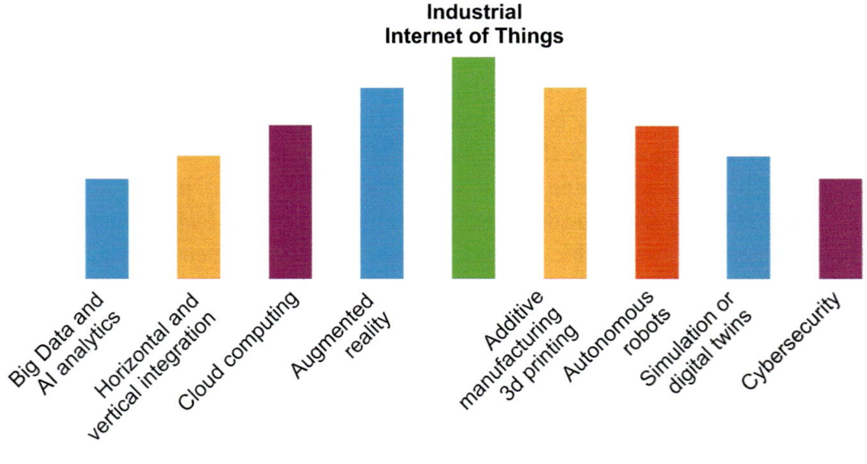

Fig. 1 Technological pillars of Industry 4.0.

1. **Big Data and AI analytics:** Industry 4.0 collects Big Data from numerous sources, including Internet of Things (IoT) sensors and equipment in factories, enterprise resource planning (ERP) and customer relationship management

(CRM) systems, mobile applications for monitoring the weather and traffic, and many others. Planning, logistics, manufacturing, R&D/engineering, enterprise asset management (EAM), and purchasing may all benefit from real-time data fuelled by artificial intelligence (AI) and machine learning (ML).

2. **Horizontal and Vertical Integration:** Horizontal and vertical integration are at the heart of the Industry 4.0. The "field level" of horizontal integration refers to how the shop floor, several manufacturing sites, and the whole supply chain work together as one cohesive unit. Vertical integration allows for unhindered communication between the production floor and the executive office, and vice versa. The elimination of data and information silos is a direct outcome of the tight integration between production and other functions of the company, such as R&D, quality assurance, sales, and marketing.
3. **Cloud Computing:** Cloud computing is the "great enabler" of Industry 4.0 and digital transformation. Cloud computing as it exists now is capable of so much more than simply being scalable, cost-effective, and rapid. It lays the groundwork for future technologies like AI and ML and the IoT, and gives them the tools they need to develop. The cloud not only stores the data that powers the technologies at the heart of Industry 4.0, but it also plays a crucial role in the communication and coordination of the cyber-physical systems at its core.
4. **Augmented Reality (AR):** Industry 4.0 is based on the idea of AR, which is the casting of digital information onto a real scene. When a worker looks at a piece of equipment or an actual product, the AR system may show real-time IoT data, digitized parts, repair or assembly directions, training materials, and more on smart glasses or mobile devices. AR is still in its early stages, but it already has big effects on the fields of repair, service, and quality assurance, as well as on the safety of workers and how they are trained.
5. **Industrial Internet of Things (IIoT):** Because of how fundamentally both the IIoT and the IoT are to the new manufacturing paradigm known as Industry 4.0, it is not uncommon to hear the two phrases used interchangeably. Industry 4.0 relies heavily on sensors and RFID (radio frequency IDentification) tags to provide real-time information on the operating state or location of devices, robots, machines, equipment, and products. Potential corporate uses for this technology include streamlining operations throughout the supply chain, shortening the time it takes to design and implement new products, lowering the number of times machinery breaks down, responding quickly to changing consumer tastes, and monitoring inventory levels.
6. **Additive Manufacturing/3D Printing:** Additive manufacturing, or 3D printing as it is more often known, is another important technology fueling the Fourth Industrial Revolution. Modern uses of 3D printing go well beyond its prototype beginnings, with applications including mass customization and

decentralized manufacturing. Reduce the need for transit by keeping digital design files for products and components in a digital warehouse and printing them out as needed via 3D printing, for instance.

7. **Autonomous Robots:** Industry 4.0 has led to the development of a new category of completely autonomous robots. Drones that scan stock inventories and mobile robots that do pick-and-place tasks are just two examples of the various kinds of autonomous robots now on the market. These robots are able to recognize, evaluate, and respond to information they receive from their environment because of sophisticated software, AI, sensors, and machine vision.
8. **Simulation/Digital Twins:** A digital twin is an electronic representation of a real-world product, service, or system, built from data collected by IoT sensors. This vital component of Industry 4.0 enables firms to get insight into, evaluate, and optimize the operation and upkeep of their industrial systems and products. An asset manager, for instance, may use a digital twin to pinpoint a faulty component, foresee possible problems, and increase availability.
9. **Cybersecurity:** Industry 4.0's dependence on interconnected systems and massive amounts of data makes it more crucial than ever to take precautions to keep sensitive information safe. Organizations may be able to lessen the likelihood of network-wide data breaches and production delays by automating threat detection, prevention, and response with Zero Trust architecture and technologies like ML and blockchain.

1.1 Benefits of Industry 4.0

1. **Intelligent Products:** Build objects that are aware of their surroundings, can communicate with one another, and can relay data about their usage and storage circumstances. You may enhance product quality, customer service, logistics, and R&D (research and development) using the data collected by these cutting-edge tools. Thus, they might theoretically foresee service requirements, get updates remotely, and pave the way for cutting-edge service-based business models.
2. **Intelligent Factories:** Manage fully digitalized, highly automated workplaces that use cutting-edge technologies like Big Data, AI, robots, analytics, and the IoT. Factory 4.0, also known as self-correcting factories, uses smart manufacturing 4.0 processes to enable the mass production and delivery of customized items at cheap cost.
3. **Intelligent Assets:** Sensors incorporated into today's deployed physical assets, in conjunction with analytics and the IIoT, are revolutionizing the way businesses manage their possessions. Intelligent assets help workers streamline the incorporation of machinery into operational procedures,

undertake dynamic and predictive maintenance, lessen the frequency of unscheduled repairs, and monitor asset performance in real time.

4. **Empowered individuals:** No matter how advanced a system becomes, it will always need human oversight. Help them stay on top of advancements on the factory floor by equipping them with real-time sensor data and cutting-edge technology like AI. AR apps and wearable technology might aid in problem solving, health tracking, and security.

Industry 4.0 technologies are many and can help millions of organizations rethink production, prioritize their customers, and integrate their operations. Here are some benefits companies are experiencing.

1. **Ability to quickly adapt to new market or economic conditions:** Adaptability and fortitude in the face of market and economic fluctuations: In order to mold the digital supply chain of the future, businesses are using state-of-the-art planning techniques.
2. **Assurance to test out novel approaches of doing business and capture promising openings:** Businesses may now combine their ocean, land, and air supply chains while saving money and increasing their competitiveness thanks to Industry 4.0 technology.
3. **Profitability with environmental responsibility:** Customers that make the switch to digital platforms save time and money while also making strides towards their environmental and other business goals, such as profitability and scalability.

2. Background

The IoT and Industry 4.0 have begun to merge, which has cleared the way for the creation of smart services that can be utilized in industrial settings. Establishing a data interchange that is both safe and effective amongst the many organizations that make up the industrial ecosystem is one of the most significant problems in this area. The idea of an IDS has recently surfaced as a potential answer to this problem. Existing research on IDS and its role in integrating IoT and smart services in Industry 4.0 and beyond includes a variety of areas of IDS, including its architecture, security measures, data governance, and its potential influence on industrial processes. This study was carried out by a number of different researchers.

The concept of the IDS as a solution to address the challenges faced by the industrial sector in data exchange, data sovereignty, and trustworthiness has been introduced by [1]. The study by [2] focuses on the architectural aspects of the IDS) and provides a comprehensive analysis of the key components necessary for the successful implementation of IDS in industrial environments. The paper by [3] proposes a reference architecture for the IDS that can serve as a blueprint for designing and implementing IDS solutions in industrial environments, whereas [4] explores the role of the IDS in enabling smart services within the manufacturing sector. It emphasizes the importance of utilizing IDS to leverage Industry 4.0 technologies and transform traditional manufacturing processes into intelligent and data-driven operations. The crucial aspects of security and data governance within the IDS are focused in [5]. It highlights the importance of robust security measures and effective data governance mechanisms to ensure the confidentiality, integrity, and availability of data exchanged within IDS. The article by [6] addresses the crucial aspects of authorization and access control within the IDS [7]. A comprehensive review of the existing research on data governance within the IDS is presented by [7]. The paper "A Data Governance Concept for the Industrial Data Space" [8] introduces a tailored data governance framework for the IDS to address the challenges of effective data management, quality, and security within the industrial context. The paper emphasizes the significance of data governance in supporting data-driven decision-making, ensuring data quality, and enabling efficient data sharing and collaboration in IDS. The research by [9] explores the role of the DS in facilitating new business models within the context of Industry 4.0. The authors emphasize IDS's potential as a key enabler for innovation, value creation, and collaboration in the industrial sector. The study by [10] explores the role of the IDS in fostering digital ecosystems within the context of Industry 4.0. It emphasizes how IDS acts as a key enabler for data exchange, collaboration, and value creation in the industrial domain.

3. Linkage of IOT with Industry 4.0

Connectivity, automation, ML, and real-time data are at the forefront of today's Industrial Revolution. The goal of Industry 4.0 is to help businesses who want to stay competitive by radically increasing their efficiency by combining traditional manual processes with digital automation to create "smart manufacturing" with a comprehensive ecosystem.

Traditionally, when new technologies become widely available, businesses lead the way in adopting them, and industrial automation soon follows. Using IoT technology makes sense since it may increase productivity and allow closer engagement with business infrastructure. Common IoT technologies will have a similar effect on hardware, software, and manpower costs as the PLC. In the present economic situation, IoT technology is transforming companies into "intelligent factories" via the use of, among other things, network connectivity, automation, real-time data monitoring, and predictive analytics to aid in decision making.

IoT and Industry 4.0 have taken a giant step forward.

There are four basic steps in the development of the IoT and Industry 4.0:

- connecting devices,
- gathering data,
- improving workflow, and
- coming up with new ideas.

3.1 What is IIoT (Industry 4.0)?

When applied to manufacturing, the term "Industry 4.0" refers to the notion of merging smart manufacturing machines, AI-powered automation, and sophisticated analytics to increase productivity for all industrial workers. This is known as the "Industrial Internet of Things" (IIoT). It fundamentally alters the potential for processing data. In order to increase productivity, businesses need IIoT devices tailored to their requirements, which can gather, monitor, and analyze data to provide actionable insights.

For instance, using IOT-enabled devices to detect flaws in leather can make Leather industry a smart factory by giving more visibility throughout the manufacturing process and can deliver higher quality products.

3.1.1 The Benefits of IIoT on Manufacturing and Beyond

1. By continually gathering and processing data, to make robots smarter and more efficient than their human counterparts. With the help of AI and ML, large amounts of data can be handled quickly by smart sensors, allowing for improved prediction of machine and production line performance. It improves output without sacrificing quality.
2. ML may teach systems to be vigilant for recurring trends. It has the ability to foresee potential failure, notify it automatically, and take timely corrective action. Companies may save millions or even billions of dollars with the aid of apps like these.
3. The IIoT unifies all employees, information, and procedures into a single system. Decisions made by business executives may be quick and precise because of the abundance of data available to them.
4. Observability of Production. In the manufacturing sector, IIoT makes machines, tools, and sensors accessible to engineers. This transparency allows for speedier tracking of all components, detection of bottlenecks, and localization of the causes of any issues.
5. Quicker iterations of enhancement. On the other side, an IoT-enabled business may save time and effort by automating data collecting.
6. Make quality management methods more affordable. To save money, IIoT uses automated sensors to simplify all activities instead of pricey Quality Management Systems.
7. Maintain high standards by always keeping an eye on things. It is possible to utilize smart sensors to track and record data about industrial processes and environmental factors in real time. The end result will be higher grade goods. Temperature might be very important in the pharmaceutical industry. Temperature and humidity sensors linked to the IoT allow administrators to keep tabs on these factors and get quick notifications if they fluctuate outside of acceptable ranges.

8. Higher rates of machine use. Sensors aid engineers in keeping tabs on all machinery and forecasting when they could break down. This allows the engineer to intervene before any machine breakdown occurs. This improves the sector's overall machine utilization.

3.1.2 What are the Present IIoT Challenges?

1. Interoperability is the first major obstacle to overcome. Machines and gadgets have diverse protocols and architectural foundations. Using an API-driven environment to facilitate API calls between devices and machines is one approach to solving this problem.
2. The issue of safety is the second difficulty. Many businesses will be wary of making the move to the cloud that IIoT necessitates because of the risks of downtime in operations and production as well as those posed by cyber-attacks and data theft. To reduce the potential for harm from internet-connected intelligent equipment, businesses, stakeholders, and legislators will need to collaborate.

4. IIoT Examples

Numerous sectors may already see the advantages of IIoT.

1. Healthcare
2. Energy and Utilities
3. The Production Sector
4. Property
5. Transportation and Supply Chain
6. The Public and Government Sector
7. Commercial

5. Survey and Analysis

To identify and establish connections between IoT and smart services in Industry 4.0, the following survey was conducted and analysis was done.

1. Which of the following best describes your familiarity with the concept of Industry 4.0?

The provided responses reflect the degree to which people are aware of Industry 4.0. Responses include "Very familiar" to "Somewhat familiar". If one is to analyze the response, it shows that a larger percentage of respondents were "Very familiar" with the concept at 55.6%. This would indicate that the sample population is relatively knowledgeable about Industry 4.0.

On the other hand, 44.4% said they are "Somewhat familiar" with Industry 4.0. Although this segment is a minority of the respondents, the answers suggest that they are fairly aware and know about the term. However, it is also important to point out that there were no choices for those to say they do not know much or anything about the concept at all in the survey.

From these answers, we can deduce that the respondents are generally familiar with Industry 4.0. Most of the respondents claimed to be "Very familiar", which means that the concept is widely known among the sample population. This is in line with the increasing importance of Industry 4.0 as a transformational framework in different industries, which means that professionals and researchers are actively engaged with the concept. However, the presence of a smaller group of respondents who reported being only "Somewhat familiar" suggests that there may still be individuals who have a limited understanding of Industry 4.0 and could benefit from further education or exploration of the topic.

2. What do you consider as the primary benefit of linking IoT and smart services in Industry 4.0?

To this, respondents answered that 44.4% claimed improved operational efficiency, 22.2% claimed better decision-making abilities, and 33.3% claimed improved product quality and customization.

The three most significant advantages of IoT and smart services in Industry 4.0 can be summarized as improved decision-making abilities, improved product quality and customization, and improved operational efficiency. These three advantages have been repeated several times in the responses given above.

The linking of IoT and smart services is an important step to enhance decision-making capabilities. Through real-time data and advanced analytics, organizations are able to make informed decisions on the basis of the most accurate and up-to-date information. Proactive and timely actions enhance operational performance and the allocation of resources.

Another important advantage is enhanced quality and customization. IoT devices, along with smart services, aid in the capture and analysis of product data from its lifecycle stage. This is helpful for manufacturers because it allows them to identify aspects that need improvements, optimize procedures, and present products that address customer needs better. The association of IoT with smart services empowers the provision of predictive maintenance in order to produce quality products while minimizing downtime.

The dominant theme of increased operational efficiency underlies the responses. IoT connectivity allows for the tracking, control, and management of many processes and systems, thus efficiently allocating resources, reducing waste, and streamlining operations. The automation of routine tasks and remote management can give an organization a higher possible level of productivity, with cost savings as well.

In summary, the linking of IoT and smart services in Industry 4.0 presents various benefits to organizations, such as enhancing decisions, quality, and product customization and increasing operational efficiency. Such gains assist in optimizing the processes of organizations in pursuit of competitive advantages and high-quality products that should be achieved in light of shifting customer demand in the digital age.

3. How would you rate the current level of adoption of IoT and smart services in your industry?

For this, the respondents answered that 11.1% claimed to have high adoption of IoT and smart services, 55.6% claimed to have moderate adoption of IoT and smart services, and 33.3% claimed low adoption of IoT and smart services.

The responses achieved reflect mixed levels of adoption of IoT and smart services in the industry in question. Of the eight responses, four reported moderate adoption, three low adoption, and one high adoption.

Moderate adoption means that the adaptation of IoT and smart services has penetrated the industry somewhat but is yet to saturate or become extensive in its application. This may be because of the issues related to the complexity of integrating IoT devices and systems, data security and privacy concerns, and additional investments needed in infrastructure. The industry seems to be at a transition point where a high number of organizations are experimenting and implementing IoT and smart services for improving their operations.

There is at least one answer that reflects very high adoption, which indicates there are some pioneers in the market who have been successful in adopting IoT and smart services, thereby perhaps gaining benefits in terms of higher efficiency, better decision-making, and improved customer experiences. The domination of answers suggesting moderate and low adoption, on the other hand, suggests that a greater portion of the industry is still far from embracing the technologies. Factors such as cost considerations, limited awareness, and possibly resistance to change may be preventing the widespread adoption of IoT and smart services.

There is a mixed landscape of IoT and smart service adoption within the industry based on the responses. Some organizations have highly ranked adoption levels, but most are in categories with moderate and low adoption. This leaves a tremendous amount of room for further IoT and smart service implementation and integration in the industry. Overcoming the problems with the cost, security, and awareness can bring about an expedited adoption process that unlocks the potential benefits offered by these technologies.

4. What do you see as the main challenge in implementing IoT and smart services in Industry 4.0?

To this, respondents replied that 33.3% believe that there is a lack of standardized protocols and interoperability, 44.4% believe that there is a limited availability of skilled personnel, and 22.2% believe that there are high implementation costs involved, which are the major challenges in implementing Industry 4.0.

According to the answers in the context of Industry 4.0, there are several critical challenges in the deployment of IoT as well as smart services. Lack of trained personnel, the high implementation costs, and the lack of standard protocols and interoperability were typically posed as one of the most stated challenges.

The lack of skilled workers is the first of these significant challenges. It is not just about smart services and IoT but also the kind of people needed for this purpose, including experts with competencies in data analytics, cybersecurity, and system integration. However, recruitment and retention are challenging. The demand from various industries often outstrips supply. In such a situation, it would take much time to implement and slow down the successful integration process of Industry 4.0.

High implementation costs are often cited as a major challenge. It is costly to deploy IoT infrastructure, upgrade existing systems, and integrate various devices and technologies. The hardware, software, networking, and data management costs can be barriers to many organizations, especially the smaller ones. Without sufficient financial resources, widespread adoption of IoT and smart services may be limited, and the full potential of Industry 4.0 may not be realized.

Another major challenge includes the standardization of protocols and interoperability. In an integrated ecosystem of IoT devices and smart services, seamless communication and interoperability are imperative. However, without standardized protocols and compatibility problems across systems and different equipment, it could hinder integration. Such problems can induce inefficiencies, inconsistency in data, and problems in working and scaling up IoT deployments; ultimately, it affects the process involved in a smooth working of Industry 4.0.

The major constraints hindering the easy implementation of IoT and smart services in Industry 4.0 include a lack of skilled labor, high costs of implementation, and a lack of established protocols and compatibility. All this calls for substantial effort in several different areas such as formulating industry-wide standards and norms, investing in education and skill development, and research to provide innovative cost-effective solutions for implementation. In the success of Industry 4.0, the integration of IoT and smart services must be done correctly. Only then will businesses unlock the full benefits of automation, data-driven decision-making, and increased operational efficiency.

5. Which of the following IoT devices or sensors are you currently using or planning to use in your organization?

To this, respondents responded that 55.6% are using RFID tags, 77.8% are using temperature sensors, 77.8% are using motion sensors, 55.6% are using GPS trackers, and 66.7% are using smart meters.

From the number of answers given, the spread of usage, or plans for different IoT devices and sensors is varied. The items listed here include GPS trackers, RFID tags, temperature sensors, motion sensors, and smart meters. These have numerous purposes and applications in quite diverse fields.

GPS trackers are widely used for fleet management and asset tracking. They are useful in reporting real-time location data, allowing the implementation of transportation and logistics operations. With GPS trackers, route planning, fuel consumption, security, and overall efficiency can be improved.

RFID tags are very popular in inventory management and asset tracking. It helps track and identify items in real time, automates the process of inventory, and gives increased visibility to the supply chain. The RFID technology makes the process efficient by making quick and accurate identification of objects possible, thereby reducing manual labor and errors.

In healthcare, food storage, and manufacturing industries, important temperature sensors are used. These help monitor and maintain the optimum levels for the temperature and assure product quality, compliance with regulations, and safety. An organization can identify possible issues and fix them right away to prevent possible spoilage, damage, or non-compliance by providing real-time temperature data.

Motion sensors are very popular for security and energy management applications. They sense movement in a given area and work on the basis of that. Motion sensors can be used for intruder detection, lighting control, occupancy sensing, and even smart automation in buildings, which would result in increased security, efficient use of energy, and cost-effectiveness.

Smart meters are widely used in utilities for electricity, water, and gas consumption monitoring. They help to provide real-time consumption data that enables proper billing, demand management, and resource optimization. Smart meters make energy conservation easier, enable the empowerment of consumers through detailed usage information, and facilitate the implementation of smart grid initiatives.

The type of organizations that are using or will use various types of IoT devices and sensors, such as GPS trackers, RFID tags, temperature sensors, motion sensors, and smart meters, is what the answers state. Such devices bring great benefits, such as real-time monitoring, higher efficiency, better security, and better resource management. Organizations can, therefore, optimize their operations, minimize costs, and make data-driven decisions to achieve both operational excellence and overall success.

6. How do you envision IoT and smart services transforming your industry in the next 5–10 years?

To this, respondents responded regarding how IoT and smart services transform your industry in the next 5–10 years that 66.7% claimed improved predictive maintenance and reduced downtime, 66.7% claimed streamlined supply chain management, 77.8% claimed advanced analytics for data-driven insights, and 55.6% claimed increased automation and autonomous systems.

The above answers outline the transformative influence that IoT and smart services may have on many sectors within the next 5–10 years. Changes include advanced analytics to drive decisions, streamlined supply chain management, predictive maintenance minimizing downtime, and automation and autonomous systems.

There seems to be a common thread that runs through comments: the application of advanced analytics to infer from data. Even though IoT lets one connect devices and sensors, collect large volumes of data, and then extract useful knowledge from it by applying tools like ML and AI, such insights can spur well-thought decisions, improve operations, identify trends, increase productivity, and unveil hidden commercial opportunities.

Streamlined supply chain management has also been termed as a major transformation. IoT and smart services allow monitoring in real time, from procurement to logistics and delivery. Therefore, with better visibility, organizations will be able to optimize their inventory management, asset tracking, demand forecasting, and eliminate all kinds of inefficiencies. Hence, streamlined supply chain management brings about cost efficiency, faster deliveries, customer satisfaction, and in general, business excellence.

Some answers show improvement in predictive maintenance and reduced cases of downtime. IoT devices and sensors allow the monitoring of equipment and machinery for any impending faults or need for maintenance, which are essential for such machinery. Predictive maintenance solutions can help organizations prevent problems, reduce their downtime, optimize schedules, and extend asset lifespans. It might even lead to more productivity, saving costs, and better operation reliability.

Finally, the responses predict an era of more automation and autonomy. Robotics, artificial intelligence, and IoT technologies will advance to make most of the current processes and activities automated. Organizations expect their autonomous vehicles, smart factories, and intelligent systems to achieve precision and safety, and to attain efficiency in most operations.

For instance, advanced analytics that deliver data-driven insights, optimized supply chain management, improved predictive maintenance with reduced downtime, increased automation and autonomous systems, and many more are part of the expected transformations from IoT and smart services in the next 5–10 years. This sets an oncoming shift in industries, unlocking levels of effectiveness and agility never previously known. However, to fully realize its benefits, organizations need to invest in adopting technology, establishing data infrastructure, developing talent, and strategic planning to successfully ride the waves of change in IoT and smart services.

7. Which of the following risks or security concerns do you associate with the widespread use of IoT and smart services in Industry 4.0?

In relation to this, the respondents shared with regard to the risks or security issues as presented by the general adoption of IoT and smart services in Industry 4.0 that "77.8% were afraid of data breach and cyberattack", "66.7% fear the unauthorized access of sensitive information", "33.3% feared lack of regulation on data privacy", and "55.6% feared integration and legacy systems".

These answers point out the risks and security issues related to the extensive adoption of IoT and smart services under the umbrella of Industry 4.0. Among the concerns that have been brought up are data breaches and cyber attacks, unauthorized access to sensitive information, no data privacy laws, and legacy system integration.

Data breaches and cyber attacks are among the most cited risks in various responses. Such inherent interconnectedness in their appliances and the humongous volume of data produced by IoT devices opens them to evil activities. Organizations have to deal with securing data transmission, storage, as well as processing, with the threat of unauthorized access, hacking, and data breach. If there is a breach into an IoT system, then the consequences would be dreadful, such as leakage of sensitive information, compromised business, loss in the form of financial revenue, as well as reputational damage.

Another significant concern is unauthorized access to confidential data. IoT devices often manage and transmit sensitive data, including trade secrets, private data, and personal information. Without proper security measures, unauthorized individuals or malicious actors may gain access to this data and compromise people's privacy, steal their intellectual property, or commit other wrongs. To mitigate this risk, there should be robust authentication, access control, and encryption processes.

Some of the answers state the lack of data privacy regulations as an area of concern. Because IoT and smart services are becoming very common, more and more personal data are being collected and processed. Without comprehensive data privacy regulations and standards, the chances of inadequate protection and misuse of personal information are high. It would therefore establish and enforce clear data privacy regulations and practices, addressing a big concern to bring users and organizations to the trust side.

Legacy system integration is a type of risk pointed out in responses. Most of the organizations had existing legacy systems before IoT technologies could be integrated with them. Thus, integrating such devices with those systems can present compatibility issues and vulnerabilities along with possible points of failure. In this regard, the need for thorough security assessments is paramount, including all necessary updates and patches to address potential risks posed by legacy system integration.

Thus, the risks and security concerns for the widespread adoption of IoT and smart services in Industry 4.0 are data breaches and cyber attacks, unauthorized access to sensitive information, lack of data privacy regulations, and integration with legacy systems. It is a multi-disciplinary effort that requires comprehensive robust security measures, encryption protocols, authentication mechanisms, and compliance with data privacy regulations. Legacy system integration is also to be considered. Continuous monitoring of the environment in conjunction with proactive risk management could mitigate potential risks and ensure a secure and trustworthy way of implementing IoT and smart services in Industry 4.0 environments.

8. Are you currently leveraging data analytics and AI technologies to drive insights from IoT-generated data?

Regarding using data analytics and AI technologies to drive insights from IoT-generated data, 44.4% respondents replied in the affirmative, 22.2% respondents in the negative, and 33.3% respondents replied that they are planning to use AI in the near future.

The responses show levels of adoption as well as the plans to employ data analytics and AI technologies for extracting insights from IoT-generated data. Some companies are already using such technologies, and others have plans to adopt the same in the near future but have not applied them yet.

The organizations that have already adopted data analytics and AI technologies are likely to be benefiting from valuable insights that are derived from the massive amount of data generated by the IoT devices. Advanced analytics techniques and AI algorithms make it possible for these organizations to unveil patterns, correlations, and trends within data. This allows them to make data-driven decisions, optimize operations, improve efficiency, and gain a competitive advantage in their respective fields.

Organizations that have yet to leverage the power of data analytics and AI technologies but expect to do so in the not-too-distant future understand and are aware of the value and the benefits these technologies bring. It is probable they know how IoT-generated data relates to their relevance and importance as well as extracting meaningful insights. By planning for the adoption of data analytics and AI, these organizations are positioning themselves to unlock the full potential of their IoT deployments and capitalize on the transformative power of data-driven decision-making.

The responses indicate a mix of current adoption and future plans to leverage data analytics and AI technologies to derive insights from IoT-generated data. Organizations that have already adopted these technologies are likely enjoying the benefits of better decision-making and operational efficiency. Organizations with future plans for adopting these technologies are recognizing the importance of data analytics and AI in their IoT strategies to unlock valuable insights and gain a competitive edge. As IoT is on the increase, producing more enormous amounts of data, organizations are going to rely more and more on the amalgamation of data analytics and AI.

9. How do you see the role of IoT and smart services in improving operational efficiency and reducing costs in Industry 4.0?

44.4% of respondents argued that real-time monitoring and optimization of processes about the role of IoT and smart services in enhancing the efficiency of operations while controlling the costs of Industry 4.0, 22.2% of respondents said that predictive maintenance minimizes failure by equipment about the role of IoT and smart services in improving the efficiency of operation and, thereby reducing costs in Industry 4.0, and 33.3% of respondents responded that IoT and smart services play an important role in improving operational efficiency and controlling costs concerning efficient inventory management and supply chain optimization in Industry 4.0.

In such a context, Industry 4.0 benefits provided through IoT and smart services would ensure efficiency and cut down the operational cost. Those mentioned advantages comprise effective inventory control and optimized supply chain management; predictive maintenance reduces the risk of machine breakdowns, and in the processes, the actual-time monitoring and optimization takes place.

More than one response mentions effective inventory control and supply chain optimization. With the use of IoT and smart services, organizations can gain real-time visibility into their inventories, track their assets, and optimize their supply chain. This enables them to make better demands, reduce the costs of carrying inventory, improve the fulfillment of orders, and prevent stockouts. Improved efficiency in inventory management and supply chain operations also lead to cost reduction for the organization, streamline processes, and more customer satisfaction.

Another important function of IoT and smart services is predictive maintenance. Organizations can spot abnormalities, anticipate future problems, and proactively plan maintenance tasks by continually monitoring machinery and equipment with IoT-enabled sensors. This method reduces downtime, prevents unexpected malfunctions, and improves maintenance plans. Organizations may switch from reactive to proactive maintenance strategies with the help of predictive maintenance, which reduces equipment failure and improves operational dependability and lifetime.

Real-time monitoring and optimization of processes are also mentioned as a significant benefit of IoT and smart services. With the collection and analysis of data from various IoT devices and sensors, organizations can gain real-time insights into their operations. This allows for immediate identification of bottlenecks, inefficiencies, and deviations from optimal performance. With real-time monitoring, organizations can make timely adjustments, optimize processes, and improve overall efficiency. This leads to reduced waste, improved productivity, and cost reductions.

The responses highlight the role of IoT and smart services in Industry 4.0 in increasing operational efficiency and reducing costs. These innovations are primarily the outcome of efficient inventory management, supply chain optimization, predictive maintenance, and real-time process monitoring and optimization. Organizations can achieve streamlined operations, efficient resource utilization, reduced downtime, and enhanced operational performance through the use of IoT technology. With the integration of IoT and smart services, organizations can make data-driven choices, optimize their operations, and maintain competitiveness in the rapidly changing environment of Industry 4.0.

10. Do you believe that collaboration and interoperability among different IoT devices and platforms are important for the success of Industry 4.0?

100% of the respondents said 'yes' that for the success of Industry 4.0, collaboration and interoperability between different IoT devices and platforms were important. The responses given point out unanimously the belief that collaboration and interoperability among different IoT devices and platforms are important for the success of Industry 4.0. The consensus established clearly points out the critical role of integration and compatibility in achieving the full potential of IoT and actualizing the vision of Industry 4.0.

The collaboration among various IoT devices and platforms would ensure smooth communication, data exchange, and coordination among the different interconnected systems. In an Industry 4.0 environment with multiple devices, sensors, and systems, the collaboration ensures an efficient information flow and provides an overall view of the operations for better optimization and decision-making as well as the interoperability in the ecosystem.

Interoperability is important for Industry 4.0 because it enables different types of IoT devices and platforms to work together despite their manufacturers and specific technical specifications. Interoperability ensures that different components, systems, and technologies can be used in interaction, data exchange, and execution in a standardized manner. It promotes scalability, flexibility, and adaptability and allows organizations to integrate new technologies, expand IoT deployments, and leverage the full potential of connected systems. Interoperability makes the implementation and maintenance processes much easier, decreases complexity, and fosters innovation through the use of diverse solutions.

In summary, the responses demonstrate that the most significant agreement about the success of Industry 4.0 is its need for cooperation and interoperability. This can allow communication between various IoT devices and platforms by sharing data as well as proper coordination. In such a connected ecosystem, by facilitating collaboration and interoperability, organizations can build systems that will foster optimization, scalability, and innovation. Collaboration and interoperability enable the various technologies involved to work synergistically together and form a coherent system–a must for the successful implementation of Industry 4.0.

6. Conclusion

The IDS is a breakthrough concept that works as a transformational bridge connecting the I IoT with intelligent services within Industry 4.0 and beyond. It is a standardized framework and infrastructure to ensure secure, controlled data interchange, collaboration, and interoperability between diverse stakeholders working in industrial environments. The Industrial Data Space enables such seamless integration and exploitation of the data generated in IoT, leading to its full potential for maximizing operational efficiency with reduced costs while driving innovation and growth.

Leveraging data analytics in real time, and AI, and some of the smartest algorithms designed for the processing of IoT with smart services deployed within the IDS. This integration supports optimized inventory management, streamlined supply chain operations, predictive maintenance, and continuous monitoring and enhancement of industrial processes in real time. Therefore, organizations can streamline their operations, minimize downtime, boost productivity, and elevate overall performance levels.

Besides this, the IDS encourages collaboration and synergy among different participants in an industry by providing a secure and trustworthy environment for sharing data. It upholds data sovereignty, privacy, and security, enabling organizations to retain ownership of their data assets while allowing for seamless data exchange and collaboration. The IDS also promotes interoperability by creating standardized protocols and interfaces, making it easier to connect various IoT devices, platforms, and systems.

The IDS is a core foundation in the dynamic landscape of Industry 4.0 that enables the efficient use of IoT and intelligent services in industrial environments. The adoption and implementation of this hold vast potential to drive innovation, competitiveness, and the discovery of new business opportunities. IDS adoption by an organization empowers it to tap into the capabilities of interconnected systems, data analytics, and collaborative ecosystems in order to address the intricacies of Industry 4.0 and lead the next industrial revolution.

References

[1] Broy, M., Steinbuch, M. and Verl, A. (2017). Industrial Data Space: Creating a Sovereign Data Ecosystem for the Industry. In: *Proceedings of the 2017 ACM International Joint Conference on Pervasive and Ubiquitous Computing and Proceedings of the 2017 ACM International Symposium on Wearable Computers* (pp. 407–412). ACM.

[2] Mader, A. and Schumann, M. (2019). Architectural Components for an Industrial Data Space. In: *Proceedings of the 14th International Conference on Availability, Reliability, and Security* (pp. 1–10). IEEE.

[3] Bauer, M., Mädche, A. and Reinhartz-Berger, I. (2019). Towards a Reference Architecture for Industrial Data Space. In: *Proceedings of the 51st Hawaii International Conference on System Sciences* (pp. 3769–3778). IEEE.

[4] Hepp, M. and Kolbe, L. M. (2018). The Industrial Data Space as a Foundation for Smart Services in Manufacturing. In: *Proceedings of the 51st Hawaii International Conference on System Sciences* (pp. 4161–4170). IEEE.

[5] Mammoliti, G., Buonanno, G., De Mola, F. and Rak, M. (2019). Security and Data Governance in the Industrial Data Space. In: *Proceedings of the International Workshop*

on Enterprise Modeling and Information Systems Architectures (pp. 1–16). CEUR Workshop Proceedings.

[6] Buchmann, R., Broens, T. and Lobov, A. (2018). Decentralized Authorization and Access Control for Industrial Data Space. In: *Proceedings of the International Workshop on Enabling Real-Time Business Intelligence* (pp. 70–84). Springer.

[7] Kärkkäinen, M., Ogonowski, C. and Sjöklint, M. (2020). Data Governance in Industrial Data Space: A Systematic Literature Review. In: *Proceedings of the 28th European Conference on Information Systems* (pp. 1–19). AIS Electronic Library.

[8] Eggert, M., Höfig, T. and Neuhaus, J. (2018). A Data Governance Concept for the Industrial Data Space. In: *Proceedings of the International Conference on Advanced Information Systems Engineering Workshops* (pp. 22–37). Springer.

[9] Krechel, M. and Kreutzer, M. (2018). Industrial Data Space - An Enabler for New Business Models in Industry 4.0. In: *Proceedings of the 9th CIRP IPSS Conference: Circular Perspectives on Product/Service-Systems* (pp. 341–347). Elsevier.

[10] Wunderlich, N., Hofer, C. and Kreutzer, M. (2019). Industrial Data Space: Enabler for Digital Ecosystems in Industry 4.0. In: *Proceedings of the 52nd Hawaii International Conference on System Sciences* (pp. 1853–1862). IEEE.

9 Resilience and Reliability in Industrial IoT Networks

Namit Singh[1], Keyur Amit Chauhan[2*], Soham Sharad Kadam[3] and Akanksha Manoj Singh[4]

Industrial Internet of Things (IIoT) networks are expanding quickly, which highlights the need to improve resilience and dependability as system complexity and data volumes rise. How businesses can create and maintain resilient IIoT infrastructures that can resist malfunctions, hacks, and other disturbances is still mostly unknown. This chapter addresses the important relationship between resilience and IIoT network dependability from operational and technology perspectives. The three main claims are that first it is important to understand the different building blocks of IIoT networks and how they make IIoT networks multi-layered, dynamic, and emergent. Secondly, the literature, which provides an overview of how a well performed and designed network plays a significant role in managing the abundant and unpredictable data generated from these IIoT systems. Third, how will IIoT networks need to integrate dependability to foster creativity and sustainable development, which raises challenging questions of the role of redundancy and proactive design in such an evolving environment.

1. Introduction

The Industrial Internet of Things (IIoT) is changing industries to gain good connectivity among machines, devices, and systems and increase innovation and operational efficiency. Guaranteed durability and stability of the grid and handling huge amounts of data continuously entered into network devices are the basis of Internet of Things (IoT) development. As with IIoT networks, the more complex these systems become, the more complex they become. It's even harder to stay networked. Creative momentum and weakness protection organizations must build fault-tolerant networks. Prompt recovery and continuity of service guarantees must deliver on all the promises of IIoT. Operational downtime and security risks can result from an unreliable network and there are more cyberattacks due to this.

[1] *MSc Business Analytics*, University of Warwick - Warwick Business School in Coventry, England.
[2-4] *Internet of Things*, Thakur College of Engineering and, Technology, Mumbai, India.
Email: namitt.singh@gmail.com[1]; sohamkadam0012@gmail.com[3]; akankshadsingh02@gmail.com[4]
* Corresponding author: keyurchauhan15@gmail.com

DOI: 10.1201/9781003587903-9

Of course, this happens when there are no strict protection measures in place. Reliability and flexibility are essential improvement situations to fully utilize IIoT technology.

1.1 Network Resilience

In the context of IIoT, network resiliency refers to the ability to continue providing services in the face of unforeseen events such as system failures, cyber attacks or natural disasters. This involves promoting networks that need to be agile, adaptable, and have the ability to recover from issues quickly. Flexibility in context provides IIoT refers to a guarantee or ecosystem because everything continues to work the same in any case. Individual components of the network fail at times. Due to their extreme diversity, IIoT networks are challenging to build resiliently. IIoT systems are often open in a variety of contexts including city centers, rural areas, and industrial plants. Each system has its own operational requirements and communication format. Therefore, flexibility is guaranteed. Hence, it is necessary to use a variety of strategies including automation or actual fail over Monitoring time and redundancy to reduce and minimize downtime is important. Impact of Network Resilience Disruption can mean the difference between continuity of operations without interruption and the resulting high costs in sectors such as the transformation or electricity sector, where IIoT systems manage critical infrastructure.

1.2 Network Reliability

Network reliability guarantees that these systems work properly. Plan all the time without suffering. Trust in the context of IIoT means the consistent performance of both the hardware and software components. No real-time data processing meeting. As IIoT devices are frequently used in challenging industrial environments, a more reliable network is thus necessary. They are essential to operational security, efficiency, and economy. IIoT distribution characterizes a system in which the equipment operates under various conditions, and produces many types of dice. It is one of two systems. The main obstacles to ensuring reliability is because of this complexity. Therefore, strict fault tolerance techniques are available. Proactive inspection, predictive maintenance, and strong system projects through identifying possible failure locations before system failure are necessary so that various companies can improve the reliability of IIoT networks and reduce downtime, increase efficiency, and guarantee safety. A trusted network helps build a solid foundation for future innovation, helping companies scale up their IIoT operations with confidence and low risk data governance and reactive analytics in IIoT. Because there are two sides of the same coin, it is therefore imperative to ensure that the use of both dice is effective and ethical. Industrial Environment is Responsive

Analytics and Data Governance in IIoT to dice governance related to administration security and editing data. Responsible analysis is guaranteed. Conclusions from these data are drawn in a fair, open, and transparent manner. IIoT context, where data is used to make critical decisions related to operations, safety, and real-time production and flow performance from various sources, the intersection of these two concepts is special. Significantly the reliability, security, and consistency of this data are guaranteed with legal standards. The dice governance system also makes ethical and unbiased decisions based on the information received. These controlled data are warranted by further analysis. Data governance and responsive analytics work together to promote innovation and reduce risk in IIoT, more automated and dice-focused platforms. Enterprises that effectively establish robust data governance and conscientious analytics methodologies will have an advantage in realizing the complete potential of the IIoT, enhancing operational efficacy and upholding regulatory compliance.

There are five basic skills to thrive in the digital business environment which includes: analytics, business, technology, architecture, and applications. All these are connected and aligned with the digital business strategy. These skills are extremely important. In data-intensive industries, such as IIoT, these companies are leveraging technology architecture and data analytics. They make informed decisions and stay competitive. Understanding and training them is important to optimize data governance structures and ensure more responsive analytics practices in IIoT environments.

Analytics: You don't know any other context as a result of IIoT. Analytics is essential to gain insights from large amounts of data. Powered by sensors and machines to help organizations predict equipment failures, productivity, and identify trends, successful analysis can guarantee data-driven decision-making, based on appropriate regulatory frameworks to guarantee the quality and accessibility of both data.

Business: The complexity of running a business is crucial to the leverage of IIoT. Emerging technologies can improve processes and reduce costs. With careful application it is ensured that it increases and does not interfere with the main purpose with careful data governance and insights-driven protocols, companies will explore new value drivers and the path to expanding profits at the same time maintaining the trust of customers. Progress requires constant re-evaluation. How can we adapt our methods to the changing landscape and at the same time remain true to our founding principles? When managed judiciously, Industry 4.0 opens the door to prosperity. And it reveals itself to be a means, not an end.

Technology: IIoT systems are based on a variety of technologies. From cloud computing to edge computing and IoT devices ensuring the availability, security, and integration of these technologies is critical to IIoT success. Data governance helps control how data is collected, stored, and processed, and maintain system security technology infrastructure aligned with business objectives.

Architecture: System architecture plays an important role in structuring IIoT systems. The architecture allows for scalability, data integrity, and soft data flow across devices, e.g., Network Adequate data governance guarantees that all architectural components are compliant to the maximum extent possible.

Applications: IIoT applications such as predictive management, intelligent manufacturing, and the chain of repression management relies on real-time data and analytics to function optimally. You can guarantee that those apps are built on a controlled database; it is extremely important to provide more reliable and accurate insights that can drive.

1.3 The Flexibility and Reliability of the Two IIoT systems is Very Important

1. Accelerate Network Performance and Stability

The most important benefit of ensuring resilience and reliability in IIoT networks is the transformation of network performance to a faster and more stable performance. This happens because IIoT systems deal with huge amounts of real-time data from various sensors and devices. Ensuring that these networks remain robust and responsive is the key to continued operation of a more reliable network. This minimizes the chances of communication failures. System breakdown and data loss are ideas for real-time decision- making and efficient operations. This stability ensures that there is uniform performance across industrial processes.

2. Improved Network Safety and Security

As the IIoT-enabled devices increase in numbers, proper security of the network becomes necessary to save against cyber attack. The flexible and reliable architecture of the network provides multilayer protection. This ensures the confidentiality of information from unauthorized access, breach, and adulteration. This, thereby, helps protect critical industrial systems. By developing more reliable networks, organizations can minimize their exposure to vulnerability and thus keep their IIoT environment secure.

3. Regulatory Compliance and Risk Management

For sectors that operate under strict regulations, such as energy, manufacturing, or healthcare, a flexible and reliable IIoT network is required to meet regulatory standards. A robust network architecture helps organizations comply with data privacy, security, and operational security regulations. This reduces the risk of being penalized for non-compliance. In addition, to proactively manage risks that may occur.

4. Better Operational Efficiency and Cost Reduction

A more flexible and reliable IIoT network guarantees operational efficiency. It involves reducing and improving the accuracy of real-time data transmission idle time. These networks ensure that your devices are always connected. These systems are faster and have a more informed decision-making process including reliable communication between each other. These machines can improve equipment inspection and production and accelerate resource management. Consequently, reduced errors, fewer delays, and lower maintenance costs all contribute to significant cost savings and improved operational efficiency.

5. Transparency and Responsibility in Network Management

A robust and reliable IIoT network promotes transparency and accountability in delivery, or burial and use in industrial environments can be guaranteed to be correct or accurate. Avoid interruptions, but organizations can clearly monitor and record network performance. Transparency makes operational processes traceable and nothing happens in terms of interference or errors. Validation results in greater responsibility and trust in all stakeholders involved.

6. Innovation and Competitive Advantage

More flexible and reliable IIoT networks are the base for innovation. Helping multiple setters can be explored. New technologies and applications are guaranteed by a more reliable network that allows for perfect integration of advanced technologies such as machine learning, predictive analytics, and autonomous steering to develop innovative solutions that provide a competitive advantage. Flexible networks allow remote locations to experiment and expand new processes without having to work or risk frequent system failures. These companies are leaders in a rapidly evolving market.

7. Informed Decision-Making and Predictive Maintenance

In an IIoT environment, timely and accurate decision-making is critical to maintaining operational continuity. A reliable network that guarantees real-time dice flow, helping companies have information to make decisions quickly. This is because flexible networks also support predictive management strategies. They provide consistent, up-to-date information that is needed to predict equipment failures or operational problems. This reduction in booking time was not planned. It reduces maintenance costs and extends the shelf life of industrial assets. That is something that has been improved around the world.

8. Clear Ownership and Responsibility of the Network

Flexible and reliable IIoT networks provide a clear path for network ownership which guarantees that stakeholders understand their responsibilities in the maintenance and operation of these systems. Clarity promotes teamwork, reduces confusion related to network management, and prevents its misuse or increased management of the network resources. Moreover, reliable networks will help enterprises survive. Responsibility for the operation of the system to ensure that performance goals are achieved with compliance requirements with confidence.

2. Challenges in Ensuring Resilience and Reliability in IIoT Networks

This is because IIoT networks must be flexible and reliable to succeed. These goals are therefore difficult to achieve. IIoT environments are complex due to the large number of interconnected systems. This makes it difficult to maintain network scalability and performance. As the IIoT ecosystem matures, the companies face challenges related to operational, financial, ethical and technical constraints. Here are the main obstacles facing companies. Must be overcome to guarantee robustness and reliability in IIoT networks:

2.1 Managing High Data Volume, Velocity, and Variety

A key issue is the sheer volume, speed, and variety of data that IIoT devices handle: large amounts of real-time data coming from multiple devices. Each device has different types of data. It must be processed by the network. If not treated properly, the two "3V" challenges include volume, speed, and variety. It may cause the traditional grid infrastructure to fail, being overworked. A variety of information sources including industrial control systems, machines, and sensors, make it difficult to create well-formulated protocols, or cause interoperability issues and additional complexity. A robust network architecture that can withstand high levels of traffic but at the same time, it is important to maintain flexibility and reliability. This is because inefficient management of these data flows can result in poor performance, interruption in operations, or even a collateral breach.

2.2 Ensuring Network Security in Distributed Systems

This is because IIoT networks often cover diverse geographic areas and depend on devices with different security requirements. Due to the large attack surface this distribution creates, IIoT networks are highly vulnerable to cyber attacks, such as malware breaches: Distributed Denial of Service (DDoS) attacks and unauthorized data access. As networks become larger and more complex, securing the network without compromising speed and scalability has become increasingly difficult. Ensuring the integrity of the dice and protecting connections between devices represents additional challenges. Building a strong network that is end-to-end secure is important. This is because every new IIoT device added to a network is a potential entry point for cyber attacks. Organizations must use cutting-edge solutions such as blockchain-based security, Zero Trust Architecture, and IA-Based Cybersecurity where traditional cybersecurity measures are often not enough.

2.3 Maintaining Data Integrity and Privacy

Massive amounts of operational and confidential data are collected and transmitted over IIoT networks. It is extremely difficult to guarantee that this information is accurate, secure, during or in transit. This is especially true when dealing with mission-critical rates that need to be completed in real time. Violations of data integrity can result in incorrect decisions, device failures, or disruptions. In addition, IIoT systems must ensure compliance with local privacy laws, such as CCPA in California or GDPR in Europe due to increasing concerns about data privacy. The challenge of protecting multiple legal systems and controlling cross-border data flows adds another level of complexity to the task of ensuring network stability. IIoT network projects are becoming more complex with privacy requirements. This requires strict data encryption, deleting personal data and how to use information ethically.

2.4 Complexity of Network Ownership and Responsibility

End users, system integrators, and equipment manufacturers are among the stakeholders involved in maintaining network performance in the IIoT ecosystem. This frequently results in a disjointed environment where ownership and accountability are unclear. To blame a fixed network performance can be challenging. This delays response times and increases downtime. Conflict can occur in complex IIoT situations where multiple parties may have access to living systems. If their duties and responsibilities are not clearly defined, creating a precise governance structure with differentiated allocations and responsibilities is essential for flexibility and reliability. This reduces the chance of system downtime for long periods of time. This ensures that any performance or security issues are resolved quickly and efficiently [2].

2.5 Enhanced Security and Protection for Networks

Number of sectors continue to rely on old systems that are not compatible with contemporary IIoT technologies. These older technologies usually lack connectivity and are also unable to exchange real-time data. It can be complex and costly to upgrade them to connect with new services or meet current cybersecurity standards. Making these legacy systems work with IIoT networks to make sure that they can communicate effectively is a very complex task. Sometimes a custom interface or middleware is required. This happens because those technologies tend to fail with frequency. However, traditional technologies can reduce overall network reliability. It is very difficult to upgrade existing systems to meet current IIoT demands without disrupting daily operations. In sectors with tight margins and low tolerance for error this is true [9].

2.6 Adapting Network Resilience Strategies for Growth

Companies want to ensure their approaches can be modified as more IIoT systems are brought. The fast growth of linked devices might weigh down older community setups, leading to problems like network disasters, delays, and overcrowding. Adding new devices to the community will increase the complexity which requires trust models, governance structure, and extra superior security measures and measuring reliability techniques across the developing wide variety of gadgets in fairly fragmented sectors inclusive of strength, logistics, and agriculture is a severe task [3].

2.7 Real-Time Data Processing and Reliability Management

In IIoT situations, delays in data processing or transmission can have serious consequences. Real-time data processing is important. In the context of smart manufacturing, for example, equipment failures or security risks can result in delays in receiving real-time data from two sensors due to the complex infrastructure, such as fast data storage, real-time analysis framework, and cutting-edge processing solutions. It is necessary to analyze data in real time and ensure network reliability. Another issue is managing security, efficiency, and data quality in real time. Such networks must be built with the ability to handle large amounts of data in a rapid fashion. while maintaining consistency and accuracy at the same time. Building a flexible IIoT network is a challenging task. This is due to the difficulty of tuning real-time data processing with reliable measures such as fault tolerance and redundancy [5].

2.8 Collaboration Problems

A key obstacle in IIoT networks is ensuring that devices produced by different companies are compatible, like devices and sensors with various protocols and data formats. They are often included in IIoT networks because of serious interoperability problems caused by a lack of mapping. It is therefore a challenge for devices to interact and function smoothly on the network. Network performance issues, lag, or even sputtering can be caused by compatibility issues. This is especially important in sectors such as healthcare, where continuous operations are critical and IIoT devices need to work together to track patients in real time. The rapid advancement of technology has made it difficult for everyone to create rules and procedures that allow for quiet communication [6].

2.9 Cost and Resource Limitations

Building a reliable and robust IIoT network requires a lot of investment for businesses and skilled workers. Many businesses, especially small and medium enterprises (SMEs) find it extremely important to implement modern scalable and secure communication systems. Websites should also be invested in innovation to ensure they are updated and current. There is a shortage of qualified and knowledgeable experts in resilient IIoT networks, and cybersecurity remains another critical challenge. For many organizations, finding the right resources is a challenge. To design, deploy, and manage flexible IIoT networks is often lacking in basic communication skills. Credibility can be undermined by operational weaknesses and risks arising from human resource scarcity. Building reliable and robust IIoT networks requires a lot of investment for businesses and skilled workers. Many businesses, especially SMEs find it extremely important to implement modern, scalable, and secure communication systems. Websites should also be invested in innovation to ensure they are updated and current regarding Technologies and Emerging Threats. There is a shortage of qualified and knowledgeable experts in resilient IIoT networks and cybersecurity is another issue. For many organizations, finding the right resources is a challenge. To design, deploy, and manage flexible IIoT networks, often lacks basic communication skills. Credibility can be undermined by operational weaknesses and risks arising from human resource scarcity [4].

2.10 Using AI and Automation in an Ethical and Transparent Manner

It's very crucial to use AI and automation in a way that is fair and understandable. This is especially true because IIoT networks rely on these technologies to manage operations. Many AI systems act like "black boxes", meaning it's difficult to know how they make decisions or why they do certain things. In industrial environments, these automated systems need to be clear and easy to explain, as they can affect productivity, safety, and efficiency. A big challenge in IIoT networks is ensuring that AI systems follow ethical guidelines, avoid bias, and are easy to understand. Companies must find a balance between the advantages of automation and the need to keep operations trustworthy and transparent, and to ensure that AI decisions are fair, accurate, and ethical, strong rules, accountability, and regular checks are essential [10].

3. Literature Survey

This research aims to identify upcoming challenges in keeping IIoT systems robust and durable, as well as explore solutions, best practices, and the potential of new technologies. To overcome these obstacles from the in-depth discussion, the participants gained a lot of knowledge about the flexibility of IIoT networks. They highlighted common barriers such as technical issues.

Security vulnerabilities: Challenges in managing real-time data and the struggle to integrate legacy, outdated systems with modern infrastructure. This summary provides comprehensive information about the fun. The focus is on new trends, important challenges, and actionable strategies to overcome resistance and availability of IIoT networks. It serves as a guide for you to overcome obstacles and leverage advances to build systems that are not only robust, but also ready to support the future [1].

4. Survey Responses

In the early stages of research we conducted a survey that revealed the occupations of 19 interviewees, with 100% indicating their occupation as students.

This uniform group indicates that most research is respectful of individuals who continue their studies more than those working as industry experts, academic researchers, or government officials. As all come from the same background, the answers therefore mainly reflect the views of the students. It is based on what we have learned or experienced during internships or projects. The lack of diversity in the workforce means that the insights provided in research responses reflect the views of particular students. This is often linked to an early academic or career understanding of the research topic [7].

Question of survey: How familiar are you with the concept of data governance?

During the data governance familiarization session, it was confirmed that most of the interviewees had a moderate level of opinion and understanding of the concept. Notably, 42.1% described themselves as neutral. This shows that they do not have good knowledge or familiarity with dice management, with only 36.8% confirming that some do. Familiar or enough to say that they have basic knowledge about the subject, but he is not a concession expert. Only 10.5% know; this reflects the depth of their knowledge, and 5.3% claim not to know them, with a maximum of two participants, respecting the two-dice rule. This is completely understandable and almost no one misunderstands this concept.

Question of survey: How important do you believe responsible data analytics is in today's digital age?

The majority of two interviewees (47.4%) endorsed responsible data analysis as very important, with 42.1% perceived it of great importance; only one small plot (5.3%) is classified as medium bearing importance. The graphs provide valuable information about perceptions of responsible governance received and provided. Analysis among students indicate a high level of familiarity with dice supervision and fortress understanding the importance of responsive data analysis.

Question of survey: Which of the following data management practices do you consider most critical in IIoT environments?

The majority (88.5%) of the interviewees endorsed with a minority (7.7%) disagreeing.

Question of survey: What are the biggest challenges your organization (or you) face in implementing data governance?

This section presents the findings, asking which data management practices are considered most important in an IIoT environment. Data security is the top priority by 80.8% of interviewees, followed by data integrity (73.1%). Data availability (61.5%) comes in third, with data privacy (57.7%) and data quality (50%) trailing behind. Data compliance is considered the least critical (19.2%).

Question of survey: On a scale of 1–5, how challenging do you find the ethical considerations in data analytics?

This section is a bar chart showing the biggest challenges organizations face in implementing data governance. The biggest challenge in implementing data governance is lack of awareness followed by insufficient resources, complexity of IIoT systems, data volume and variety, and lastly regulatory compliance.

Question of survey: Do you think responsible data analytics should be prioritized over quick decision-making in IIoT?

This section is a bar chart showing how challenging people find the ethical considerations in data analytics on a scale of 1–5. The majority of the respondents (50%) find the ethical considerations in data analytics to be a challenge on a level 4 out of 5, followed by 26.9% on a level 3, and 15.4% on a level 5. Only 3.8% of the respondents found this challenge to be on a level 1 or 2.

Question of survey: Which ethical considerations do you think are most important in data analytics?

The results showed that 92.3% of the two interviewees answered ' YES', while 7.7% answered "NOT SURE", and no one answered 'NO'.

The second graph is a horizontal bar graph that shows your findings. It asks which ethical considerations are most important in dice analysis.

The research asks that interviewees be selected based on three considerations.

The results show that the most important considerations are:

- Privacy Protection (73.1%)
- Transparency (69.2%)
- Accountability (61.5%)
- Fairness (57.7%)
- Bias Prevention (34.6%)

5. Conclusion

In conclusion, for IIoT networks to work well and last long, they need to be strong and stable. There's a growing need for secure and reliable systems that can handle problems without failing. As IIoT systems grow bigger, more complex, and handle more data, it's important to build networks that can adapt and stay flexible. These networks must also be able to resist cyberattacks. Since needs and processes change quickly, it's crucial to focus on backup systems and predictive maintenance to keep everything running smoothly and encourage innovation. Many organizations are aware of the challenges in making IIoT networks stronger. Because of this, more research and practical solutions are needed to close the gap between what's needed and what's available [8].

References

[1] Jeschke, S., Brecher, C., Song, H., et al. (2020). *Industrial Internet of Things: Cyber Manufacturing Systems*. Springer Series in Wireless Technology.

[2] Sarmento, L.F.G. and Melodia, T. (2020). Resilient Network Design for the Industrial Internet of Things. *IEEE Transactions on Network and Service Management*, *17*(1), 145–159. doi:10.1109/TNSM.2020.2979846.

[3] Gringoli, F. and Bononi, L. (2022). Reliability and Resilience in Industrial Internet of Things: Challenges and Solutions. *ACM Computing Surveys*, *54*(4), 1–35. doi:10.1145/3455415.

[4] Zhang, Y. and Wu, H., et al. (2021). *Handbook of Computer Networks and Communications* (Volume 3). Networking Technologies and Applications. Springer.

[5] Xie, L., Lee, G. M. and Kim, H. J. (2020). A Survey on Industrial Internet of Things: Architecture, Security and Privacy, and Applications. *IEEE Access*, *8*, 126–140. doi:10.1109/ACCESS.2019.2955848.

[6] Bandyopadhyay, S. and Sen, J. (2011). Internet of Things: Applications and Challenges in Technology and Standardization. *Future Generation Computer Systems*, *29*(7), 1244–1252. doi:10.1016/j.future.2013.01.010.

[7] Atzori, L., Iera, A. and Morabito, G. (2010). The Internet of Things: A Survey. *Computer Networks*, *54*(15), 2787–2805. doi:10.1016/j.comnet.2010.05.010.

[8] Miorandi, D., Sicari, S., Pellegrini, F. and Chlamtac, I. (2012). Internet of Things: Vision, Applications and Research Challenges. *Ad Hoc Networks*, *10*(7), 1497–1516. doi:10.1016/j.adhoc.2012.02.016.

[9] Bertino, E. and Islam, N. (2017). Botnets and Internet of Things Security. *Computer*, *50*(2), 76–79. doi:10.1109/MC.2017.62.

[10] Yang, Y. and Wu, Y. (2017). A Survey on Security and Privacy Issues in Internet of Things. *IEEE Internet of Things Journal*, *4*(5), 1250–1258. doi:10.1109/JIOT.2017.2701238.

10 Collaborative Innovation for Responsible IIoT

Hybrid Approach of Seed Spreading and Weed Plucking for Enhanced Crop Yields using Machine Learning

Noopur Divekar[1*], Disha Garje[2], Anjali Priya[3] and Rupinder Kaur[4]

This review study bridges a knowledge gap in agricultural systems by describing an integrated strategy to seeding and weeding at the same time. We present a novel hybrid method based on a number of research publications that combines robotic technology for precise weed plucking and seeding with machine learning for weed detection. Powered by solar energy, the self-sufficient system includes a container for constant seed sowing and a blower for optimal seed distribution. In analysing multipurpose seed-sowing devices, the study emphasises the importance of adaptability and the possibility for an integrated agricultural management system to boost farm productivity and sustainability.

1. Introduction[1]

A significant portion of the recent changes in agricultural operations can be linked to the increased need for sustainable and efficient farming methods. Even while technology improvements have enabled several devices for weed control and seed sowing, there is still a significant paucity of solutions for integrating these two critical jobs. This gap study, which included a thorough evaluation of multiple research articles on automated seed spreading systems and weed detection, identified a large research need in the sector. Thus, the fundamental goal of this review paper is to offer a novel hybrid technique that combines the advantages of weed eradication and seed planting through the use of robotics and machine learning. The proposed method entails creating an autonomous agricultural robot

[1-3] Department of Electronics Engineering, Thakur College of Engineering and Technology, Mumbai, India.

[4] Faculty of Electronics Engineering, Thakur College of Engineering and Technology, Mumbai, India.
Email: rupider.kaur@tcetmumbai.in

* Corresponding author: noopur.div188@gmail.com

DOI: 10.1201/9781003587903-10

that uses cutting-edge machine learning techniques for weed recognition and robotic arms for precision weed removal. To ensure regular seed planting, the gadget includes a container with a moving motor and a blower for efficient seed dispersal. The system employs solar power and sophisticated printed circuit board (PCB) designs to achieve maximum energy efficiency while focussing on sustainability. Using robotics and machine intelligence in agriculture could transform farming processes, raise crop yields, and improve overall farm management. To fill these research gaps, this review paper conducts a thorough examination of existing weed detection and seed-sowing procedures before proposing a fresh, holistic strategy for agricultural automation. This study intends to demonstrate the potential of the proposed comprehensive agricultural management system, with the objective of encouraging the development of more adaptive and sustainable weed control and seed-sowing equipment evaluation methods in modern agriculture.

2. Background and Motivation[2]

Precision seeding techniques and effective weed management are critical for long-term farming operations with high yields. Nonetheless, a fundamental impediment has been the lack of complete, integrated solutions that address both processes concurrently. Conventional farming methods are known to cause environmental difficulties, increased labour expenses, and inefficiency.

Using cutting-edge technology such as robotics and machine intelligence provides a tremendous chance to revolutionise traditional agricultural operations, particularly as the world's population grows and the significance of sustainable farming methods increases. Furthermore, the need to mitigate environmental damage highlights the importance of ecologically sustainable farming practices.

Using robotics, machine intelligence, and renewable energy sources, this review paper proposes a complete and integrated strategy for weed management and seed-sowing operations. The goals include increasing agricultural output, cutting labour costs, and promoting environmentally friendly farming techniques. This review study uses robotics, machine learning, and renewable energy sources to present a comprehensive and integrated strategy for weed control and seed-sowing activities. The goals include increasing agricultural output, cutting labour costs, and promoting environmentally friendly farming techniques.

3. Literature[3]

Gabor wavelet transforms, Support Vector Machines (SVMs), and convolutional neural networks (CNNs) have all been used successfully in the field of sickness detection. These methods have achieved high accuracy rates in plant disease diagnosis, frequently exceeding 90%. There have even been situations where the accuracy exceeded 99%. Among the limits we observed are the necessity for big datasets, sensitivity to segmentation quality, and computational resource requirements. We used neural networks, CMOS (complementary metal-oxide semiconductor) digital cameras, and a variety of image processing algorithms to detect weeds. We have demonstrated precision rates of up to 99% in weed detection. Some recommended development paths include the use of advanced image processing techniques and algorithm parallelisation.

Exploration of the remaining areas of ignorance Assessment of Multi-Purpose Seed-Sowing Devices: While some progress has been made in the development of image processing algorithms for evaluating seed-sowing devices, more adaptable approaches that are appropriate for a wide range of seeders, including drill, hill, and precise seeding, are required. Modern approaches frequently focus on specific types of seed-sowing apparatus.

Sturdiness and Practical Assessment: Many of our existing methodologies have been tested in controlled settings or with little real-world data. More research should be done to determine the efficacy of these strategies in various agricultural situations, such as soil types, lighting conditions, and plant species [2].

Integration of Technologies: Combining seed-sowing equipment evaluation, disease diagnosis, and weed detection into a single, automated agricultural system could result in tremendous progress. Although little study has been conducted in this area, the development of integrated systems has the potential to improve farm management overall.

This communication describes an autonomous seed planting robot that is durable, environmentally benign, and energy efficient. Aluminium is employed in its construction, and caterpillar treads provide mobility. The proposed system has its own mobility systems and is completely autonomous. The solenoid actuator assembly is part of the seed metering system. This effectively regulates the flow of seeds into the grooves formed by the accompanying tillers. Furthermore, unlike current systems that rely on carbon-based fuels, the robot's power source is rechargeable lead-acid batteries. Our work fills these research gaps, allowing for more adaptable and practical solutions for weed management and seed-sowing equipment evaluation. Crop yields and agricultural production could be significantly boosted by examining how to incorporate these technologies into a holistic farm management system.

4. Scope[4]

Our study focus es on two areas of agricultural technology: weed management and the performance of seed-sowing equipment. Our inquiry includes the following topics:

Weed Control: We will look into cutting-edge techniques for quickly finding and removing weeds from agricultural fields. Our research will combine neural networks, complicated and cutting-edge algorithms, and image processing methodologies to achieve precise and long-term weed control results [30]. The goals are to reduce the use of herbicides and increase agricultural output.

The following hardware and software were required for the experimental investigation into image processing techniques for evaluating the performance of a seed-sowing system. An industrial computer has a Core 2 E4300 CPU, a 1 GB memory stick, and a 1.8 GHz primary frequency. The Basler A602fc is a high-speed industrial colour digital camera with a maximum resolving power of 651 by 496 pixels. Its maximum frame rate is 100 frames per second slower than its maximum resolving power [18].

It has a 12 mm focus distance, an F1.4 aperture, a Computer M1214-MP lens, an IEEE1394 output interface, and can produce images in both colour and greyscale. Two rows of eight-watt LED lights are uniformly distributed across two rectangular aluminium boards to provide lighting. The program development tool is Microsoft Visual C++ 6.0. With the assistance of Beijing Xiandai Fubo Technology Ltd.'s MIAS, an image processing software development tool for the 2-dimensional dynamic measure analysis system. The study course has ended. When the belt in the testbed obtains an image for seed-sowing detection, it does so at a pace of 0.2 meters per second. The unusually high amounts of precipitation in January and the first half of February may have influenced ant activity and seed removal rates. There were most likely more naturally existing weed seeds in January, which would have an impact on how well seed removal performed overall that month (Lobry De Bruyn, 1993) [31].

5. Objectives Weed Control

Investigate neural network-based and image processing strategies for weed identification in crops.

- To reduce false positives and negatives, boost detection precision, and accuracy. Look into parallelisation and effective approaches for increasing processing speed and real-time applicability.
- Assessment of Seed-sowing Instruments: Create image processing algorithms capable of recognising and evaluating seeding equipment in a variety of agricultural settings.
- Recognise camera coverage constraints and adapt algorithms to accommodate various seeding approaches.
- Examine the applicability of these algorithms in real-world agricultural contexts [4]. Investigate the feasibility of merging weed management and seeding equipment assessment into a single farm management system.

Consider the potential for automated crop management and fast decision-making.

- Data and Resource Accessibility: Develop and share standardised datasets in the fields of weed control and seed-sowing device evaluation to facilitate future research in these areas. Consider how to increase farmers' access to these technologies while keeping costs and processing power in mind.

In October 2001, bare ground was more prevalent in non-planted seed regions than in planted seed areas [12]. However, in October 2004, things improved, leaving less bare soil in areas where seeds had not been sowed [13]. In July and October 2002 and 2004, the percentage of bare ground was higher when cuttings remained after both cuts [3]. In areas where no seeds were sowed and both cuttings were removed, the amount of barren ground gradually decreased. Determines the fraction (I%) of unsown annual and perennial anemochorous species between 2001 and 2004 [14]. From July 2002 to October 2003, the proportion of annual unsown anemochorous species was higher in areas without seed planting [15]. However, by 2004, the values had changed little if at all. The fraction of annual unsown anemochorous species decreased throughout time, regardless of plant community or extent of herbage removal [16].

Plant community and herbage removal had no significant effect on the percentage of annual unsown anemochorous species three years after planting. The percentage of yearly unsown anemochorous species in July 2002 and the amount of bare ground in October 2001 had a positive and significant correlation with 0.58 ($P = 0.02$) [17]. By pursuing these goals, we hope to help resolve some of the most pressing difficulties plaguing modern agriculture while also increasing farm productivity, reducing resource use, and improving crop output.

6. Methodology [5]

Let's take a closer look at the machine learning-based hybrid technique for weed management and seed distribution to boost crop yields.

Collecting and Preparing Data: First, create a thorough data collection system. Throughout the growing season, use satellite imagery, drones, or stationary cameras to acquire high-quality photographs of the agricultural field. Learn more about the temperature, moisture content, soil composition, and historical crop and weed growth patterns. Preprocessing was performed on this data to eliminate outliers and balance factors.

Image Labelling and Annotation: Annotate the collected photos to highlight areas of interest, such as weed species and agricultural crops. This stage entails identifying each object in the images in order to provide a training set for machine learning models. To facilitate model training, ensure that the labelling is precise and consistent.

Training Machine Learning Models: Use the labelled dataset to train deep learning models such as CNNs. Create distinct models for categorising weeds (several species) and recognising objects (both crops and weeds). Make regular modifications to these models to improve their accuracy.

Real-time Data Integration: Develop an integrated system capable of collecting data in real time from field-deployed sensors. This includes weather data, soil moisture readings, and environmental parameters. Incorporating this data into machine learning models will help you make better decisions [7].

Weed Classification and Identification: Instal cameras or other sensors capable of capturing real-time images in the field. Implement proven object detection and weed categorisation models for real-time processing on edge devices. These models will recognise and categorise weeds as they appear in the field.

Check Crop Health: Use machine learning models to track crop health indicators over time using images. These models should evaluate leaf size, colour, and any indications of illness or stress. This knowledge is critical for early intervention.

Development of Decision Support Systems (DSS): Combine the results of weed detection, weed classification, and crop health monitoring to create a DSS. This system should use historical data, machine learning predictions, and real-time data to decide when to pull weeds and spread seeds [6].

Seed Distribution Optimisation: Use the DSS to optimise seed dispersion. Determine the optimal sowing rate for each section of the field based on crop shortfalls and soil conditions. To ensure optimal seed dispersion, program the seed-spreading machinery to target certain spots.

A Robotic Weed Pulling equipment: Create a machine learning-powered robotic weed-picking equipment. The system should identify weeds using computer vision, and robotic arms or other tools should be used to 10.

Real-time Control and Observation: Implement a control system that receives continuous feedback from sensors, cameras, and the DSS. Based on changing field conditions, this system optimises sowing and weeding efforts in real time.

Loop for Evaluation and Feedback: Compare crop yields, weed control efficacy, and resource utilisation to traditional farming methods on a regular basis to assess the system's performance. Use this input to make changes to machine learning models, algorithms, and control settings as they evolve.

Integrate and Scale: Investigate how the hybrid technique will adapt to various crop varieties and farming practices. Check the compatibility and integration of many agricultural systems, including various types of planting and harvesting equipment.

Ethical and Environmental Considerations: Monitor the hybrid strategy's impact on the environment, notably in terms of reduced pesticide use and better soil health. Address moral issues about data privacy, intellectual property rights, and the ethical use of AI in agriculture.

Knowledge and Training Transfer: Provide farmers with educational materials, seminars, and training sessions to help them understand this hybrid strategy. Collaborate with local agricultural extension services and farmer cooperatives to encourage adoption and offer technical assistance. This comprehensive methodology addresses all aspects of the project lifecycle, from data collection and model training to real-time deployment, assessment, and knowledge transfer. It focusses on integrating robots and machine learning technologies into agricultural processes to improve crop yields and sustainability while reducing weed impact [5].

7. Discussion and Analysis[6]

In the coming years, there will be a number of significant advancements in hybrid weeding and seeding strategy research for improved agricultural fields. These emerging patterns will provide useful information on the progress of agricultural technology and sustainability.

Integration of Machine Learning: The use of machine learning approaches in agriculture will only grow. Researchers researching a variety of machine learning methods, including support vector machines, neural networks, and decision trees, will improve the hybrid technology, which will eventually transform agricultural field management [19].

Precision Agriculture will continue at the forefront of R&D (research and development) efforts. Scientists and farmers alike will invest in innovative methods and cutting-edge technology that enable targeted weed control and precise sowing [20].

Sustainable Practices: Ecological farming techniques will grow more widespread. Researchers will work tirelessly to identify sustainable weeding and seeding methods that will preserve natural resources, use less chemicals, and improve ecological equilibrium.

IoT and Sensor Technologies: The Internet of Things (IoT) and sensor technologies will be increasingly integrated into agriculture in the future. Real-time monitoring and data collection enabled by these improvements will lead to better decision-making and more efficient use of available resources.

Automation and Robotics will play an increasingly important role in agricultural activities. The use of drones and self-driving automobiles will make weed management and seed sowing more productive and cost-effective in farming operations [21].

Data-Driven Insights: Big data and data analytics will have a significant impact on the development of future weeding and seeding tactics. Researchers will use enormous datasets to gain valuable insights and continuously improve the hybrid technique.

Crops Resistant to Climate Change: Priority will be placed on producing agricultural varieties that can endure climate change. Researchers will focus on breeding and genetic manipulation to develop crops that can survive in shifting environments.

Cooperation and Expertise Sharing: Agriculture groups will increasingly collaborate to trade best practices and experience. This global initiative will speed the use of more effective weeding and sowing practices [22]. Crop rotation will become more important in maintaining soil health and managing weeds. Long-term success requires the use of sustainable crop rotation systems.

Organic Farming: For chemical-free, sustainable crop production, there will be a growing interest in organic farming practices that follow sowing and weeding rules.

Conservation of Biodiversity: To support healthier ecosystems and more resilient farming systems, researchers will look into ways to boost biodiversity in and around crop fields [23].

Government Incentives: Governments will pass laws and provide incentives to encourage the use of novel weeding and sowing technology that promotes enhanced food security and sustainable agriculture.

Global Collaboration: Addressing the world's food challenges would necessitate international cooperation and knowledge exchange, with a particular emphasis on developing and implementing cutting-edge weeding and planting techniques.

Enhanced Productivity and Environmental Sustainability: Ultimately, a balance between higher crop productivity and improved environmental sustainability will be attained as long as we continue to refine the hybrid weeding and seeding technique for the benefit of both agriculture and the environment.

Weed Seed Viability: The effect of ensiling and/or rumen digestion on weed seed viability varied according to species.

The impact on grass species was greater than that of broadleaf species. According to research, grass species, particularly those in the Leguminosae family, lose more viability when passed by cattle than broadleaves. The degree of hardheadedness is associated with seed survival via animal digestion; broadleaf weed species frequently have tougher seed coats than grass weed species [8].

Particular Weed Reactions: Rumen, ensiling, and ensiling plus rumen treatments eradicated all downy brome, foxtail barley, and barnyard grass seeds. It has also been reported that cheat seeds, which are closely related to downy brome, do not germinate after passing through cattle [24].

Ensiling and ensiling + rumen treatments destroyed green foxtail seeds, while 17% survived the rumen treatment. The seeds of green foxtail were shown to be more resistant to passage through cattle than the seeds of several other grass species [29]. All three treatments reduced the viability and germination of kochia and flixweed seeds, although they were not totally eliminated. For flixweed and kochia species, the viability loss caused by rumen, ensiling, and ensiling + rumen treatments was similar.

8. Comparative Analysis of Approaches and Solutions[7]

When comparing different ways or solutions for the hybrid strategy of weed control and seed dissemination, consider the following factors:

- **Efficiency and Accuracy:** Determine the effectiveness of each method for weeding and spreading seeds. Compare their efficiency to the time and resources required to achieve the desired objectives [10].
- **Cost-effectiveness:** Determine how each strategy will impact your finances. This covers both the initial investment and ongoing maintenance and operating expenses.
- **Scalability:** Consider whether the technique can be made larger or smaller to handle fields of different sizes. Scalability is crucial in many agricultural contexts to ensure optimal application.
- **Consider the Environmental Impact of Each Practice:** Evaluate each strategy's environmental impact by taking into account pesticide use, soil disturbance, and water use [24].
- **Dependability and Robustness:** Evaluate the dependability and robustness of each solution. How well do they function under various weather and field conditions?
- **Ease of Integration:** Consider how easy it is to combine each plan with current tools and farming procedures. There are various benefits to minimizing disruptions to existing workflows.
- **Consider the Data Needs for Each Solution:** Certain approaches, such as soil quality, may be substantially reliant on data inputs, whilst other systems may require far fewer data-intensive inputs.

9. Flowcharts [8]

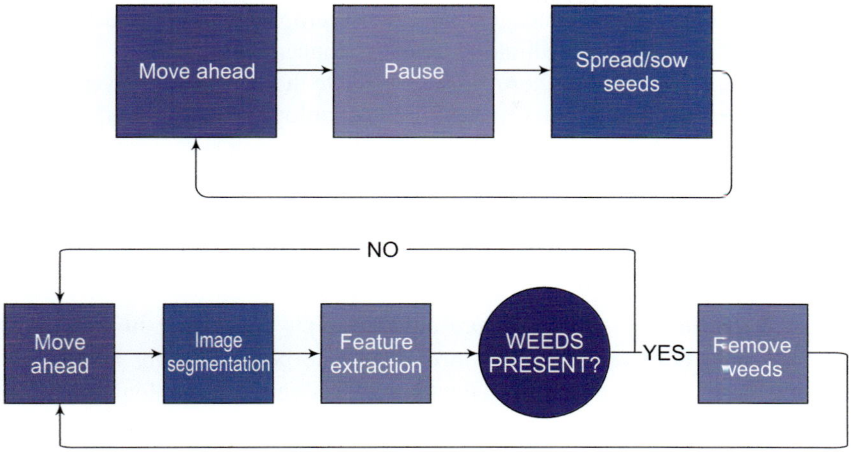

Fig. 1 Flow chart of seed spreading and weed removal.

10. 3D RAW Design [9]

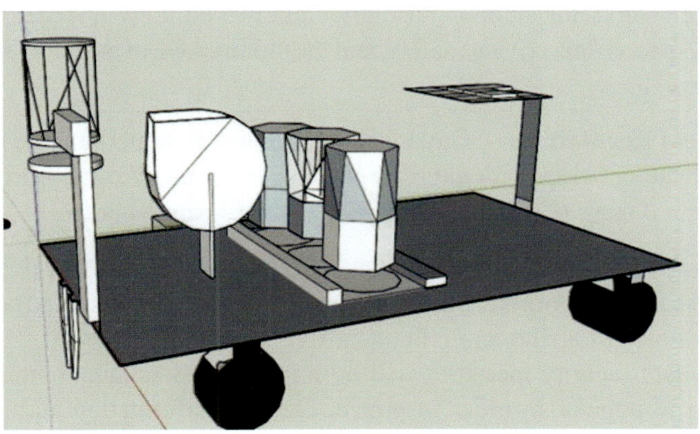

Fig. 2 Design model of System.

11. Conclusion

To summarise, the use of machine learning in conjunction with weeding and sowing to improve crop fields is a huge development in agricultural technology. This strategy enhances farming operations by using the power of automation and artificial intelligence, resulting in increased crop yields, resource efficiency, and sustainability. As research in this sector advances, it becomes clear that the convergence of robots, sensor technology, precision agriculture, and machine learning is altering the way we farm crops [9].

Sustainable resource management, precise crop and weed detection, and real-time decision-making will all be essential concerns to handle in the future. Furthermore, it is critical to thoroughly investigate the ethical and governmental consequences of autonomous agriculture technology. Researchers, farmers, and industry stakeholders will need to work closely together to overcome these challenges and ensure that the hybrid strategy protects the environment while preserving food security [25]. The integration of cutting-edge technologies will continue to enable real-time decision-making. This will improve our ability to quickly adjust to changing field conditions and crop management.

Better Detection Systems: Future advancements will result in even more accurate and effective techniques to discriminate between crops and weeds, utilizing machine learning, artificial intelligence, and sensor technologies to assure weeding accuracy.

Sustainable Resource Management: By implementing eco-friendly practices and technologies, resource management may reduce waste and increase resource efficiency.

Ethical Frameworks: In the coming years, it will be critical to develop strong ethical frameworks for autonomous agricultural technology in order to address challenges such as data privacy, safety, and the moral usage of artificial intelligence (AI) in farming.

Government Regulations: Governments will play an important role in creating the future of autonomous agriculture by enacting laws and rules to ensure the moral and safe use of these technologies. Multidiscplinary word is before technologies.

Collaboration: To close the technology-agriculture divide and develop solutions that improve food security and the environment, scientists, farmers, and commercial partners will work together more frequently.

Strict data security measures will be implemented as autonomous systems become more popular in order to protect personal information and ensure the integrity of agricultural operations.

Education and Training: Preparing personnel to run and manage autonomous systems will be critical to future agricultural success.

Climate Change Adaptation: Farmers can respond quickly to changing weather patterns by using autonomous farming technologies [26].

Global Accessibility: Efforts will be undertaken to guarantee that farmers around the world have access to cutting-edge weeding and seeding equipment, promoting a more equitable distribution of agricultural benefits.

Ecosystem Preservation: Future developments will prioritize preserving and increasing biodiversity in agricultural areas to enable healthier ecosystems and sustainable food production [27].

Remote Sensing: As autonomous farming progresses, satellite technology and remote sensing will be employed more frequently, enabling for larger-scale crop field management and monitoring.

Risk Mitigation: Researchers and specialists will develop plans to lessen the likelihood of equipment failures, pest infestations, and unexpected environmental changes.

Economic Viability: Efforts will be made to lower the cost of autonomous agricultural technology for farmers, ensuring that adoption is financially possible.

Increase Public Acceptance and Understanding: The effective adoption of autonomous agriculture technology will be dependent on growing public acceptance and understanding of its benefits and security measures. Given the increasing pressure on the agriculture industry to produce more with less resources, the hybrid method has the potential to transform farming techniques [28]. It demonstrates that some of today's most important challenges, such as feeding a growing global population and decreasing the environmental damage caused by agriculture, may be addressed by technological means [11].

References

[1] Smith, J.A. and Johnson, R.L. (2020). Machine learning applications in precision agriculture. *Computers and Electronics in Agriculture*, *176*, 105687.

[2] Smith, J.A. and Johnson, R.L. (2020). Machine learning applications in precision agriculture. *Computers and Electronics in Agriculture*, *176*, 105687.

[3] Panetta, F.D. and Timmins, S.M. (2004). Evaluating the feasibility of eradication for terrestrial weed incursions. *Plant Protection Quarterly*, *19*(1), 5–11.

[4] Zhang, Y. and He, Y. (2019). A review of autonomous navigation systems for agricultural vehicles. *Computers and Electronics in Agriculture*, *165*, 104942.

[5] De Cauwer, B., Reheul, D., Nijs, I. and Milbau, A. (2008). Management of newly established field margins on nutrient-rich soil to reduce weed spread and seed rain into adjacent crops. *Weed Research*, *48*(2), 102–112.

[6] Liu, H., Zhang, L. and Zhang, S. (2018). A review of deep learning in the field of precision agriculture. *Computers and Electronics in Agriculture*, *147*, 70–78.

[7] Nidhi, P. and Saini, R. (2021). An overview of agricultural robots and automation. *International Journal of Agricultural and Biological Engineering*, *14*(5), 9–22.

[8] Jacob, H.S., Minkey, D.M., Gallagher, R.S. and Borger, C.P. (2006). Variation in post-dispersal weed seed predation in a crop field. *Weed Science*, *54*(1), 148–155.

[9] Smith, C.A. and Smith, A.B. (2019). Precision agriculture and environmental quality: Challenges for research and extension. *Journal of Environmental Quality*, *48*(2), 329–341.

[10] Blackshaw, R.E. and Rode, L.M. (1991). Effect of ensiling and rumen digestion by cattle on weed seed viability. *Weed Science*, *39*(1), 104–108.

[11] Christoffoleti, P.J., Carvalho, S.J.P., Nicolai, M., Doohan, D. and VanGessel, M. (2007). *Prevention Strategies in Weed Management. Non-chemical Weed Management: Principles, Concepts, and Technology*. CABI, Oxfordshire, pp. 1–16.

[12] Swain, S. and Misra, P. (2020). Sustainable agriculture practices using machine learning: A review. *Sustainable Computing: Informatics and Systems*, *25*, 100443.

[13] Food and Agriculture Organization (FAO). (2018). *The Future of Food and Agriculture: Trends and Challenges*.

[14] Yenish, J.P. (2007). Weed management in chickpea. In: *Chickpea Breeding and Management* (pp. 233–245). Wallingford UK: CABI.

[15] Thill, D.C. and Mallory-Smith, C.A. (1997). The nature and consequence of weed spread in cropping systems. *Weed Science*, *45*(3), 337–342.

[16] Haibo, L., Qing, L., Yufeng, X. and Chuijje, Y. (2010). Research and development on the key technology of wheat single seed robot. In: *IEEE World Automation Congress*; September (pp. 339–343).

[17] Haug, S., Michaels, A., Biber, P. and Ostermannm J. (2014, March). Plant classification system for crop/weed discrimination without seg mentation. In: *IEEE Winter Conf. on Application of Computer Vision* (pp. 1142–1149).

[18] Chocron, O., Delaleau, E. and Pleureau, J.L. (2007, June). Flatness based control of a mechatronic weed killer autonomous robot. In: *IEEE Int Symp on Indust Electron,* (pp. 2214–2219).

[19] Jeon, H.Y. and Tian, L.F. (2009). Direct application end effector for a precise weed control robot. *Biosyst Eng*., *104*, 458–464.

[20] Kladivko, E.J. (2001). Tillage systems and soil ecology. *Soil Till. Res*., *61*, 61–76. [7] Liu J., Li Z., Li P., Mao, H. (2008). Design of a laser stem-cutting device for harvesting robot. In: *IEEE Int. Conf on Automation and Logistics,* September (pp. 2370–2374).

[21] Nakai, S. and Yamada, Y. (2014). Development of a weed suppression robot for rice cultivation: Weed suppression and posture control. *Int. J. Electr. Comput. Electron. Commun. Eng.*, *8*, 1736–1740.

[22] Parker, C. and Fryer, J.D. (1975). Weed control problems causing major reduction in world food supplies. *FAO Plant Prot. Bull*., *23*, 83–95.

[23] Mallikarjun Mudda, Dr. Manjunath R., Dr. Krishnamurthy, N. (2015, Oct.) Review on Brain Tumor Detection and Analysis Using MRI Multi Slice Sequences and further progress proposed. *International Journal of Multidisciplinary Educational Research (IJMER), with International Scientific Indexing Value*: 2.286, *4* (10).

[24] Mallikarjun Mudda,1 Dr. Manjunath R., , Dr. Krishnamurthy, N. (2017). Enhanced Image Registration Technique for Medical Image Segmentation. In: *International Medical Sciences Academy*, *30* 3 (July –September 2017) (*SCI*, *Scopus* Indexed).

[25] Aguiar, M.R. and Sala, O.E. (1997). Seed distribution constrains the dynamics of the Patagonian steppe. *Ecology*, *78*, 93–100.

[26] Aizen, M.A., Sabatino, M. and Tylianakis, J.M. (2012). Specialization and rarity predict nonrandom loss of interactions from mutualist networks. *Science*, *335*, 1486–1489.

[27] Anderson, J.T., Nuttle, T., Rojas, J.S.S., Pendergast, T.H. and Flecker, A.S. (2011, Mar. 23). Extremely long distance seed dispersal by an overfished *Amazonian frugivor*e. Proc. Biol. Sc., 278(1723), 3329–3335.

[28] Bacles, C.F.E., Lowe, A.J. and Ennos, R.A. (2006). Effective seed dispersal across a fragmented landscape. *Science*, *311*, 628.

[29] Bartuszevige, A.M. and Gorchov, D.L. (2006). Avian seed dispersal of an invasive shrub. *Biological Invasions*, *8*, 1013–1022.

[30] Bascompte, J. and Jordano, P. (2007). Plant-animal mutualistic networks: The architecture of biodiversity. *Annual Review of Ecology Evolution and Systematics*, *38* 567–593.
[31] Lisa Lobry de Bruyn (1990), Arthur Conacher,"The role of termites and ants in soil modification-a review", *Soil Research*, 28, 55–98.

11 | Social Impact of IIoT
Community and Workforce Considerations

Jignesh Patel[1*], Seema Jamal[2] and Swati Mahalle[3]

The IIoT is a force for transformation in shaping global industries by connecting a wide array of machines, systems, and processes through advanced analysis techniques that uncover patterns, advanced computer systems that can make decisions like humans, and big information. While IIoT presents unparalleled opportunities for operational efficiency and innovation, it also has enormous social implications, especially in terms of impacts on the workforce and communities. The chapter further discusses the twin impact of IIoT on workers and societal structures, considering how automation and machine intelligence are redefining traditional roles while creating new opportunities that require very specialized skills in areas such as information analysis techniques that uncover patterns, cybersecurity, and smart systems management. The discussion emphasizes the need for reskilling initiatives and relevant policy measures as a means of aiding workforce transitions.

On a community level, it drives the construction and expansion of smart cities and enhances public infrastructure; on the other hand, it deepens technological gaps between rural and underdeveloped areas by exacerbating social inequalities. Also, considering that IIoT systems can gather really huge amounts of data, this means a number of privacy and data security risks, which in turn bring ethical issues that demand action from regulating authorities. The chapter will discuss the opportunities and risks of IIoT, outline some case studies within the manufacturing, healthcare, and agricultural sectors, and conclude with recommendations for equitable technological access, worker rights, and community resilience in the face of industrial digitalization.

1. Introduction

1.1 The IIoT–Industrial Internet of Things and Social Impact

The Industrial Internet of Things (IIoT) is the directional and transforming technological framework that reconstitutes innovation and operational efficiency

[1-3] *Assistant Professor*, Thakur College of Engineering and Technology, Mumbai.
Email: Jamal.seema@tcetmumbai.in[2]; swati.mahalle@tcetmumbai .in[3]
* Corresponding author: jignesh.patel@tcetmumbai.in

DOI: 10.1201/9781003587903-11

across industries. Different from IoT (internet of things), which has a primary focus on consumer-based applications like smart home devices, wearables, and mobile apps, IIoT applies to large-scale industrial operations: manufacturing, logistics, energy production, healthcare, and supply chain management. IIoT integrates connected sensors, devices, and advanced analysis techniques that uncover patterns for better productivity, safety, and informed, information-driven decisions.

IIoT allows for Industry 4.0—a revolution whereby digital and physical systems will be inextricably linked—allowing machines, equipment, and processes to share information in real time. Core to IIoT is M2M communication, whereby industrial devices send information on their own initiative. This extends into collecting information via smart systems, advanced analysis techniques that uncover patterns, and cloud computing in order to optimize operations, predict maintenance needs, and streamline supply chains.

For instance, predictive maintenance powered by IIoT reduces production delays by warning operators of impending equipment failures. In healthcare, IIoT devices track critical patient data in real time, enabling better care at lower costs. Similarly, in energy, IIoT enables efficient energy monitoring, promoting sustainability and reducing environmental impact.

1.2 Motivation

While IIoT has major operational benefits, there are significant and growing societal challenges with this rapid adoption, particularly related to the workforce and greater community. Reshaping industrial processes, IIoT replaces traditional jobs and at the same time opens up new opportunities that require very specific skills in information analysis techniques that uncover patterns, for example, cybersecurity, and the management of smart systems systems. Yet, all these technological advancements push low-skilled workers, and this demands a move toward reskilling and upskilling efforts.

The chapter searches for social consequences of IIoT, which have something to do with the impact this technology makes on redefining workforce dynamics and community structures. The evolution in the workforce from manual to knowledge-based skilled employment is discussed, as well as a rise of reskilling programs helping people transition between jobs. The communal dimension in the implementation of IIoT covers a number of important ideas; it has to do with the emergence of smart cities, creation of better public infrastructure, and the benefits derived by society. On the other hand, it addresses concerns about increased digital divide, privacy challenges, and ethical issues of data use.

IIoT is a source of opportunities and risks since, while driving innovation and productivity on one side, it also produces economic and ethical concerns on the other. For example, the adoption of IIoT technologies by underdeveloped regions is so unequal that the digital gap will be increased in some populations. Further, large volumes of data emanating from IIoT systems raise a specter of some threat to privacy and security concerns that need regulatory interventions. This chapter, therefore, evaluates these challenges and opportunities and recommends how industrial digitalization can balance with fair access and community welfare.

2. Defining IIoT: A Focus on Industry Applications

IIoT is basically a network of different industrial devices that are embedded with sensors and software, thus enabling real-time data facilitation and decision-making. As opposed to IoT, which has been designed for consumers, emphasizing convenience and connectivity, IIoT has been designed to meet industrial needs like resource optimization, reduced downtime, and better efficiency in operations.

IIoT Adoption Cuts across Many Industries

- **Manufacturing:** Smart factories are one of the most numerous users of IIoT, and apply automation, smart systems, and information analysis techniques that uncover patterns to be able to monitor and optimize production lines on the fly.
- **Supply Chain and Logistics:** IIoT improves supply chain efficiency by enabling better inventory tracking, real-time asset monitoring, predictive maintenance of delivery vehicles, and end-to-end visibility of goods in transit, which helps reduce delays and minimize operational costs.

2.1 Problem Statement

This chapter underlines the fact that while IIoT greatly enhances operational efficiency and industrial innovation, it also brings about changes in workforce structure, societal norms, and communal well-being. Automation of workforce mechanisms for lowly skilled roles, creating more specialized opportunities, and disrupting traditional labor dynamics—are some of the roles that IIoT calls for reskilling programs and making policy interventions highly essential to have a smooth workforce transition.

At the community level, IIoT speeds up the process of building smart cities and public infrastructure but has further potential to increase social disparities, in particular in less-developed areas. Enormous data from IIoT systems increase critical issues about privacy, security, and ethics that raise the need for strong regulatory oversight.

The chapter discusses, through cases in manufacturing, healthcare, and agriculture, the ways in which IIoT might be responsibly leveraged toward balancing digitalization benefits with worker protection, equitable access, and community resilience.

3. Role of IIoT in Industry

The IIoT is technologies, processes, and different information analysis techniques that uncover patterns enabling innovation in manufacturing, healthcare, and logistics. IIoT has been fundamental in changing and transforming industrial functions by addressing inefficiencies and enhancing their effectiveness. Speaking of this transformative journey, at the core have been two aspects automation and efficiency, and also information and connectivity. These help derive better performance and less downtime or more effective, fact-based decisions.

3.1 Automation and Efficiency

IIoT is a key enabler of automation, offering industries the ability to monitor and control operations in real time through interconnected sensors, devices, and machines. This connectivity minimizes downtime, optimizes resource utilization, and facilitates predictive maintenance. Predictive maintenance is one of IIoT's most notable advancements. Traditionally, machinery maintenance occurred only after a breakdown, resulting in wasted time and resources. IIoT resolves this by enabling machines to communicate their operational status and potential wear through sensors. Data collected by these sensors allows industries to predict maintenance needs and address them before major failures occur. For instance, IIoT has reduced unscheduled downtime significantly in smart factories while also cutting maintenance costs. Examples of IIoT's efficiency-enhancing capabilities include smart factories where automated systems and devices operate with minimal human intervention. Companies like General Electric use IIoT platforms, such as Predix, to connect thousands of machines, enabling real-time performance monitoring. This has led to operational efficiency improvements of up to 10% and significant cost savings through proactive maintenance. Additionally, IIoT-powered robots and automated systems handle repetitive tasks with precision, allowing human workers to focus on higher-level decision-making and problem-solving. In logistics, IIoT-enabled devices optimize inventory management, track shipments, and streamline delivery routes. For instance, Amazon's automated guided vehicles efficiently transport goods within warehouses, reducing human interference and improving supply chain operations. These examples demonstrate that IIoT not only optimizes existing processes but also introduces groundbreaking automation capabilities that revolutionize industries.

4. Data and Connectivity

Beyond automation, IIoT leverages information to enable smarter decision-making. Through the integration of big information analysis techniques that uncover patterns, advanced computer systems that can make decisions like humans (smart systems), and machine learning (ML), IIoT transforms raw information from machines and systems into actionable insights. This information-driven approach enhances performance, safety, and innovation across industries. One of IIoT's significant advantages is its ability to collect and analyze large volumes of data in real time. Sensors embedded in industrial equipment monitor parameters such as temperature, pressure, and vibration, providing insights that help companies identify patterns, predict potential issues, and optimize processes. For example, in the oil and gas industry, IIoT sensors monitor drilling equipment health and environmental conditions, minimizing operational risks and improving extraction efficiency. IIoT's connectivity also enables seamless communication between machines and systems, fostering coordination and improving overall productivity. When integrated with smart systems and machine learning (ML) algorithms, IIoT systems further automate decision-making processes. For instance, in smart grids, IIoT devices monitor energy consumption patterns and, using smart systems, optimize energy distribution to reduce waste and costs.

Remote monitoring and control of industrial systems are additional benefits of IIoT's connectivity. Industries can manage operations from anywhere in the world via cloud-based platforms. In healthcare, for example, smart medical devices collect and analyze patient data remotely, providing healthcare professionals with real-time feedback and enhancing patient care while reducing hospital visits. Moreover, IIoT facilitates innovation by enabling the development of advanced technologies and services. Autonomous vehicles, for instance, leverage IIoT to communicate with road infrastructure and other vehicles in real time, improving safety and efficiency. Similarly, precision farming techniques use IIoT sensors to monitor soil conditions, optimize irrigation, and increase crop yields. These innovations demonstrate that IIoT is not just a tool for improving efficiency but also a catalyst for creating new opportunities and redefining industry standards.

4.1 Workforce Considerations in IIoT (Industrial Internet of Things)

The IIoT is transforming industries by connecting machines, sensors, and devices to enhance operational efficiency and streamline decision-making processes. However, this transformation significantly impacts the workforce, introducing challenges such as job displacement, skill evolution, and the emergence of new roles. As IIoT adoption expands, understanding its effects on the workforce is crucial for mitigating challenges and leveraging opportunities. This section explores the key considerations related to workforce dynamics, including shifts in skills, job displacement, and the emergence of new employment opportunities driven by IIoT.

5. Skill Shifts and Job Displacement

The integration of IIoT technologies into industries is reshaping skill requirements, leading to the obsolescence of traditional roles and creating demand for new, specialized expertise. Organizations must address these changes by prioritizing workforce reskilling and upskilling to manage job displacement caused by automation.

5.1 Automation and Job Redundancy

Automation driven by IIoT presents both challenges and opportunities for the workforce. Industries such as manufacturing, logistics, and healthcare are witnessing job displacement as machines take over tasks traditionally performed by humans.

In manufacturing, for instance, IIoT-enabled robotics automate repetitive tasks like assembly, inspection, and quality control. These smart machines communicate in real time, predict maintenance needs, and make autonomous decisions, thereby reducing the reliance on manual labor. Similarly, in logistics, IIoT technologies automate warehouse operations and supply chain management, streamlining inventory tracking, stock counting, and order processing. In healthcare, IIoT-enabled devices automate routine tasks like patient monitoring and diagnostics, increasing efficiency and accuracy but reducing the need for human intervention in these areas.

While automation eliminates low-skill, repetitive roles, it also frees up workers to focus on higher-value activities, such as complex decision-making and creative problem-solving. However, the challenge lies in providing displaced workers with opportunities to transition into new roles through reskilling initiatives.

5.2 Upskilling and Reskilling

To counteract the negative effects of job displacement, upskilling and reskilling the workforce is essential. IIoT demands a workforce equipped with advanced skills in digital technologies, including information analysis techniques that uncover patterns, advanced computer systems that can make decisions like humans (smart systems), cloud computing, and machine-to-machine communication. Workers with these skills are better positioned to thrive in an IIoT-driven environment.

Educational institutions, government policies, and corporate training programs play critical roles in this transition. Schools and universities must integrate IIoT-related topics into their curricula, offering courses in IoT technologies, data science, cybersecurity, and smart systems. Governments can support these efforts by incentivizing companies to invest in employee training programs and providing subsidies for workers seeking to enhance their skills.

Corporate training programs are particularly vital in industries heavily impacted by IIoT. For example, factory maintenance workers may require training to interpret IIoT sensor data to predict equipment needs rather than reacting to failures. By equipping the workforce with relevant skills, industries can ensure workers are prepared to adapt to the changing technological landscape.

6. Workforce Opportunities

While IIoT may lead to the displacement of certain roles, it also creates new opportunities centered on managing and maintaining IIoT systems, data analysis, cybersecurity, and advanced computer systems that can make decisions like humans. Additionally, IIoT enables flexible work arrangements, allowing employees to monitor and control industrial processes remotely.

6.1 New Job Roles

IIoT has given rise to new roles that did not exist a decade ago. For instance, the role of an IIoT system architect involves designing, implementing, and managing IIoT networks. This position requires an in-depth understanding of hardware, software, and networking technologies to ensure smooth and secure operations.

Another emerging role is that of a cybersecurity specialist focused on IIoT systems. As the connectivity of devices increases, so does the potential for cyberattacks. Cybersecurity experts are essential for protecting IIoT networks, securing sensitive data, and maintaining operational integrity.

Additionally, roles such as data analysts and smart systems trainers are in high demand. Data analysts interpret vast amounts of data generated by IIoT devices to derive insights that drive decision-making, while smart systems trainers develop machine learning algorithms that enable IIoT systems to learn and improve over time. These roles offer lucrative career opportunities for individuals with strong skills in data science, smart systems, and machine learning.

6.2 Remote and Flexible Work

IIoT facilitates remote and flexible work arrangements, particularly in industries that rely on real-time monitoring and predictive analysis techniques that uncover patterns. Workers can oversee operations from anywhere using cloud platforms, which provide access to information and decision-making tools.

For example, in agriculture, IIoT sensors monitor soil conditions, weather patterns, and crop health, allowing farmers to make informed decisions remotely. Similarly, in the energy sector, IIoT devices monitor pipelines for potential issues such as leaks or corrosion, enabling engineers to address problems before they escalate. These remote capabilities enhance efficiency while offering employees greater flexibility in terms of work location and hours.

The post-pandemic era has further highlighted the importance of remote work, with many industries adopting hybrid or fully remote models. IIoT supports this transition by providing the infrastructure necessary for remote operations, making it a valuable tool for industries seeking to increase flexibility and reduce on-site labor requirements.

The rise of IIoT is reshaping the workforce by automating traditional roles while creating opportunities for new, high-skill positions. Preparing the workforce for this transformation through upskilling and reskilling initiatives is essential. At the same time, IIoT offers promising opportunities in fields such as cybersecurity, data analysis, and smart systems, while enabling more flexible work arrangements.

As industries continue to embrace IIoT, prioritizing workforce development will be key to ensuring a smooth transition into the next industrial era.

7. Community Considerations in the Context of IIoT

The IIoT has immense potential to enhance productivity, streamline operations, and improve quality of life, particularly through its applications in smart cities and industrial advancements. However, the widespread adoption of IIoT also poses significant social implications for communities. These include challenges such as digital inequality, impacts on social infrastructure, privacy concerns, and environmental sustainability. Understanding these implications is critical to ensuring that IIoT benefits communities equitably and responsibly.

7.1 Digital Divide and Inequality

7.1.1 Access to Technology

While IIoT promises transformative benefits for industries and societies, its effectiveness depends heavily on equitable access to technology. Communities that lack access to IIoT systems, particularly in economically disadvantaged areas, risk being excluded from these advancements. This digital divide exacerbates disparities in economic opportunities, education, and essential services.

In underdeveloped regions, where basic technological infrastructure is often inadequate, communities are unable to leverage IIoT innovations. This disparity hinders economic growth and restricts job opportunities. For instance, as industries adopt IIoT solutions, regions without the necessary infrastructure experience slower progress. Studies, such as those by Baccelli [1], highlight that while IoT technology drives economic development, its implementation must prioritize inclusivity to ensure broad societal benefits.

Urban areas with advanced technology infrastructures attract investments and talent, further widening the gap between urban and rural communities. Without intervention, this imbalance could lead to uneven development, leaving economically weaker regions behind in the race for industrial digitalization.

7.1.2 Rural vs Urban Divide

The divide between rural and urban areas is a significant facet of digital inequality, further emphasized by IIoT adoption. Urban centers, with their higher population density, access to skilled labor, and established infrastructure, are more likely to receive the investments needed to implement IIoT technologies. This urban bias accelerates the development of smart cities, efficient transportation systems, and improved public utilities, while rural communities often struggle to keep up.

For example, agriculture, predominantly a rural industry, can benefit from IIoT technologies such as automated irrigation systems and soil monitoring sensors. However, without adequate technological infrastructure, rural farmers may find it challenging to adopt these advancements. As a result, rural areas risk economic stagnation while urban centers thrive.

Bridging this rural-urban divide requires investments in technological infrastructure for rural communities. Governments and industries must work collaboratively to ensure equitable access to IIoT, fostering a more inclusive digital economy that benefits all communities.

8. Social Infrastructure

8.1 Smart Cities and Communities

One of the most visible applications of IIoT is in the creation of smart cities, where connected technologies enhance urban infrastructure and improve quality of life. IIoT-powered systems optimize resources and provide solutions in areas such as traffic management, waste disposal, and public safety.

For instance, IIoT-enabled traffic systems use real-time data to adjust signal timings, reducing congestion and improving mobility. Similarly, smart grids optimize energy distribution, minimizing waste and lowering costs. These systems empower public administrators with data-driven insights for better decision-making.

While smart cities represent a model for how IIoT can improve urban living, their reliance on interconnected networks introduces vulnerabilities, particularly in terms of security and privacy. Addressing these challenges is vital to maintaining trust in IIoT technologies.

8.2 Privacy and Security Concerns

As IIoT becomes more integrated into communities, concerns about privacy and data security grow. IIoT systems collect large volumes of data, including sensitive information such as location, operational details, and personal identifiers. This makes them attractive targets for cyberattacks and raises questions about data ownership and ethical usage.

Critical sectors, such as healthcare or public utilities, face severe consequences in case of data breaches. This will affect safety and trust in technology. According to Salunke and Mahajan [5], "the increased connectivity of IIoT devices also increases the number of entry points for cyberattacks, thereby emphasizing the need for robust cybersecurity measures".

Communities must be assured that their data is being handled securely. Governments and industries must ensure stringent data protection regulations and ethical standards to safeguard privacy, ensure transparency, and build trust in IIoT technologies.

9. Environmental Impact

9.1 Sustainability Initiatives

One of the most important benefits of IIoT is its ability to promote sustainability. IIoT technologies allow industries and communities to optimize resource use, reduce waste, and minimize their environmental footprint. For instance, smart grids balance energy loads, thereby reducing the generation of excess power and greenhouse gas emissions. In community settings, IIoT supports efficient utility management. Smart water meters detect leaks in real time, which reduces the amount of wasted water, while smart energy meters help households monitor and decrease their energy consumption. These initiatives empower communities to align with broader environmental goals, such as carbon reduction and resource conservation.

Another facet through which IIoT impacts supply chain efficiency includes optimizing the movement of goods, reducing fuel consumption, and also minimizing emissions. For instance, IIoT devices allow monitoring the condition of perishable goods during transportation, which reduces their spoilage and thus waste. Nonetheless, the sustainability of IIoT technologies can be compromised by its environmental drawbacks. Manufacturers of IIoT devices add to electronic waste due to their disposal. To ensure long-term sustainability, industries must adopt environmentally responsible practices for device production and recycling.

10. Ethical and Policy Considerations

The rapid expansion of the IIoT involves not only advancing technology but raises a host of ethical and policy challenges. As industries embraced automation, big data, and smart systems-powered systems, critical job security, data ownership, data privacy, and regulatory frameworks raised concerns. It is in relation to these critical issues that are associated with three primary ethical policy issues of the IIoT: job security and worker rights, data ownership and privacy, and the necessity of robust regulatory frameworks. This will be addressed as a vital part of making the transition to an IIoT-driven future both fair and sustainable.

10.1 Job Security and Worker Rights

Among the most serious ethical issues associated with IIoT implementation is the displacement of jobs. Through IIoT-enabled devices and systems, industries can replace human labor with machines to perform jobs such as assembling products, inventory control, and logistics management. These improvements enhance efficiency and save money but also bring up questions on the future of work and the rights of workers.

IIoT is fast-tracking the automation trend, mainly impacting industries that are heavily reliant on manual labor. As noted by Baccelli [1], the sectors most affected include manufacturing and agriculture, where IIoT-enabled systems automate repetitive tasks. This leaves the workers in those fields facing uncertainty over their job security and the need to acquire new skills to remain employable. The corporations and governments must take proactive measures to address these challenges. Companies should offer reskilling and upskilling opportunities, fair severance packages, and create pathways for displaced workers to move into new jobs. Governments must also enact policies such as job retraining programs, unemployment benefits, and incentives for companies' worker development. Other regulatory measures include requiring businesses to disclose the impact of automation on the workforce and outline strategies to mitigate the effects, which could help protect the rights of workers.

The ethical debate extends beyond job displacement to broader questions about how society values human labor in an increasingly automated world. Policymakers and businesses must collaborate to ensure the benefits of automation are shared equitably and do not come at the expense of workers' livelihoods.

11. Data Ownership and Privacy

IIoT systems produce a huge amount of data, which raises very important questions about data ownership and privacy. In industrial settings, IIoT devices monitor machine performance, environmental conditions, and worker activities. While this data is invaluable for optimizing operations and improving safety, it also raises concerns about surveillance and the misuse of personal information.

Ethically, there is a need to set the framework about data ownership and usage. Employees have the right to know what data is being recorded about them, how it is being used, and who has access to it. In some areas, it would be proper if workers have ownership rights on particular data, especially if this refers to performance or activities. According to Baccelli [1], ethics frameworks to guide human ethical behaviour must exist to prevent companies from exploiting other workers' data.

The implications of data ownership extend beyond individuals to entire communities. Public IIoT systems, such as smart city infrastructure, collect data to improve services and quality of life. However, communities should have a say in how their data is used and whether it is shared with third parties. Transparent policies prioritizing consent and accountability are essential to ensure that IIoT benefits society without compromising privacy.

Balancing corporate interests, individual rights, and community welfare in the management of IIoT data requires well-defined ethical and regulatory frameworks. These must safeguard privacy while enabling the responsible use of data to drive innovation.

11.1 Regulatory Frameworks

The rapid evolution of IIoT technologies has outpaced the development of regulatory frameworks, leaving gaps in areas such as worker rights, data privacy, and cybersecurity. Addressing these gaps is essential to managing the societal impact of IIoT responsibly. Policies regarding job displacement, worker monitoring, and reskilling should be amended in line with the realities of an IIoT-driven workforce. Current labor laws often cannot consider the IIoT-related issues that come along. For example, the governments must develop policies to safeguard the interests of the workers who get affected by the adverse consequences of automation, such as compulsory training for workers whose jobs get automated and compelling the companies to inform about the workforce implications of the automation [4]. There are existing regulations such as the General Data Protection Regulation (GDPR) in Europe, but these regulations might not address the IIoT issues at a level that can ensure its smooth working. For instance, continuous and real-time data collected by IIoT devices often do not let the individuals control their information. The policymakers must frame regulations, especially tailoring to the needs of IIoT. This includes transparency, data minimization, and informed consent [5].

Cybersecurity is another area of major concern in the case of IIoT. The complex web of interconnected devices can be an avenue for cybercrime. There should be regulations to set up security standards for the devices, impose high security on organizations, and also provide means to deal with breaches. Ethical and policy challenges of IIoT are better addressed by the collaboration of governments, industries, and communities. Through the regulation of matters such as job security, data privacy, and cybersecurity, society can pave a way for a fair and well-balanced evolution into smart systems.

12. Case Study

AWS Industrial IoT Motor Monitoring and Control System for a Smart Factory

1. Overview

A manufacturing facility was looking to establish a system for the remote monitoring and control of its motors. The requirements included operational efficiency and timely maintenance. It was desired to set up an IoT-based architecture that would be able to report the status of the motors, manage the power levels, and securely control several devices using a centralized application. Using the AWS Industrial IoT services, it developed a scalable, cloud-based solution that allows real-time communication between the factory's motors and its supervisors.

2. Requirements

The factory supervisor required a remote application—accessible via desktop or web—that could: Monitor motor status, including power state, power levels, and revolutions per minute (RPM). Enable remote control of motors, including turning them on or off and adjusting power levels. Safely control many motors with unique identities, and ensure that access was limited to only authorized users.

3. Solution Architecture

The architecture used AWS IoT services to scale and secure the solution: Factory Motors (IoT Devices): Each motor had a connectivity module that reported its status (on/off state, power level, RPM) and accepted commands over the internet.

- ***AWS IoT Core:*** It acted as the central MQTT message broker, enabling secure communication between the motor control application and the motors. Each motor was represented as a ' thing' in AWS IoT, with its operating status stored in a virtual 'shadow'.
- ***Motor Control Application:*** This application served as an interface between the factory supervisor and the motors. It communicated with AWS IoT Core to exchange messages on motor statuses.

4. Implementation Steps

- ***Things and Shadows Creation:*** Each motor was registered as a virtual device ('thing') in AWS IoT, with an associated shadow object to store its operational state.
- ***The shadow maintained two key states:***
  - ***Reported State:*** The current operational data of the motor.
  - ***Desired State:*** The target operational settings requested by the application.
- ***MQTT Communication:*** Motors communicated with AWS IoT Core using MQTT topics. For example, a reserved 'update' topic was published for the status updates. State change requests (e.g., adjusting power levels) were sent to the same topic, with responses indicating whether the changes were accepted or rejected.
- ***Security Considerations:*** Each motor has a unique certificate to ensure secure communication with AWS IoT Core. A strict security protocol was followed to enforce authentication and authorization for all devices and applications involved in the system.

5. Use Case: Real-Time Monitoring and Control

In a typical scenario, the motor control application received real-time status updates from the motors. If the supervisor wanted to adjust a motor's power level, the application would send a desired state change request via the MQTT 'update' topic. The motor would respond accordingly, and the updated state would be reflected in the application's interface. This bidirectional communication ensured seamless control over the motors' operations.

6. Results

The AWS IoT-based motor monitoring and control system delivered the following outcomes: Enabled real-time monitoring and remote control of motors, significantly improving operational efficiency. Established secure communication channels, ensuring the integrity of all connected devices. Provided a scalable solution capable of managing multiple motors centrally, streamlining factory operations of AWS IoT.

13. Conclusion

The integration of IIoT technologies into industries and communities presents both opportunities and challenges. IIoT has the potential to revolutionize operations by bridging social divides, enhancing public infrastructure, and driving sustainability. It enables industries to optimize processes, reduce inefficiencies, and improve decision-making through real-time information analysis techniques that uncover patterns and automation. Similarly, its applications in smart cities and public utilities contribute to better resource management and improved quality of life.

However, these advancements come with significant challenges, including the risk of deepening digital inequalities, exacerbating the rural-urban divide, and raising concerns about privacy and security. The ethical and social implications of IIoT demand careful consideration, particularly with regard to job displacement, how jobs and skills are changing, and the responsible use of data.

To ensure the benefits of IIoT are widely shared, policymakers, industries, and governments must collaborate on equitable technology access, comprehensive reskilling programs, and robust regulatory frameworks. Such measures will help address digital disparities, safeguard privacy, and promote environmental sustainability. By prioritizing these considerations, society can harness the transformative power of IIoT while fostering inclusivity, resilience, and responsible innovation.

Ethical and Policy Considerations

The rapid adoption of the IIoT presents not only technological advancements but also a range of ethical and policy considerations. As industries embrace automation, big data, and smart systems-powered systems, questions about job security, data ownership, privacy, and regulatory frameworks become increasingly critical. This chapter explores three key ethical and policy challenges related to IIoT: job security and worker rights, data ownership and privacy, and the need for robust regulatory frameworks. Addressing these concerns is essential to ensuring a fair and just transition to an IIoT-driven future.

Job Security and Worker Rights

One of the most pressing ethical concerns surrounding the widespread implementation of IIoT is the potential for job displacement. Automation, facilitated by IIoT-connected devices and systems, has the potential to replace a significant number of human workers in industries such as manufacturing, logistics, and agriculture. For instance, robots and automated systems can now perform tasks that were once the domain of human labor, from assembling products in factories to managing inventory in warehouses. While these advancements lead to increased efficiency and reduced costs for companies, they raise ethical questions about the future of work and the rights of workers.

The displacement of workers due to automation is not a new issue, but IIoT accelerates this trend by enabling greater levels of automation and data-driven decision-making. According to Baccelli [1], industries that heavily rely on manual labor are particularly vulnerable to job displacement as IIoT-enabled machines and systems take over repetitive tasks. As a result, workers in these sectors face uncertain futures, with many needing to acquire new skills to remain employable.

Corporations have a responsibility to ensure that workers are adequately supported during this transition. Ethical considerations include offering reskilling and upskilling opportunities, providing fair severance packages, and creating pathways for workers to transition into new roles within the company. However, the responsibility should not rest solely on corporations. Governments must also play an active role in ensuring that workers are not left behind. Policy interventions such as job retraining programs, unemployment benefits, and incentives for companies to invest in worker development are crucial. Governments could also consider implementing regulations that require companies to disclose the potential impact of automation on their workforce and outline plans for mitigating job displacement [2].

The ethical debate extends beyond immediate job displacement to the broader question of how society values human labor in an increasingly automated world. As industries continue to adopt IIoT, policymakers and corporations must work together to ensure that the benefits of automation do not come at the expense of workers' rights and livelihoods.

Data Ownership and Privacy

The proliferation of IIoT devices generates vast amounts of data, raising important questions about data ownership and privacy. Who owns the data generated by IIoT devices, particularly in workplaces where workers may be monitored through sensors and connected systems? This question has significant implications for both individual workers and communities.

In an industrial setting, IIoT devices collect data on various parameters, including machine performance, environmental conditions, and even the activities of workers. This data can be invaluable for optimizing operations, improving safety, and enabling predictive maintenance. However, the collection of data about workers' activities raises concerns about surveillance and privacy. Workers may feel that their privacy is being invaded if their every movement is tracked and analyzed by their employer.

From an ethical perspective, it is important to establish clear guidelines regarding data ownership. Workers should have a right to know what data is being collected about them, how it is being used, and who has access to it. In some cases, it may be appropriate for workers to have ownership rights over certain types of data, particularly if it pertains to their personal activities or performance. Baccelli [1] argues that ethical frameworks for data ownership should be developed to ensure that workers are not unfairly exploited by companies seeking to extract value from their data.

Other than the issue of individual privacy, the question of data ownership also bears broader implications to communities. For instance, smart city infrastructure has IIoT devices that collect data in public spaces, hence improving services and quality of life. However, communities should be involved in whether the data should be shared or not with third parties. Data use ethical frameworks in public spaces should focus on transparency, consent, and accountability so that data is used for the public good rather than for corporate profit [3]. Ultimately, the ethical management of data generated by IIoT devices requires a balance between the interests of companies, workers, and communities. Policymakers must develop regulations that protect individuals' privacy while allowing for the responsible use of data to drive innovation and efficiency.

Regulatory Frameworks

The rapid pace of IIoT technology development has overtaken the regulatory framework development process, creating gaps in workforce rights, data privacy, and cybersecurity. While some regulations exist to deal with these issues, more work is needed at the policy level to ensure that the social impact of IIoT is managed appropriately. Regarding workforce rights, the existing labor laws do not provide for the special challenges posed by automation and IIoT. For instance, the law on job displacement, worker monitoring, and reskilling should be updated to accommodate the new realities of an IIoT-driven workforce. Governments should consider enacting regulations that protect workers' rights in the face of automation, such as requiring companies to provide training and support for workers affected by IIoT implementation [4]. Another area where regulatory frameworks need to catch up with IIoT developments is data privacy. While data protection laws such as the General Data Protection Regulation (GDPR) in Europe provide some safeguards for individuals' privacy, they may not fully address the unique challenges posed by IIoT devices. For instance, IIoT devices often collect data continuously and in real-time, making it difficult for individuals to have control over their data. Policymakers should consider coming up with regulations that particularly deal with the privacy aspects of IIoT, which include transparency, data minimization, and informed consent [5]. Workforce rights and privacy aside, the other important concern in the IIoT scenario is cybersecurity. The extensive network of interconnected devices opens new avenues for vulnerabilities that cybercriminals can take advantage of Regulatory frameworks should include measures to ensure that IIoT devices and the data they generate are secure. Such measures may be in the form of standards set for device security, requirements to implement appropriate cybersecurity measures in companies, and procedures for reporting and addressing security breaches.

References

[1] Baccelli, E. (2021). *Internet of Things (IoT): Societal Challenges & Scientific Research Fields for IoT*.

[2] Nizetic, S., Solic, P., Lopez-de-Ipina Gonzalez-de-Artaza, D. and Patrono, L. Internet of Things (IoT): Opportunities, issues, and challenges towards a smart and sustainable future. *IPE Journal of Management*, *14*(18).

[3] Kumar, S.N. (2024). Connecting the World: IoT's Influence on Society and Industry. *IPE Journal of Management*, *14*(18).

[4] Kaushik, N. and Bagga, T. (2020). Internet of Things (IoT): Implications in Society. *3rd International Conference on Innovative Computing and Communication* (*ICICC*).

[5] Salunke, V. and Mahajan, S. (2017). The Evolution and Implications of the Internet of Things. *ICADEMS Conference Proceedings*.

12 Development and Implementation of State of Charge Algorithm for Battery Management System

Archana Talhar Belge[1*], Pankaj Shelar[2], Dipesh Tare[3], Abhishek Patil[4] Jitendra Chavan[5] and Kshama Shukla[6]

The Battery Management System (BMS) plays a pivotal role in ensuring the optimal performance and longevity of rechargeable batteries. One critical parameter monitored and managed by the BMS is the State of Charge (SOC), which represents the amount of energy stored in a battery as a percentage of its total capacity. Accurate SOC estimation is essential for efficient utilization of battery energy and preventing undesirable outcomes such as overcharging or deep discharging, which can lead to performance degradation and reduced battery life. This abstract focuses on the development and implementation of a sophisticated SOC estimation algorithm within the BMS. The algorithm employs a combination of advanced mathematical models, data-driven approaches, and real-time sensor feedback to enhance the precision and reliability of SOC calculations. By leveraging state-of-the-art techniques, the proposed SOC estimation method addresses challenges associated with non-linear battery behavior, temperature variations, and aging effects. The BMS architecture integrates various components, including voltage and current sensors, temperature sensors, and a microcontroller unit (MCU) responsible for executing the SOC estimation algorithm. Matlab Simulink uses a BMS to manage state of charge. The algorithm utilizes battery characteristics and historical performance data to dynamically adapt to changing operating conditions. Additionally, machine learning techniques are employed to continuously improve SOC prediction accuracy based on real-world usage patterns. Furthermore, the BMS incorporates a feedback control mechanism to adjust charging and discharging strategies based on the calculated SOC. This ensures that the battery operates within optimal operating ranges, minimizing stress on the cells and maximizing overall system efficiency. The SOC information

[1-6] Department of Mechanical & Mechatronics Engineering, Thakur College of Engineering and Technology, Mumbai, India.
Email: shelarpankaj95@gmail.com[2]; dipesh.tare@tectmumbai.in[3]; abhishek.patil@tcetmumbai.in[4]; jitendra.chavan@tcetmumbai.in[5]; kshama.shukla@tcetmumbai.in[6]
* Corresponding author: archanabelge@gmail.com

DOI: 10.1201/9781003587903-12

is communicated to external systems or user interfaces, enabling transparent monitoring of the battery's energy status. In summary, this abstract outlines a comprehensive approach to SOC management within a BMS. By combining advanced algorithms, sensor technologies, and adaptive control strategies, the proposed system aims to enhance the overall performance, reliability, and lifespan of batteries.

1. Introduction

Battery Management System (BMS) technology continues to advance, especially in the context of electric vehicles. Advanced BMS helps optimize the performance, lifespan, and safety of the batteries used in electric vehicles. BMS algorithms are becoming more sophisticated, allowing for better state-of-charge (SOC) and state-of-health (SOH) estimation. This is crucial for accurately predicting battery life and performance [1]. Some BMS designs incorporate wireless communication capabilities, making it easier to monitor and manage batteries remotely. Research continues on new battery materials and chemistries. BMS needs to adapt to the specific characteristics of these new materials to ensure optimal performance and safety [2]. Some BMS designs incorporate wireless communication capabilities, making it easier to monitor and manage batteries remotely. BMS monitors individual cells within a battery pack. This involves measuring voltage, temperature, and sometimes current for each cell. Monitoring helps ensure that all cells are operating within safe limits. BMS calculates the SOC, indicating how much energy is left in the battery. It also estimates the SOH, providing insights into the battery's overall condition and remaining lifespan [3]. In multi-cell battery packs, cells may have slightly different capacities or performance characteristics. The BMS helps balance the charge across cells to prevent overcharging or over-discharging of any individual cell, thus extending the overall battery life [4]. BMS monitors for any abnormal conditions or faults in the battery system, providing diagnostics and alarms when necessary. This proactive approach helps identify and address issues before they escalate. BMS incorporates safety features such as temperature sensors, fuses, and disconnect switches to minimize the risk of fire or other safety hazards associated with battery operation. By actively managing various parameters, the BMS helps optimize the performance and efficiency of the battery system, contributing to longer lifespan and better overall energy utilization [5].

One of the primary functions of a BMS is to ensure the safety of the battery pack. It monitors and manages the battery's temperature, voltage, and current, preventing conditions that could lead to overcharging, over-discharging, overheating, and short circuits. By actively managing these parameters, a BMS helps to minimize the risk of fire, explosion, or damage to the battery [6, 7]. A BMS helps improve the overall efficiency of the battery system by optimizing charging and discharging processes. It can implement strategies such as charging at the most efficient voltage and current levels, reducing energy losses, and improving overall energy utilization [8]. BMS often includes communication interfaces to interact with external systems, controllers, or user interfaces [9, 10]. This communication capability allows for remote monitoring, diagnostics, and control, enhancing the overall management and integration of the battery system into larger energy storage or power systems. BMS accurately estimates and manages the state of charge in a battery, providing users with information about the remaining capacity. BMS ensures that individual cells within a battery pack are balanced in terms of voltage. This helps in maximizing the overall capacity and life of the battery [11].

SOC estimation is inherently challenging, and the accuracy of BMS in estimating the state of charge may vary, leading to potential errors in predicting available capacity. While BMS performs cell balancing, it may not always be able to achieve perfect balance due to factors such as cell aging, differences in manufacturing tolerances, and other external influences. BMS can control temperature to some extent, but extreme operating conditions or rapid temperature changes may pose challenges for effective thermal management. High-quality BMS solutions can be expensive, and this cost can be a limitation, particularly in applications where budget constraints are a concern [12]. The complexity of BMS systems can be a challenge, especially in large battery packs, requiring careful design and implementation to ensure reliable performance. BMS may have limitations in controlling certain aspects of battery chemistry, which can impact the overall performance and life of the battery [13]. Periodic maintenance and calibration may be required to ensure the continued accuracy and effectiveness of the BMS. In summary, while BMS plays a vital role in managing and protecting batteries, it's important to be aware of its limitations and work within its scope to maximize the benefits it provides. Advances in technology and ongoing research may address some of these limitations over time.

2. Literature Review

According to Global EV Outlook 2023, Battery demand for Electric Vehicles continues to rise. The demand for automotive lithium-ion (Li-ion) batteries saw a notable uptick, registering a 65% increase to reach 550 GWh in 2022, compared to the 330 GWh recorded in 2021. This surge is primarily attributed to the expanding market for electric passenger cars, evidenced by a 55% rise in new registrations in 2022 compared to the preceding year. In China, the demand for vehicle batteries witnessed a robust growth of over 70%, accompanied by an 80% increase in electric car sales during 2022 relative to 2021. The escalation in battery demand was somewhat moderated by the rising prevalence of Plug-in Hybrid Electric Vehicles (PHEVs). Similarly, the United States experienced an 80% surge in battery demand for vehicles, surpassing the 55% increase in electric car sales during 2022. Despite a modest 7% growth in the average battery size for battery electric cars in the United States in 2022, the typical battery size for these cars remains approximately 40% higher than the global average. This discrepancy is influenced, in part, by the greater market share of SUVs in U.S. electric car sales compared to other major markets. Additionally, manufacturers' strategies to provide extended all-electric driving ranges contribute to this variation. On a global scale, the sales of Battery Electric Vehicles (BEVs) and PHEVs are eclipsing those of Hybrid Electric Vehicles (HEVs). The larger battery sizes of BEVs and PHEVs contribute to a further escalation in battery demand [14]. Figure 1 shows the battery demand overall in the world and highlighted most prominently used countries where as Figure 2 shows the battery demand in various modes of transportation [15].

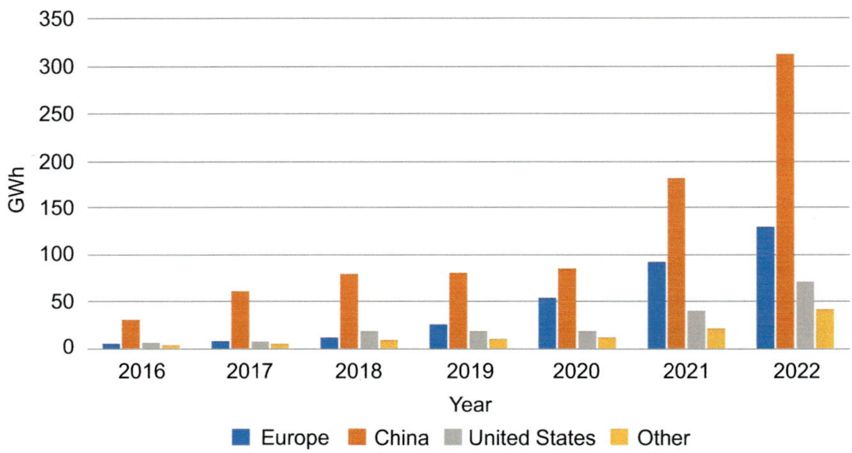

Fig. 1 Battery Demand by Region [15].

Batteries find extensive use in numerous high-power applications, notably in electric vehicles (EVs) and HEVs, where the implementation of a suitable battery management system (BMS) plays a crucial role in ensuring the batteries' safe and reliable operation. Paper [16] seeks to provide a concise overview of several pivotal BMS technologies, encompassing battery modeling, state estimation, and battery charging. It initiates with a survey of prevalent battery types employed in EVs, followed by an introduction to key BMS technologies. The review delves into various battery models, including the electric model, thermal model, and coupled electro-thermal model. Subsequently, it comprehensively examines battery state estimations for parameters such as state of charge, state of health, and internal temperature. In conclusion, the paper discusses several essential and traditional battery charging approaches, along with associated optimization methods. Papers [17–19] present the BMS which plays a pivotal role in EVs and

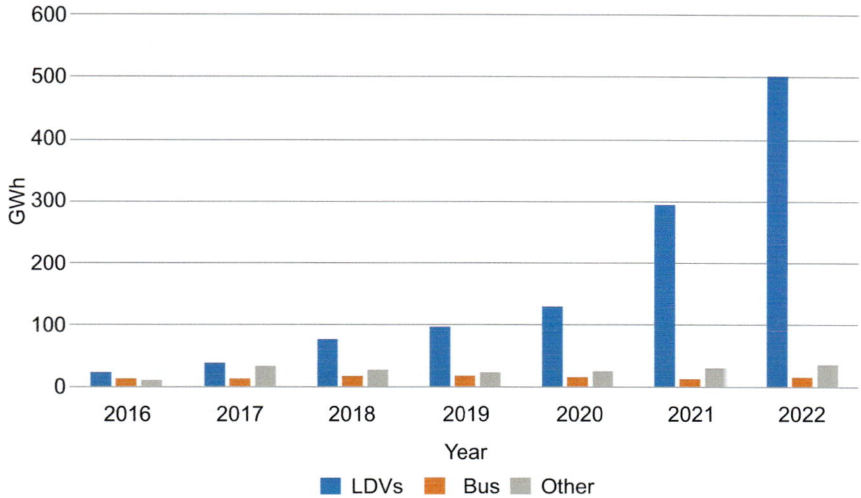

Fig. 2 Battery Demand by Mode [15].

PHEVs ensuring the safe and reliable operation of batteries. Its primary objective is to uphold the safety and dependability of the battery. Key functionalities, including state monitoring and evaluation, charge control, and cell balancing, have been incorporated into BMS to achieve this goal. Given the varying operational and environmental conditions affecting a battery's behavior as an electrochemical product, maintaining safety and reliability poses a challenge. This paper addresses current concerns regarding BMS functionality. The critical task of state evaluation for a battery, encompassing SOC, SOH, and state of life, is explored. By reviewing the latest methodologies for battery state evaluation, the paper outlines future challenges for BMSs and proposes potential solutions. Paper [20] presents a design methodology for an innovative, generalized BMS that has been introduced, leading to the development of an experimental, cell-based, modular, and fault-tolerant system. This system is proficient in providing charge and discharge equalization along with battery protection. It has the capability to communicate with a PC via a serial port, and a dedicated computer program was devised and tested in the laboratory. All challenges encountered during the testing and operational procedures were successfully addressed, ensuring that measurement errors remained within acceptable limits. The module exhibited the ability to perform more than six measurement sets per second, indicating a swift response to battery issues. The battery discharge curves illustrate the advantages of the discharge equalization schemes implemented. In papers [21, 22], the author concentrates on the hardware facets of BMSs designed for both EV and stationary applications. The objective is to provide an overview of existing concepts in state-of-the-art systems, allowing the reader to assess the factors to be taken into account when developing a

BMS for a specific application. Following a brief analysis of general requirements, the paper explores various potential topologies for battery packs and delves into their implications for the complexity of the BMS. Four battery packs sourced from commercially available EVs are presented as illustrative examples. Subsequently, the paper delves into implementation considerations related to the measurement of essential physical variables (voltage, current, temperature, etc.), along with addressing balancing issues and strategies. Finally, safety considerations and reliability aspects are thoroughly examined.

3. Issues and Challenges Faced in BMS [23 27]

1. **Battery Degradation:** Batteries in EVs experience degradation over time due to factors like cycling, temperature, and charging patterns. Managing battery health and predicting degradation accurately is crucial for optimizing performance and extending battery lifespan.
2. **Temperature Management:** Temperature extremes, both high and low, can significantly impact battery performance and longevity. Efficient thermal management is necessary to prevent overheating, reduce internal resistance, and ensure optimal charging and discharging.
3. **State Estimation Accuracy:** Accurate estimation of the SOC and SOH is essential for effective battery management. Inaccurate SOC or SOH estimations can lead to suboptimal performance, affecting range predictions and overall battery efficiency.
4. **Cell Balancing:** Variations in individual cell capacities can lead to imbalance issues, affecting overall battery pack performance. Effective balancing ensures uniform charging and discharging of cells, maximizing energy utilization and extending pack life.
5. **Fault Detection and Diagnosis:** Identifying and diagnosing faults in the battery system promptly is crucial for preventing catastrophic failures. Reliable fault detection methods improve safety and reduce the risk of system malfunctions.

3.1 Different State of Charge (SOC) Estimation Method

SOC estimation is a critical aspect of BMSs in EVs and other applications. Accurate SOC estimation ensures efficient and safe use of the battery. Various methods are employed for SOC estimation, and they can be broadly categorized into three main types: direct, model-based, and data-driven methods.

Integration methods like Coulomb Counting integrate the current over time to estimate SOC. It requires accurate current measurement but may suffer from cumulative errors over time. A Throughput Method Estimates SOC by integrating current over time and considering the ampere-hour efficiency of the battery. Model-based methods use physics-based models describing electrochemical reactions

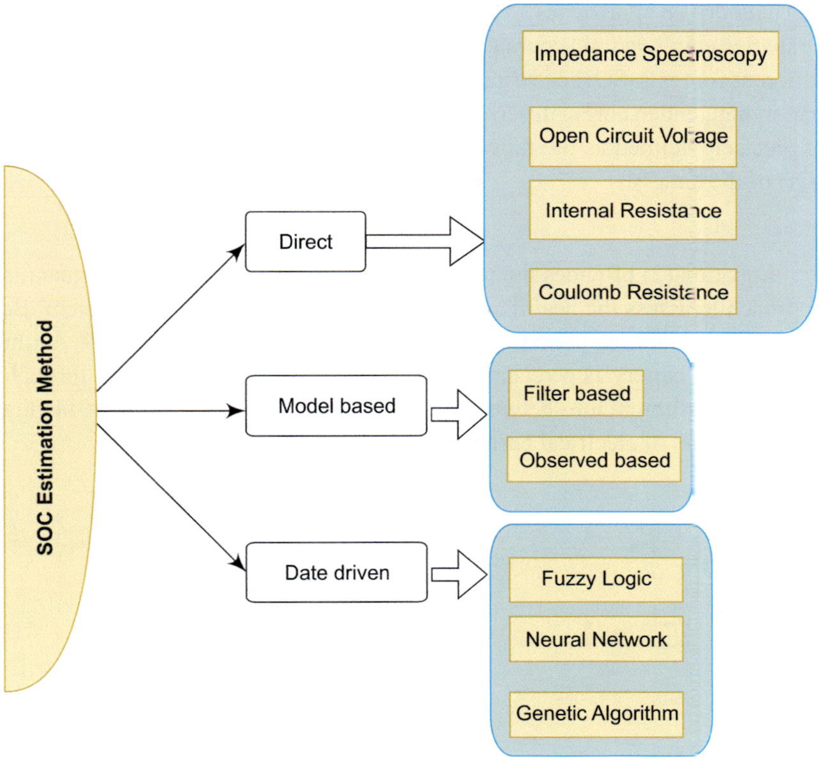

Fig. 3 Different State of Charge (SoC) Estimation Method.

within the battery. Examples include the Newman model, Doyle-Fuller-Newman (DFN) model, etc. Neural Networks are trained using historical data to predict SOC. They can capture complex relationships but require sufficient training data.

4. Design and Simulation Results

Accurate estimation of SOC is vital for effective energy management in EVs , given that SOC serves as a key indicator of available battery capacity. This parameter is defined as the ratio of the current capacity of a cell to its nominal capacity, where the nominal capacity corresponds to the maximum capacity specified by the battery manufacturer. Equation (1) outlines the formula for SOC estimation, emphasizing its crucial role in optimizing energy utilization in EVs.

$$SOC = QC/QN \qquad ...(1)$$

The various techniques utilized for SOC estimation fall into distinct categories: direct, model-based, or data-driven. In these methods, QC represents the current capacity, and QN denotes the nominal capacity. Direct, model-based, and data-driven approaches all play a role in SOC estimation, reflecting the diverse strategies employed to accurately assess the current state of a battery's charge in different contexts. SOH ensuring safety and preventing potential battery failures is paramount, underscoring the critical need for a thorough evaluation of SOH, which is precisely defined as the ratio of discharge capacity (Qd) to the rated capacity (Qr) of the battery.

$$SoH = Qd/Qr \tag{2}$$

Equation (2) provides the means for estimating SOH, offering a quantitative approach to assess the health and overall condition of the battery system. Below represents lithium-ion battery simulation to manage their state of charge. A lithium-ion (Li-ion) battery is a rechargeable type of battery that operates through the migration of lithium ions between its positive and negative electrodes, facilitating the generation of electrical current.

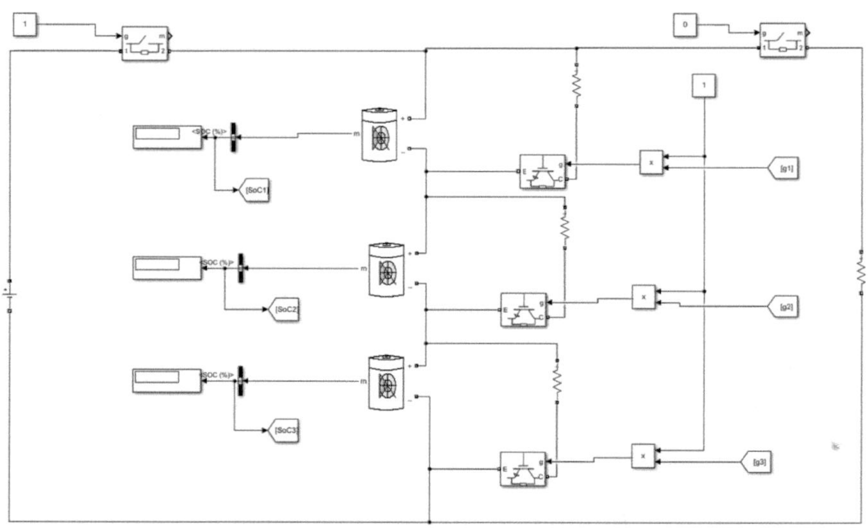

Fig. 4 Lithium-ion battery management system power supply applied condition.

Figure 4 shows the BMS system with power supply applied condition. An IGBT (Insulated Gate Bipolar Transistor) switch is a type of power semiconductor device used for switching and amplifying electrical signals in electronic circuits, especially in power electronic applications. In BMS, the IGBT (Insulated Gate Bipolar Transistor) switch plays a crucial role in controlling and managing the flow of electrical energy within the system. Here five IGBT switches are used to connect the power supply, resistor and gate output. Ideal switches are used to start and end power supply using binary digit 1 and 0 same for load application.

Once power supply is applied from the ideal switch then simultaneously all batteries get equally charged to prevent overcharge. When applied load through the resistor it discharges equally because of the battery management algorithm that manages the state of charge of the battery. In MATLAB, the 'Scope' is often associated with tools or functions that allow you to observe and analyze signals graphically. Display reflects the exact value of SOC. Figure 5 shows the SOC for sample battery, similarly the respective batteries gets charged.

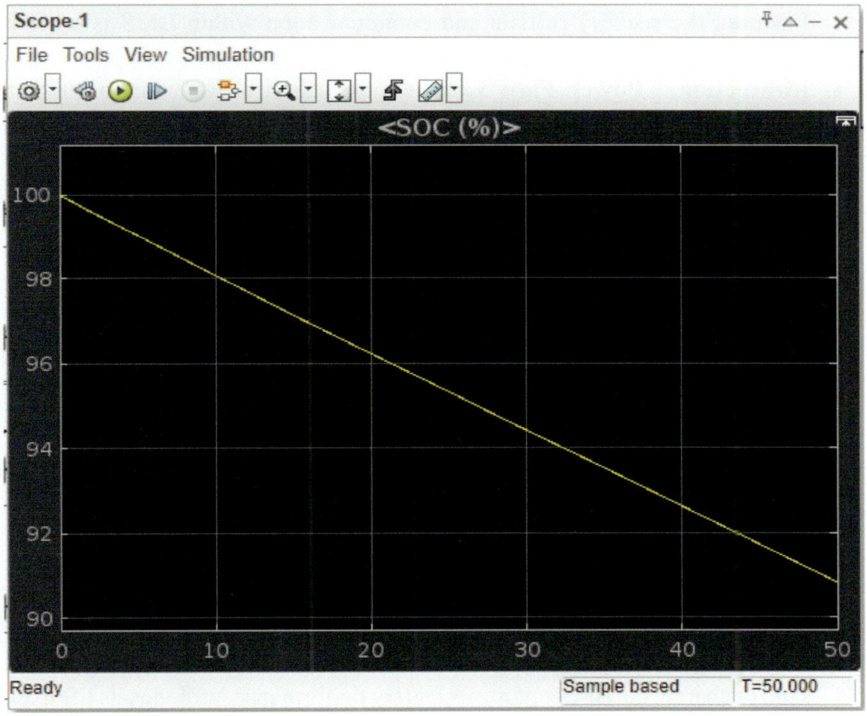

Fig. 5 State of charge condition for battery 1.

This simple MATLAB code simulates a charging process for a battery. It uses a basic charging control logic where the charging continues until the SOC reaches 80%. In a real BMS, you would need to incorporate more sophisticated control strategies, safety checks, and potentially implement features such as cell balancing, temperature monitoring, and fault detection.

5. Recent Trends in BMS

1. **Integration with Energy Storage Systems (ESS):** BMS are increasingly being integrated with energy storage systems, allowing for better control

and optimization of energy storage in applications such as renewable energy storage and grid management.

2. **Advanced Algorithms for SOC and SOH Estimation:** Research and development in algorithms for accurate SOC and SOH estimation are ongoing. Advanced methods, including machine learning algorithms, are being explored to enhance the precision of battery state estimation.
3. **Cybersecurity in BMS:** With the rise of connected devices and the Internet of Things (IoT), there is an increased focus on the cybersecurity of BMS. Ensuring the security of data and communication within BMS is crucial, especially in applications like EVs.
4. **Bidirectional Power Flow Control:** BMSs are evolving to support bidirectional power flow, enabling functionalities such as vehicle-to-grid (V2G) communication in EVs and allowing batteries to contribute power back to the grid when needed.
5. **Thermal Management and Cooling Systems:** Thermal management is a critical aspect of BMS. New trends involve the development of more efficient cooling systems to maintain optimal operating temperatures and extend the lifespan of batteries.
6. **Standardization and Regulatory Compliance:** There is an increasing push towards standardization in BMS to ensure interoperability and compatibility across different battery technologies. Compliance with safety and regulatory standards is also a growing trend.
7. **Advancements in Battery Technologies:** As new battery chemistries and technologies emerge, BMSs are adapting to cater to the specific requirements of different types of batteries, such as lithium-sulfur and solid-state batteries.

6. **Conclusion and Future Scope:** A lithium-ion BMS plays a crucial role in ensuring the safe and efficient operation of lithium-ion batteries. Through the monitoring and control of parameters like voltage, current, temperature, and state of charge, the BMS optimizes battery performance, prolongs its lifespan, and prevents safety hazards such as overcharging, over-discharging, and overheating. The integration of an effective BMS is indispensable in diverse applications, including EVs, renewable energy storage systems, and portable electronics. Matlab Simulink simulates the BMS effectively in the future as the number of battery cells increases and then needs the assistance of AI and ML tools to manage the number of batteries. In this study extended research is to determine the SOH, and temperature condition of each battery cell. The role of BMS in maintaining battery health and safety becomes increasingly crucial.

References

[1] Engelhardt, J., Zepter, J.M., Gabderakhmanova, T. and Marinelli, M. (2022). Energy management of a multi-battery system for renewable-based high power EV charging. *eTransportation*, *14* (August), 100198. doi: 10.1016/j.etran.2022.100198.

[2] Li, J., Zou, W., Yang, Q. and Bao, H. (2022). Towards net-zero smart system: A power synergy management approach of hydrogen and battery hybrid system with hydrogen safety consideration. *Energy Convers. Manag.*, 263 (May), 115717, 2022, doi: 10.1016/j.enconman.2022.115717.

[3] Xie, K., et al. (2022). A method for measuring and evaluating the fault response performance of battery management system. *Energy Reports*, *8*, 639–649. doi: 10.1016/j.egyr.2022.01.223.

[4] Behera, S. and Choudhury, N.B.D. (2022). Modelling and simulations of modified slime mould algorithm based on fuzzy PID to design an optimal battery management system in microgrid. *Clean. Energy Syst.*, *3* (August), 100029. doi: 10.1016/j.cles.2022.100029.

[5] Hamed, M.M., El-Tayeb, A., Moukhtar, I., El Dein, A.Z. and Abdelhameed, E.H. (2022). A review on recent key technologies of lithium-ion battery thermal management: External cooling systems. *Results Eng.*, *16* (July), 100703. doi: 10.1016/j.rineng.2022.100703.

[6] Yang, S., et al. (2021). Implementation for a cloud battery management system based on the CHAIN framework. *Energy AI*, *5*, 100088. doi: 10.1016/j.egyai.2021.100088.

[7] Szumanowski, A. and Chang, Y. (2008). Battery management system based on battery nonlinear dynamics modeling. *IEEE Trans. Veh. Technol.*, *57* (3), pp. 1425–1432. doi: 10.1109/TVT.2007.912176.

[8] Youssef, R., Hosen, M.S., He, J., Al-Saadi, M., Van Mierlo, J. and Berecibar, M. (2022). Novel design optimization for passive cooling PCM assisted battery thermal management system in electric vehicles. *Case Stud. Therm. Eng.*, *32* (February), 101896. doi: 10.1016/j.csite.2022.101896.

[9] Balasingam, B., Ahmed, M. and Pattipati, K. (2020). Battery management systems-challenges and some solutions. *Energies*, *13* (11), 1–19. doi: 10.3390/en13112825.

[10] Li, W., Rentemeister, M., Badeda, J., Jöst, D., Schulte, D. and Sauer, D.U. (2020). Digital twin for battery systems: Cloud battery management system with online state-of-charge and state-of-health estimation. *J. Energy Storage*, *30* (April), 101557. doi: 10.1016/j.est.2020.101557.

[11] Lelie, M., et al. (2018). Battery management system hardware concepts: An overview. ” *Appl. Sci.*, *8*, (4). doi: 10.3390/app8040534.

[12] Xu, H., Duo, Y. and Tang, T. (2022). Development of an edge-cloud collaboration framework for fission battery management system. *Int. J. Adv. Nucl. React. Des. Technol.*, *4* (4), 177–186. doi: 10.1016/j.jandt.2022.11.002.

[13] Zadeh, P.G., Gholamalizadeh, E., Wang, Y. and Chung, J.D. (2022). Electrochemical modeling of a thermal management system for cylindrical lithium-ion battery pack considering battery capacity fade. *Case Stud. Therm. Eng.*, *32*, (February), 101878. doi: 10.1016/j.csite.2022.101878.

[14] IEA. (2023). *Global EV Outlook 2023*. IEA, Paris https://www.iea.org/reports/global-ev-outlook-2023, License: CC BY 4.0.

[15] IEA. (2022). *Battery Demand by Mode, 2016-2022.* IEA, Paris https://www.iea.org/data-and-statistics/charts/battery-demand-by-mode-2016-2022, IEA. Licence: CC BY 4.0.

[16] Kailong, Liu, Kang, Li, Qiao Peng, Cheng Zhang. (2019). A brief review on key technologies in the battery management system of electric vehicles. *Front. Mech. Eng.*, *14*(1): 47–64. https://doi.org/10.1007/s11465-018-0516-8.

[17] Yinjiao Xing, Eden W. M. Ma., Kwok L. Tsui and Michael Pecht. (2011). Battery Management Systems in Electric and Hybrid Vehicles. *Energies*, *4* 1840–1857. doi:10.3390/en4111840.

[18] Ranjith Kumar, R., et al. (2023). Advances in Batteries, Battery Modeling, Battery Management System., *IEEE Access*, e *11*, 105761–105809.

[19] Liu, K., Li, K., Peng, Q. and Zhang, C. (2019, Mar.). A brief review on key technologies in the battery management system of electric vehicles. *Frontiers Mech. Eng.*, *14* (1), 47–64.

[20] John Chatzakis, Kostas Kalaitzakis, Nicholas C. Voulgaris and Stefanos N. Manias, (2023, Oct.). Designing a New Generalized Battery Management System. *IEEE Transactions on Industrial Electronics*, *50*, (5)

[21] Markus Lelie, Thomas Braun, Marcus Knips, Hannes Nordmann, Florian Ringbeck, Hendrik Zappen and Dirk Uwe Sauer, (2018). Battery Management System Hardware Concepts: An Overview", *MDPI Open Access*, *Appl. Sci.*, *8*, 534. doi:10.3390/app8040534.

[22] Zhao, Y., Pohl, O., Bhatt, A.I.. Collis, G.E., Mahon, P.J., Rüther, T. and Hollenkamp, A.F. (2021). A Review on Battery Market Trends, Second-Life Reuse, and Recycling. *Sustain. Chem.*, 2, 167–205. https://doi.org/10.3390/suschem2010011.

[23] Li, P., et al., (2022, Mar.). A Comprehensive Review on the Management of Battery Degradation in Electric Vehicles. *Journal of Power Sources*, *367*, 123–138.

[24] Wang, Y., et al. (2020). Recent Advances in Thermal Management of Battery Systems in Electric Vehicles. *Energy Reports*, *6*, 265–278.

[25] Zhang, J. et al. (2020). State of Charge Estimation Methods for Battery Management System in Electric Vehicles: A Review. *Energy Reports*, *6*, 223–235.

[26] Liu, W. et al. (2016, Mar.). A Review of Active Cell Balancing Technologies for Lithium Ion Battery Packs in Electric Vehicles. *Journal of Power Sources*, *324*, 160–180.

[27] Lu, X. et al. (2018, Sep.). Fault Diagnosis and Prognosis in Lithium-Ion Batteries: A Review. *Energies*, *11* (9), 2346.

13 Strategies for Overcoming Ethical Challenges in IIoT Adoption

Vidyadhari Singh[1*], Archana Kulkarni[2] and Sony Jha[3]

Industrial Internet of Things (IIoT) is an industrial transformation involving internet-connected devices, advanced analytics, and automation. While using IIoT results in operational efficiency, productivity, and innovation across all sectors, it introduces serious ethical and cybersecurity challenges to its adoption. These are some of the most pressing issues related to data privacy, lack of transparency, environmental sustainability, and ethical disposal of IIoT hardware. This paper delves into these challenges, emphasizing the need for robust ethical frameworks, cybersecurity measures, and sustainable practices in a way that allows for responsible IIoT deployment. Privacy-by-design, zero-trust architectures and sustainable supply chain initiatives foster a balance between technological advancement and ethical integrity.

1. Introduction

The Industrial Internet of Things (IIoT) is revolutionizing industries by integrating physical devices, sensors, and software with digital systems to enable seamless data exchange and automation. From manufacturing and transportation to energy and healthcare, IIoT has the potential to improve decision-making, maximize the use of resources, and stimulate creativity. However, this rapid integration raises critical ethical and security concerns. Issues such as data privacy breaches, cybersecurity weaknesses, lack of transparency, and environmental impact have become a challenge for the responsible use of IIoT. As IIoT systems grow, so does the need to address the challenges through wide-ranging ethical frameworks, robust cybersecurity strategies, and effective ways to ensure the durability and adoption of IIoT technology.

[1] *Associate Professor*, Department of Computer Science and Engineering (Cybersecurity), Thakur College of Engineering and Technology, Mumbai, Maharashtra, India.

[2] Department of Computer Science Engineering (Cybersecurity), Thakur College of Engineering and Technology, Mumbai, Maharashtra, India.

[3] Thakur College of Engineering and Technology, Mumbai, Maharashtra, India.
Email: archana.kulkarni@tcetmumbai.in[2]; sony.jha@tcetmum bai.in[3]

* Corresponding author: vidyadhari.singh@tcetmumbai.in

DOI: 10.1201/9781003587903-13

1.1 Industrial Internet of Things (IIoT) Overview

The IIoT is an interconnected network of industrial devices, machinery, sensors, and software applications that enables the instantaneous collection, exchange, and analysis of data. Its goal is to enhance production processes, boost operational efficiency, and support more informed decision-making across sectors like manufacturing, energy, transportation, and utilities. Unlike traditional automation systems, the IIoT integrates advanced technologies directly with industrial machinery and processes. This allows IIoT systems to automate processes, predict maintenance requirements, track the condition of equipment, and increase overall productivity.

At the heart of the IIoT is the concept of cyber-physical systems (CPS), in which physical devices such as sensors, actuators, and control units are integrated with digital systems for efficient data flow and communication. For example, sensors in a manufacturing plant can monitor temperature, pressure, or humidity and transmit the information to an analytics platform. Such real-time feedback allows organizations to identify anomalies in advance, optimize resource allocation, and correct issues before they get out of hand, thus minimizing downtime and costs.

The primary intent of the IIoT is to create "smart factories" in which machines and systems can operate autonomously and adapt to changing production requirements and environmental conditions. Often, this is achieved using digital twins; virtual models of physical systems that simulate real-world scenarios and allow manufacturers to test and troubleshoot processes virtually before implementing them in production.

Although IIoT has several excellent benefits, some challenges arise specifically in the contexts of security and privacy when installing it. Mostly, IIoT systems are used in enterprise networks and the internet, and hackers can use different cyber threats to disrupt important functionalities or steal sensitive data; hence, significant security mechanisms with data availability and integrity, maintaining confidentiality, ought to be established to reduce such a risk.

2. Importance of Ethics in the Adoption of IIoT

The IIoT, which connects systems, sensors, and devices to maximize efficiency, lower expenses, and improve output, is transforming industries. However, the adoption of IIoT brings forward significant ethical considerations because it involves massive data collection, monitoring, and potential impacts on workers, stakeholders, and the environment. Ethical frameworks in the adoption of IIoT are crucial to ensure that trust is safeguarded, and transparency is guaranteed to mitigate risks on data privacy, security, and fair labor practices [6].

The adoption of the IIoT presents a myriad of ethical considerations, which are fundamental to ensuring that technology is integrated responsibly, securely, and sustainably in industrial settings. Ethical issues in IIoT include data privacy, security, transparency, accountability, and the possibility of bias in automated decision-making systems. Ethical frameworks are necessary to address concerns about data ownership, informed consent, and the prevention of misuse or unauthorized access to sensitive industrial data. Ethical adoption also fosters trust among stakeholders, enhances regulatory compliance, and reduces risks associated with worker safety, environmental impact, and economic inequalities. Effective ethical policies ensure that the benefits of IIoT are maximized while minimizing harm, thereby fostering a balanced approach to technological advancement and societal well-being in industrial applications.

2.1 Common Ethical Challenges in IIoT Deployment

In deploying IIoT systems, several ethical challenges commonly arise:

1. **Data Privacy and Ownership:** The IIoT systems generate voluminous amounts of sensitive industrial and personal data, creating questions regarding the ownership of this data and the usage thereof. Without well-defined regulations, unauthorized access and data misuse can become a significant issue while breaching privacy for both the individual and organizations.
2. **Cybersecurity and Data Protection:** IIoT networks are highly interconnected, which exposes them to cyberattacks that can disrupt operations, compromise safety, and expose confidential data. Their protection is a significant ethical priority for both organizational and employee information.
3. **Transparency and Informed Consent:** Often, employees or clients may be unaware of how their data is collected, used, or analyzed within IIoT systems. A lack of transparency and insufficient consent processes can lead to data misuse and a loss of trust.
4. **Bias and Fairness in Decision-Making Automation:** The algorithms of the automated IIoT systems may inadvertently imprint biases, causing unfair treatment of some individuals or groups. These have to be addressed to achieve fair decision-making in industrial conditions.
5. **Worker Safety and Job Displacement:** While IIoT can automate numerous industrial tasks, reducing human intervention, this shift can also lead to job displacement and safety concerns. Organizations must navigate the ethical implications of automation on the workforce, ensuring that workers' safety is not compromised.
6. **Environmental Impact:** IIoT devices and systems consume energy and resources. Therefore, their impact on the environment is significant. Ethical deployment should focus on sustainability by ensuring minimal energy consumption, proper management of electronic waste, and minimal carbon emissions.
7. **Accountability and Liability:** There are several stakeholders involved in an IIoT system, including device manufacturers, software vendors, and others, which makes it very difficult to define accountability in the event of a

system failure, accident, or breach. Ethical issues concerning responsibility and liability arise within the IIoT ecosystem.

2.2 Ethical Issues in IIoT

The IIoT integrates traditional industrial systems with internet-enabled devices, sensors, and data analytics and drives innovation in manufacturing, healthcare, transportation, and other sectors. While IIoT adoption promises operational efficiency, enhanced productivity, and improved decision-making, it also raises some significant ethical concerns, primarily centered on data privacy and cybersecurity risks. The chapter addresses such challenges and provides strategies for overcoming them, thereby ensuring the responsible and sustainable growth of IIoT in various industries.

3. Data Privacy Issues

The most pressing ethical issues facing the adoption of IIoT are the protection of data generated by linked devices and systems from privacy invasion. IIoT systems generate much data, most of which are sensitive, including ambient conditions, staff activity, and equipment utilization. The amount and sensitivity of such data create risks associated with unauthorized access, surveillance, and misuse.

3.1 Lack of Data Collection Transparency

Because many IIoT devices require very little user interaction to operate, there is sometimes a lack of transparency regarding the data being collected. Users or staff might not be completely aware of the types of data being collected, how they are stored, and how they are used. This calls into question informed consent and may cause organizations and their stakeholders to become distrustful of one another [2].

3.2 Inadequate Data Governance

IIoT systems often operate across multiple jurisdictions, each with its own data privacy regulations. While other regions may be less strict, the European Union's General Data Protection Regulation imposes strict data handling rules. To stay compliant, organizations implementing IIoT must navigate this complex regulatory landscape; otherwise, there could be ethical and legal implications [4].

4. Cybersecurity Risks and Vulnerabilities

IIoT networks consist of many interconnected devices, sensors, and machines, which increases the attack surface for malicious actors. The complexity of IIoT systems, along with the limited security capabilities of some devices, amplifies the risk of cyberattacks, which may cause severe financial and reputational damage [1].

4.1 Device Vulnerabilities

Most of the IIoT devices have been designed to have low computation power and without high-end security features, so they are very prone to attack through DDoS and malware penetration. This situation is exacerbated by the existence of legacy systems throughout industries, which are not designed for cybersecurity [5].

4.2 Lack of Standardization

The IIoT environment is quite fragmented with many vendors developing devices that operate by different protocols and standards. The extent to which certain parts are more susceptible than others makes standardization hard to achieve in most cases and difficult to impose uniform security mechanisms across the network [5].

4.3 Insider Threats

The majority of IIoT systems involve shared management from several stakeholders such as staff, third-party vendors, and contractors. An inside threat either inadvertent or malicious is considered the most important potential source of attack on security.

4.4 Controls over Cyber Security Risk

Security breaches from insiders usually arise due to misconfigured devices, insecure access points, or inadequate training of the user.

- **Zero Trust Architecture:** The organizations must adopt a zero-trust security model where no user or device, even inside a network, is trusted automatically. Instead, they need to continuously monitor, verify, and authenticate.
- **Endpoint Security:** Use strong encryption, firmware update, and protect the communication channels between the devices to protect IIoT endpoints.
- **Fostering a Culture of Cybersecurity Awareness:** To reduce the danger of insider threats, regular training and awareness initiatives for contractors and employees are essential. Security can be further improved by promoting safe habits like multi-factor authentication and strong password policies.

4.5 Privacy-by-design in IIoT Systems

Modern industrial initiatives and developments focus on "connecting the unconnected." Applications range from critical infrastructure to contemporary automobiles and industrial control systems, all of which rely on millions of embedded devices today for safety and security. Traditional production engineering, automation, and sophisticated computer systems have combined to form the

IIoT over the last few decades. Production lines, factories, and industrial control systems are becoming more and more equipped with computational components. Programmable logic controllers are being replaced gradually by developments in CPS, which are freely programmable embedded devices that regulate physical processes. Although CPS usually function through closed industrial communication networks, they frequently have Internet connections as well [3].

The development of automated production facilities driven by steam and water indicates the first industrial revolution. The second industrial revolution led to the development of division of labor-based electrically driven mass manufacturing. The third industrial revolution enhanced factory automation by integrating electronics and information technology into production systems. Traditionally, production facilities were composed of a number of separately functioning systems with little to no communication between them. A new trend is the incorporation of more sophisticated electronics into production processes and their connection to traditional company IT systems. The outcome of this has been the rise of the Industrial IoT, thereby setting the basis for a new level of industrial value chain management and organization that will facilitate flexible, resource-effective production and better operations.

5. Security for Industrial IoT

Adapting traditional information security concepts to CPPS is particularly challenging. Classical enterprise IT systems differ from conventional IT systems in many ways. Integrity and confidentiality are the key security goals in classical enterprise IT systems, with protection against cyberattacks often having to balance availability. This cannot be applied in CPPS as availability is important.

Many industries also require traceability of production steps to provide provenance and accountability. As CPPSs become increasingly connected and the possibility of applying Big Data analytics to CPPS data arises, privacy also becomes an important issue. For example, Big Data analysis may infringe on employee privacy or make sensitive customer information available to CPPS manufacturers or service providers.

5.1 The Dimensions of Privacy

Privacy, in the industrial sector, particularly in the new digital age of the IIoT, may become hard to put into words and define since many traditional boundaries of services are washed out by the mere integration of service. Privacy will vary with cases, relationships, and even individual products. Hence, technologists must investigate which privacy aspects have to be created the most for setting IoT infrastructure privacy settings.

Following the aforementioned aspects, we can define privacy:

a. **Identity privacy:** The protection of an entity's identity details is closely related to authentication and authorization. Much of the data collected by I IoT systems is intended for protected user categories. As a result, authentication and authorization become very important, especially while dealing with copyright, patents, and other sensitive issues that determine whether an industry lives or dies.

b. **Location privacy:** Locating an entity may infringe on personal or business privacy, which is a problem in information detection, identification, storage, processing, and dissemination within legal or technical frameworks.
c. **Footprint privacy:** Smart Energy Grids, which rely on two-way, real-time communication, are one example of this. For network users, managing and safely recovering energy data provided with third parties poses a serious privacy concern.
d. **Multidimensional privacy:** It refers to the complicated identification elements of an entity, which can encompass several of the above dimensions. The processes needed to solve these privacy issues are complex, integrated, or tailored solutions that take care of each of the elements within a customized framework.

6. Countermeasures Against Cybersecurity Risks

6.1 Strong Encryption Standards

Encryption safeguards data against unauthorized people for confidentiality and unavailability. Data in IIoT environments is at rest and in transit, and sensitive data is protected by high standards of encryption [7].

- *Symmetric Encryption:* With algorithms like AES, it becomes constantly used for its speed and security.
- *Asymmetric encryption:* The key exchange for secure keys is done with the use of RSA and ECC.
- *TLS (Transport Layer Security):* Provides end-to-end encryption between devices and servers.
- *End-to-End Encryption (E2EE):* Encrypts data from the sender to the receiver without decrypting at intermediate points.
    1. Implement industry-standard encryption protocols such as AES-256 and TLS 1.3.
    2. Use HSMs or cloud-based key management services for secure key management.
    3. Periodically update encryption algorithms to fix vulnerabilities and keep up with standards such as NIST.

6.2 Regular Vulnerability Assessments and Penetration Testing

Vulnerability assessments and penetration testing (VAPT) are proactive approaches to identify and remediate security weaknesses in IIoT systems before they are exploited.

- *Vulnerability Assessments:* Automated tools like Nessus or OpenVAS scan systems for known vulnerabilities.
- *Penetration Testing:* Simulates attacks to find undiscovered vulnerabilities and assess how well security measures are working.

6.3 Multi-Factor Authentication (MFA) for IIoT Devices

Multi-factor authentication (MFA) makes it harder for attackers to gain unauthorized access to IIoT systems [8].

- Components of MFA:
  - Something you know (password or PIN).
  - Something you have (hardware token, mobile device).
  - Something you are (biometric authentication like fingerprint or facial recognition).
- Applications in IIoT: MFA allows only authenticated users or systems to interact with devices.

6.4 Network Segmentation and Secure Communication Protocols

Network segmentation limits the potential spread of breaches by breaking a network up into smaller parts. Secure communication methods ensure the confidentiality and integrity of data while it is being transmitted [9].

- Network Segmentation:
  - Separates critical systems (e.g., SCADA) from less secure systems.
  - Reduces attack surface and lateral movement opportunities.
- Secure Communication Protocols:
  - OPC UA (Open Platform Communications Unified Architecture): Designed for secure, encrypted communications in industrial environments.
  - MQTT with TLS: Lightweight encryption protocol, good for IIoT.

7. Responsible Environment

It refers to the promise of conducting sustainable activities and conscious decision-making to ensure reduced environmental effects. It includes moral, scientific, and innovative ways of reconciling economic growth, social well-being, and environmental conservation.

7.1 Energy-efficient IIoT Devices

Energy-efficient IIoT device integration is crucial to the sustainable and social industrial transformation of industries like manufacturing, transportation, and energy, which are being revolutionized by the IIoT . The purpose of energy-efficient IIoT devices is to reduce power consumption while ensuring reliable connectivity, data processing, and operational capabilities. However, their application raises a number of ethical challenges, especially with regard to resource conservation and environmental sustainability.

Integrating energy-efficient devices into IIoT networks addresses the urgent requirement to reduce carbon footprints in industrial operations. Such gadgets use cutting-edge technologies that make reductions in energy usage possible, such as those applied in edge computing, energy-harvesting sensors, and low-power wireless communication protocols such as ZigBee and LoRaWAN. Industries are thus enabled to increase productivity while holding on to environmentally friendly practices.

Still, there are still some issues in maintaining a balance between performance and energy efficiency. Industries often face the dilemma of device longevity, data privacy, and cost trade-offs with respect to gadget longevity. Giving importance to energy efficiency could have an influence on the devices' processing powers or real-time responsiveness, affecting overall productivity in general. The recycling and disposal processes of IIoT devices at the end of their life-cycle pose ethical dilemmas.

The only way to overcome these challenges is to have a holistic approach that balances long-term sustainability goals with energy-efficient technologies by industries. It involves evaluating the IIoT devices at every step of their life cycle, starting from manufacturing and deployment to disposal, while ensuring adherence to ethical and environmental standards. With the engagement of stakeholders in developing standards and best practices, there should be an opportunity to promote energy-efficient devices without compromising the moral principle of data integrity and the ethical responsibility of protecting the environment.

7.2 Sustainability Aspects in Supply Chains of IIoT

IIoT supply chain integration offers unprecedented opportunities for better decision making, operational transparency, and efficiency. However, the sustainability of supply chains driven by IIoT brings its own set of moral dilemmas that need planning and careful deliberation. The objectives of green supply chains are to minimize the adverse impacts they have on the environment, maximize resource efficiency, and ensure the welfare of all the parties involved. The IIoT can help waste reduction and resource optimization by enabling automation, real-time monitoring, and predictive analytics. To achieve long-term sustainability, however, issues including the energy consumption of IoT devices, the environmental cost of IIoT hardware production, and e-waste management must be resolved.

Organizations must strike a balance between innovation and moral sustainability principles in order to integrate IIoT technologies ethically. This means taking into account the carbon footprint of IIoT operations, ensuring that IIoT systems are energy-efficient, and adopting circular economy concepts for the management of device lifecycles. Increased monitoring and data collection in IIoT supply chains also raises ethical issues and requires careful management of labor conditions and privacy issues to ensure human rights are maintained. Such strategies can help organizations align the implementation of IIoT with the broader sustainability goals in a way that the supply chain ecosystem becomes more ethically and environmentally responsible.

7.3 Responsible Disposal of IIoT Equipment to Reduce Electronic Waste

Adoption of the IIoT has transformed industries through efficiency, automation, and data-driven decision-making. However, the issue of ethical disposal becomes an essential issue as IIoT technology advances rapidly, significantly increasing the amount of electronic trash (e-waste). Since IIoT devices have shorter life spans due to technological enhancements, improper disposal may lead to hazardous components causing health hazards and environmental damage.

7.4 Key Practices for Responsible Disposal

Sustainable Design and Manufacturing: Responsible disposal starts with design. IIoT manufacturers should use sustainable practices, including the use of eco-friendly materials, reducing the use of toxic substances, and producing long-lasting and recyclable products. Modular designs that allow for easy component changes minimize e-waste by reducing the need for frequent hardware repairs.

(a) **Circular Economy Initiatives:** The adoption of a circular economy strategy is necessary to reduce e-waste. This involves establishing a system that allows IIoT devices to be remanufactured, refurbished, or reused instead of being discarded. Industries can reduce waste output and the need for additional resources by recovering valuable materials and returning them into the manufacturing cycle.

(b) **Manufacturers** are subject to EPR legislation, which includes taking back their products at the end-of-life time. Businesses that develop IIoT devices should have take-back programs where they collect and recycle retired equipment with assurances that it is disposed of in an environmentally friendly way. In addition, it encourages producers to design products that are easy to demanufacture and recycle.

(c) **Responsible Recycling Ethics:** During the disposal of IIoT devices, facilities intended for recycling them need to be selected. Recycling processes that comply with environmental and ethical norms can help retrieve valuable materials, such as rare earth metals, reduce their environmental impact, and limit toxic chemicals from filling the landfills. To ensure the responsible dismantling of old IIoT devices, industries should engage with companies that are authorized by relevant authorities to recycle e-waste.

(d) Consciousness and Education: Educating the customer and the organizations should be done regarding the ethical consequences of disposing of IIoT equipment. Training sessions must be held about the benefits of responsible electronic equipment disposal, dangers of improper handling of e-waste, and benefits of recycling. Awareness drives could make the companies take a much more sustainable life cycle of their IIoT devices.

8. Conclusion

The Industrial Internet of Things (IIoT) has the potential to revolutionize industries by streamlining processes, increasing efficiency, and promoting innovation. However, its adoption is accompanied by inherent challenges, especially in terms of ethical compliance, data security, and environmental responsibility. These concerns need to be addressed through a multi-faceted approach that includes robust ethical guidelines, advanced cybersecurity measures, and sustainable lifecycle management of IIoT devices. Through focus on transparency, stakeholder trust, and sustainability in the environment, industries may ensure that they balance the benefits of IIoT while lessening its downsides. Future shaping of a wholesome ethical and secure ecosystem of IIoT will primarily depend on collaborative and proactive measures among stakeholders.

References

[1] Ahmad-Reza Sadeghi1, Christian Wachsmann and Michael Waidner. (2020, May). *Security and Privacy Challenges in the Industrial Internet of Things*. ResearchGate.

[2] Vasiliki Demertzi, Stavros Demertzis and Konstantinos Demertzis. (2023, Jan.). *A Summary of Industrial Internet of Things Privacy Considerations (IIoT)*. arXiv.2301.06172.

[3] Cavoukian, A. (2010). *Privacy by Design: The 7 Foundational Principles: Implementation and Mapping of Fair Information Practices.* Information and Privacy Commissioner of Ontario, Canada.

[4] Mantelero, A. (2014). The EU proposal for a general data protection regulation and the roots of the right to be forgotten. *Computer Law & Security Review*, *30*(3), 229–235.

[5] Roman, R., Lopez, J. and Mambo, M. (2018). Mobile edge computing, Fog et al. A survey and analysis of security threats and challenges. *Future Generation Computer Systems*, *78*, 680–698.

[6] Lee, I. and Lee, K. (2015). The Internet of Things (IoT): Applications, investments, and challenges for enterprises. *Business Horizons*, *58*(4), 431–440.

[7] Bhargavi, H. and Thakur, D. (2022). Implementing Encryption Standards in IIoT for Data Security. *International Journal of Secure Systems*, *8*(3), 245–261.

[8] Dutta, A. and Singh, P. (2023). MFA Implementation in IIoT Environments: Challenges and Solutions. *IEEE Transactions on Industrial Informatics*, *19*(4), 569–581.

[9] Kothari, R. and Zhang, L. (2022). Network Segmentation and Communication Protocols for IIoT Security. *Journal of Industrial Cybersecurity*, *7*(2), 87–102. *(Provides a detailed analysis of segmentation and protocol security in IIoT.)*

14 Security and Privacy in Industrial IoT Environments

Prutha Trivedi[1*], Aman Verma[2], Nandini Seth[3] and Jalpaben Pandya[4]

Technological advancements have transformed industrial operations, underscoring the importance of security and privacy in Industrial Internet of Things (IIoT) ecosystems. Real-time surveillance and rapid response mechanisms are essential to counter challenges such as unauthorized access, equipment tampering, and system vulnerabilities. This chapter outlines a security framework built around an ESP32 camera module, functioning as the main controller, paired with a PIR sensor for motion detection. The framework ensures comprehensive monitoring of industrial sites by securely transmitting data to a cloud-based server, reinforced with firewalls and Java socket programming to address potential security breaches. The PIR sensor activates the ESP32 camera when motion is detected, prompting data capture and alert generation for potential threats. Experimental evaluations confirm the system's effectiveness in delivering reliable surveillance and maintaining privacy, emphasizing its role in improving security across industrial applications.

1. Introduction

In Industrial IoT (IIoT) environments, addressing security and privacy concerns has emerged as a crucial priority [1]. As industries adopt interconnected systems to optimize their operations, they simultaneously face increased exposure to security threats. Risks such as unauthorized access, data leaks, and system interruptions can severely impact industrial processes, underscoring the need for robust security measures to safeguard these advanced infrastructures[2]. To address these challenges, it is essential to develop robust systems that combine real-time monitoring with secure data transmission, ensuring privacy while maintaining functionality[9].

Security has grown essential, especially for home security, given the variety of scenarios that arise, such as crimes and resident negligence that can cause

[1-3] Dept. of Electronics Engineering, Thakur College of Engineering and Technology, Mumbai, India.
[4] Faculty of Electronics Engineering, Thakur College of Engineering and Technology, Mumbai, India.
Email: jalpa.pandya@thakureducation.org[1]
* Corresponding author: Prutha19121@gmail.com

DOI: 10.1201/9781003587903-14

problems like security issues. In this case, a remote visual status reporting system for the room is necessary. This helps with the surveillance procedure by enabling the system to notify a room's condition via a smartphone in the event of a crime occurring there [10].

In recent times, camera technology has become a vital tool for room monitoring, as it provides visual data that is simple and intuitive for users to interpret. Additionally, microcontroller technology plays a significant role in the monitoring and control aspects of security systems [5].

Although room monitoring systems are widely implemented, their application in residential settings, where residents often leave their belongings unattended, introduces distinct challenges that require innovative solutions. One way to address these issues is to create a basic home security system that can monitor a room's condition and deter crime.

Studies have to be performed to construct and install a Esp32 Cam room security monitoring device that enables photographs from photo seizure to be visible on mobile devices like computers and smartphones which can be linked to the community, so that you can improve home protection for the owner [4]. Certain security device customization is required in the current home security systems in order to perform room monitoring and remote control of the system and prevent cyber threats such as ransomware and malware assaults [5].

2. Design Framework

The development of a home security system begins with a comprehensive literature review to understand existing technologies and methodologies. This study provides the foundational knowledge required to proceed with system planning. During the planning phase, the key design elements, functionalities, and system requirements are outlined.

Following this, the hardware and software components of the system are developed based on the established plan. Once the system components are ready, a series of system tests are conducted to evaluate the performance and functionality of the security system [12]. If the test results indicate that the system is not suitable or does not meet the requirements, necessary adjustments or improvements are made, and the testing process is repeated.

When the system is deemed suitable, data retrieval processes are carried out to collect relevant information from the system's operation. This data is then analyzed to derive meaningful insights and conclusions regarding the system's effectiveness and potential areas for further improvement. The process concludes with a comprehensive evaluation of the system's overall performance and its readiness for deployment.

The design approach relies on leveraging the insights from prior research involving PIR sensors, firewalls, ESP32-CAM microcontrollers, and GSM modules. These components have been extensively studied for their effectiveness in enhancing system functionality, security, and communication. By referencing these studies, the design aims to establish a robust framework that combines motion detection, real-time surveillance, and secure data transmission. The literature review plays a pivotal role in validating the integration of these technologies and identifying potential improvements [9]. Following this, system planning becomes the next critical step, ensuring seamless implementation and optimization of the proposed design.

At this point, planning activities will be carried out to begin creating an overview of the system that will be constructed. The AM312 PIR sensor, firewall, Esp32 cam micro controller, GSM module, and the necessary software—Java, Arduino IDE, Eloquent motion, AWS, and the notification app on a smartphone to show the results—are the required tools and supplies [7].

The subsequent phase focuses on system development, where Arduino programming is employed alongside the integration of mechanical tools and notification applications [4]. These elements are carefully incorporated into the device to ensure functionality and alignment with the desired objectives, paving the way for achieving the intended results efficiently.

The tool will be tested as a whole in the next step, called system testing, which is done to determine whether or not the tool is operating as intended. Repairs and rearranging will take place whether the errors persist or disappear [12]. The most distance that the AM312 PIR sensor will not detect is five meters. Even though records collection is needed to set up this limit, the object's distance at the sensor may additionally have an effect on the Passive Infrared (PIR) sensor's reaction time to occur.

The system architecture integrates multiple components for efficient operation and communication. A PIR sensor detects motion and sends signals to the microcontroller, which acts as the central processing unit [3]. The microcontroller is connected to a relay to control the solenoid, enabling specific mechanical actions. A buzzer is also connected to the microcontroller, providing audible alerts when triggered [13]. The microcontroller is powered by a power adapter and is linked to an access point for internet connectivity. This connection allows seamless communication with the Internet cloud, enabling remote monitoring and control via a smartphone. Wired connections are utilized between the PIR sensor, relay, solenoid, buzzer, and microcontroller, while wireless communication is established between the microcontroller, access point, and smartphone, ensuring reliable and flexible system operation. Initially, the program will use the motion sensor as a detector to alert the user to a potential threat in the room by sending a text message to their Telegram application. The PIR sensor on the Esp32 cam has the capabilities as an energetic digicam that when an item moves, it activates the camera to take photographs and ship the statistics to the cloud database [7] in which the dispatched statistics can be seen remotely, in actual time, the use of the Telegram app.

2.1 AM312 PIR Sensor

The AM312 PIR sensor is designed to detect motion by sensing infrared radiation changes in its environment [3]. When the sensor detects motion, it transitions to a high state (logic 1), triggering the connected buzzer to emit an alarm sound. The PIR sensor has three pins: the input voltage pin, the output pin, and the ground pin. In contrast, the buzzer is equipped with two pins: the positive pin and the ground pin. The output pin of the sensor is connected to the input of the buzzer to enable the alarm functionality [13]. This simple configuration allows for efficient motion detection and immediate alert generation. The pin out of the device is linked to the microcontroller's GPIO 12 for data, the Vin pin is connected to the microcontroller's 5V VCC, and the ground pin is connected to the microcontroller's GND pin. The device requires 5V as a power supply. The GND pin of the Esp32 cam micro controller is attached to the negative (–) buzzer pin, whilst its GPIO pin 2 is attached to the high quality (+) buzzer pin. The GND pin of the Esp32 cam micro-controller is hooked up to the buzzer pin that is negative (–) [14].

2.1.1 Planning Telegram's Data Message Display

The setting of the notification of message content that shows up on the Telegram utility while the property owner uses the far flung manipulate. In addition, homeowners get real-time access to the surveillance room's condition and the state of the door that is currently locked or open in the house [9].

3. Setting up Firewall Using Java Socket Programming

1. Define Firewall Rules

A firewall developed using Java socket programming involves designing a system that regulates and monitors network traffic based on predefined criteria. These criteria, known as firewall rules, determine the conditions under which connections are allowed or denied. Rules may be based on factors such as IP addresses, port numbers, protocols, or other attributes of data packets. For instance, access may be granted or restricted to specific ports like port 80 for HTTP (hypertext transfer protocol) traffic, while certain IP addresses may be blocked or permitted depending on the security requirements [11]. Additionally, the firewall logic evaluates incoming connections by comparing the source and destination IP addresses, port numbers, and protocols against the defined rules. A dynamic rule configuration may also be implemented, allowing administrators to modify the rules—such as adding or removing allowed IP addresses—without needing to interrupt the firewall's operation. This approach ensures that the system can adapt to changing security needs in real time, providing flexibility and enhanced control over network access.

2. Set Up a Server Socket

Creating a server application that listens for incoming client connections involves setting up a "ServerSocket" in Java, a key component of the java.net package. The "ServerSocket" facilitates the creation of a server that binds to a specific port and accepts connections from clients. To begin, it is crucial to import the necessary Java packages and select an appropriate port number, which serves as the logical endpoint for client communication. Ports are often chosen based on convention, such as port 80 for HTTP services [10]. After deciding on a port, a 'ServerSocket' object is instantiated to bind the server to the host machine's specified port. Developers must account for potential exceptions like 'IOException' during this process to ensure robust error handling. Once initialized, the server enters a loop using the 'accept()' method, which blocks until a client connects, subsequently returning a 'Socket' object that represents the connection. For handling multiple clients simultaneously, it is common practice to create a new thread or use an executor for each connection, enabling efficient and concurrent communication between the server and its clients [11].

3. Accept Incoming Connections

Once a 'Server Socket' is set up and bound to a specific port, it continuously waits for incoming connection requests. When a client sends a connection request to the server, the 'Server Socket' accepts the request and returns a 'Socket' object that represents the communication channel between the server and the client.

The 'accept ()' function, a method within the 'ServerSocket' class, is designed to handle incoming connection requests. This function operates by halting execution until a connection is successfully established. Once a connection is accepted, it provides a 'Socket' object that facilitates the communication channel between the client and the server. Commonly, this process of accepting connections is implemented inside an infinite loop to ensure continuous handling of client requests.

This allows the server to continuously listen for and accept incoming connections, providing the ability to handle multiple clients concurrently [1]. Once a connection is accepted, you can pass the 'Socket' object to a separate thread or class (e.g., "Connection Handler") to manage the communication with the client. This approach is essential for handling multiple clients simultaneously. Handle exceptions may occur during the acceptance of connections. The 'accept()' method can throw an 'IO Exception', and proper error handling is essential for robust server applications.

4. Implement Connection Handling

Connection handling involves defining the behavior of the server once a connection is established with a client. This can include tasks such as reading and writing data, processing client requests, and managing the lifecycle of the connection. Define a class, often named 'Connection Handler', that implements the 'Runnable' interface. This class will encapsulate the logic for handling an individual connection.

In the 'run ()' method of the 'Connection Handler' class, the core functionality for handling client communication can be implemented. This includes utilizing input and output streams to read data from the client and send responses back. Additional logic, such as interpreting client requests, executing required actions, or managing connection lifecycles, can be incorporated based on the application's needs. Since each instance of the 'Connection Handler' typically operates in its own thread, this enables the server to manage multiple client connections simultaneously. Proper thread management is crucial to maintaining the server's scalability and ensuring efficient operation.

5. Implement Firewall Rules in Connection-Handler

Firewall rules specify the criteria for permitting or denying a connection. Within the 'Connection Handler' class, these rules are implemented to evaluate incoming connections before further processing [10]. The process involves creating an instance of the firewall logic directly within the 'Connection Handler' class. The IP address of the connected client can be retrieved using the method clientSocket.getInetAddress(). By integrating the firewall logic, the system determines whether to approve or block the connection based on the client's IP address, ensuring only authorized access is granted while maintaining security. Within the 'if [firewallRules.isAllowed (clientIpAddress)]' block, implement the logic for handling allowed connections. This can include reading and writing data, processing client requests, etc.

6. Handle Allowed Connections

When the firewall rules permit a connection, the responsibility of managing communication with the client falls to the 'Connection Handler' class. This involves several tasks, such as reading incoming data, handling client requests, and sending appropriate responses. Input and output streams are established to facilitate data exchange over the connection. The input stream is utilized to receive data sent by the client, enabling the server to interpret and process the received information or requests [11]. Based on the data processed, the output stream is employed to transmit suitable responses back to the client. Proper handling of the connection lifecycle is crucial, which includes closing the connection once the requests have been addressed and responses have been delivered, ensuring efficient resource management and system reliability.

7. Run and Test

To test the functionality of a firewall program, it is essential to compile and execute the code, ensuring it operates as intended. Once the program is running, connect to it from various client systems to evaluate its behavior. During this process, observe whether the configured firewall rules are enforced accurately by monitoring the responses to different connection attempts. This approach verifies the effectiveness of the rules in controlling access and ensures that the firewall performs its intended function of filtering and managing network traffic securely.

3.1 Connecting ESP32 cam to Wifi and Display Results on Eloquent Motion Server

To develop a fully functional smart home automation system with ESP32 integration, several key steps are involved [4]. The first step is to securely connect the ESP32 to your home Wi-Fi network. This requires programming the ESP32 with the correct Wi-Fi credentials, utilizing strong encryption protocols such as WPA2 or higher to safeguard the connection against unauthorized access [6]. Once the ESP32 is connected to the Wi-Fi, the next task is to integrate various sensors and actuators. This involves writing custom firmware for the ESP32 to enable it to gather data from sensors like motion detectors or temperature sensors and control devices like smart lights or locks. Libraries and communication protocols should be chosen carefully to ensure seamless and secure data transfer between the ESP32 and the connected devices [4].

Next, the system can be extended to the cloud for remote control and monitoring. By integrating the ESP32 with a cloud platform, device registries need to be set up, along with security configurations to prevent unauthorized interactions. Communication protocols such as MQTT (message queuing telemetry transport) or HTTPS are commonly used to facilitate secure and reliable communication between the cloud and the ESP32 [10]. Additionally, establishing a cloud database to store sensor data and device statuses enables data persistence and easy retrieval, which enhances the scalability of the smart home system. To further strengthen security, implementing end-to-end encryption for all communications between the ESP32 devices and the cloud is essential [8]. Secure user authentication and access controls must also be in place to restrict control to authorized users only.

For the end user, creating a mobile app or web interface is crucial to provide a convenient and secure way to interact with the smart home system. This interface will communicate with the cloud platform, allowing users to control and monitor their devices remotely [8]. It is important to ensure the interface is intuitive and user-friendly, providing a seamless experience. Cloud-to-device communication must also be established so the cloud can send commands to the ESP32 devices and receive updates, enabling real-time synchronization of the system. User authentication and authorization protocols should be applied throughout to maintain security by ensuring only authorized users can access and control the devices.

Finally, the smart home system must undergo comprehensive testing to verify its functionality, security, and reliability. Identifying and resolving any issues during testing is crucial to ensure the system operates smoothly. A system for regular updates and monitoring should also be implemented, allowing the ESP32 firmware to be updated regularly to address any security vulnerabilities or performance issues, while also monitoring system logs for anomalies or potential threats. This approach ensures the long-term effectiveness and security of the smart home automation system [4].

To create a comprehensive smart home automation system with ESP32 integration, let's break down each step into a detailed paragraph explanation:

4. Results and Discussion

The system's performance was observed under two conditions: day and night, with varying distances between the PIR sensor and the object. During the day, the sensor detected motion at all tested distances ranging from 1–4 meters. At distances of 1 meter, 2 meters, and 3 meters, motion was detected almost immediately, and the photo was successfully sent to the Telegram bot [3]. At a distance of 4 meters, the sensor responded within 1–5 seconds, and again, the photo was sent successfully. However, at a distance of 5 meters, no motion was detected, and the system failed to send a photo.

At night, the system showed similar behavior but with slightly longer response times. For distances of 1 meter and 2 meters, motion was detected, and the photo was sent, although the response time increased to 1–2 seconds for the 1-meter distance, and 1–7 seconds for the 2-meter distance [3]. At a 3-meter distance, the motion detection response was slightly delayed at 2 seconds, but the photo was still successfully sent. However, at distances of 4 meters and 5 meters, no motion was detected, and the system failed to send the photo [5].

This analysis suggests that the PIR sensor performs well within certain distance ranges and under different lighting conditions, but its effectiveness diminishes beyond a certain threshold, especially when no motion is detected.

4.1 Results of Implementation of Telegram Application

It is possible to deduce from this test that every menu in the utilized application may effectively and efficiently carry out every task.

4.2 Using Wireshark to Measure Delay

Delay measurement is used to determine whether the system is designed to be of sufficient quality or not. Wireshark is used for testing, and the IP address of the destination is examined. It'll be capable of processing statistics for verbal exchange built from this machine primarily based at the IP that is later acquired, that is, a photograph of the Wireshark put off studying.

4.3 Packet Loss Check of Wireshark

The quantity of packets out of place on a packet network because of collisions, community capacity exhaustion, and packet drops added on by the use of TTL (Time to stay) packet expiration is referred to as packet loss [14].

The delay values recorded during testing using Wireshark highlight the time intervals between successive transmission control protocol (TCP) packets transmitted from the source IP address 192.168.2.77 to the destination IP address 149.154.167.99. The measurements reveal a range of delay durations, starting from as low as 0.000006 seconds to as high as 0.254377 seconds. Notable time differences include intervals of 0.000329 seconds, 0.115237 seconds, and 0.250652 seconds, which emphasize variations in packet transmission times. The calculated average delay for the observed transmissions is approximately 0.05974 seconds. This demonstrates that while some packet transmissions occur almost instantaneously, others exhibit slight delays, likely influenced by network conditions or processing latency, which is typical in TCP communications. Such insights are essential for understanding network performance and optimizing data flow in similar scenarios [15].

The analysis indicates that the packet loss rate observed is 10%, which places it within the medium category of packet loss based on the predefined classification criteria. This classification is derived from previously established thresholds, where a 10% loss is characterized as moderate. Such a categorization is significant as it highlights a notable level of data loss during transmission, potentially impacting network performance. Understanding this classification helps in diagnosing and addressing issues related to data transfer inefficiencies and ensuring optimal communication reliability.

5. Conclusion

The integration of an AM312 PIR sensor with the ESP32-CAM microcontroller presents a highly effective approach to enhancing security and privacy in IIoT settings [3]. This system leverages the ESP32-CAM's built-in Wi-Fi capabilities to transmit data seamlessly to connected devices, including smartphones, facilitating real-time alerts through applications like Telegram. Evaluations indicate the system performs exceptionally well, with an average service quality delay of 0.0597 seconds, classified as optimal, and a packet loss rate of 7.2%, which is within acceptable thresholds. This innovative architecture enhances industrial security by providing continuous monitoring and instant alerts while safeguarding data integrity and ensuring secure communication. By addressing key privacy challenges and optimizing performance parameters, this solution underscores the critical role of advanced IoT technologies in strengthening industrial operations against modern threats and bolstering safety and resilience.

References (APA Format)

[1] M. Rouse, "IoT security (Internet of Things security)," TechTarget, 2023. [Online]. Available: https://www.techtarget.com/searchsecurity/definition/IoT-security

[2] C. Maple, "Security and privacy in the Internet of Things (IoT): Models, algorithms, and implementations," Journal of Cyber Policy, vol. 2, no. 2, pp. 208–229, 2017.

[3] H. Suo, J. Wan, C. Zou and J. Liu, "Security in the Internet of Things: A review," in Proc. IEEE Int. Conf. Computer Science and Electronics Engineering (ICCSEE), Hangzhou, China, Mar. 2012, pp. 648–651.

[4] A. Banerjee, R. Pal and S. Mukherjee, "Design and implementation of a low cost home security system using Raspberry Pi," International Journal of Engineering Research and Technology, vol. 3, no. 12, pp. 478–481, 2014.

[5] S. Ray, "Integrating camera modules in IoT systems for surveillance," International Journal of Advanced Research in Computer Science, vol. 10, no. 2, pp. 230–236, 2019.

[6] D.A. Gajjar, "Home automation using ESP32 and Blynk," International Journal of Engineering Sciences & Research Technology, vol. 9, no. 5, pp. 155–160, 2020.

[7] A. Kumar and S. Tripathi, "A smart security system using PIR sensor and GSM module," in Proc. Int. Conf. on Computing, Communication and Automation (ICCCA), Greater Noida, India, 2016, pp. 1335–1339.

[8] D. Kaur and P. Kaur, "IoT based home automation system using ESP32," International Journal of Scientific Research in Engineering and Management (IJSREM), vol. 5, no. 7, pp. 1–5, 2021.

[9] R. Kumar, A. Singh and N. Sharma, "ESP32-CAM based real-time surveillance system with cloud integration," International Journal of Computer Applications, vol. 177, no. 9, pp. 21–25, 2020.

[10] Y. Zhang, "An overview of socket programming in Java," Journal of Computer Science and Applications, vol. 6, no. 2, pp. 45–51, 2018.

[11] A. Maheshwari, "Firewall architecture using Java sockets for IoT-based systems," International Journal of Computer Applications Technology and Research, vol. 8, no. 6, pp. 239–243, 2019.

[12] K.S. Sandhu, "Smart security using PIR sensor with IoT," in Proc. 2020 Int. Conf. on Computing, Communication and Automation (ICCCA), 2020, pp. 645–649.

[13] N. Patel and S. Dey, "Design and implementation of motion sensing alert system using PIR and ESP32," Journal of Electronics and Communication Engineering, vol. 14, no. 4, pp. 15–21, 2021.

[14] A. Sharma and V. Agarwal, "Remote monitoring using ESP32 and Telegram Bot API," International Journal of Engineering Research & Technology, vol. 10, no. 8, pp. 1–5, 2021.

[15] R. Singh, "Network packet analysis using Wireshark," International Journal of Scientific & Technology Research, vol. 9, no. 4, pp. 2008–2011, 2020.

15 Conclusion
Shaping a Responsible Technological Future

Bhushankumar Nemade[1*], Deven Shah[2], Kiran Maharana[3] and Vinayak Bharadi[4]

The rapid expansion of the IIoT raises both transformative opportunities and significant challenges related to ethical, social, and environmental responsibility. This concluding chapter synthesizes key insights developed in the preceding chapters and puts forward a vision for responsible development of the IIoT with an emphasis on integrating ethical considerations throughout the entire lifecycle of an IIoT system, its design and implementation with sustainability at the core, and robust security, privacy, and data governance. The chapter focuses on human-centric and community-oriented IIoT solutions, highlighting the role of transparency, accountability, and regulatory compliance in building trust. Furthermore, it emphasizes the need for cross-industry collaboration to foster innovation and ensure that IIoT advancements are aligned with societal needs. Embracing collective responsibility—of developers, policymakers, businesses, and end-users alike—IIoT can be leveraged toward a more sustainable, ethical, and equitable technological future.

1. Introduction

The Industrial Internet of Things (IIoT) represents a revolutionary shift in industrial processes, offering unprecedented levels of efficiency, connectivity, and innovation [1]. But there are significant societal, ethical, and environmental issues raised by this development. This last chapter summarizes the main ideas from the earlier talks and provides a road map for creating a responsible technological future. Our goal is to direct the creation of IIoT systems that benefit society as a whole by emphasizing ethical issues, sustainability, security, human-centric design, cooperation, and regulatory frameworks.

Collective responsibility can shape the future of IIoT, wherein developers, engineers, business leaders, policymakers, and end-users engage actively with it

[1-2] Shree L.R. Tiwari College of Engineering, Mumbai, India.
[3] Researcher, ICICI Lombard GIC Ltd, Mumbai, India.
[4] Finolex Academy of Management and Technology, Mumbai University, India.
Email: sir.deven@gmail.com[2]; kiranmaharanak@gmail.com[3]; vinayak.bharadi@outlook.com[4]
* Corresponding author: bnemad@gmail.com

DOI: 10.1201/9781003587903-15

[2, 3]. Embracing this collective responsibility and agreeing upon the principles within this book means that we'll be making technology enabled through IIoT advance a better future for this planet-being more sustainable, ethical, and equitable.

2. Integrating Ethical Considerations Across the IIoT Lifecycle

2.1 The Ethical Imperative in IIoT Development

The ethical imperative in the development of IIoT is not a mere theory but rather a practical need. As IIoT systems become an integral part of industrial operations, decisions made by developers and engineers will have far-reaching consequences in society [4]. Ethical considerations must be incorporated into every phase of the IIoT lifecycle, from design to deployment and beyond.

It goes without saying that the core of this ethical requirement lies in the balance of technological innovation and societal values. Developers and engineers must consider how IIoT may impact the lives of employees, customers, communities, and the environment. They must ask critical questions: How will this technology affect people's lives? What are the potential risks and benefits? How can harm be minimized and positive outcomes maximized?

According to a survey by Cisco, 84% of consumers are concerned about the privacy of their data when using IoT devices [5]. Another study found that 70% of IoT devices have security vulnerabilities, with each device averaging 25 flaws [6]. These weaknesses contributed to a 300% surge in cyberattacks on IoT devices in the first half of 2019 [7]. Attackers exploit these vulnerabilities in various ways, including intercepting personal information, forming large-scale botnets [8], executing distributed denial of service (DDoS) attacks against vital infrastructure [9], distributing phishing and spam emails [10], and stealing valuable intellectual property [11].

These imperfections can be corrected by IoT manufacturers to minimize the risk of cyberattacks and avoid potential harm to consumers and organizations. One of the major difficulties in incorporating ethical considerations is the complexity of IIoT systems. In fact, IIoT systems involve multiple stakeholders with diverse interests and concerns, so a more subtle approach to ethical decision-making should be considered for all perspectives and values.

2.2 Navigating Ethical Challenges in IIoT Implementation

Implementing IIoT systems raises a number of ethical considerations, particularly with regard to unanticipated consequences. For instance, although IIoT may increase efficiency and production, it may also lead to job displacement and changes in workforce dynamics. Similarly, the potential of the IIoT to increase data collection and analysis raises privacy and security concerns.

To address these ethical challenges, there is a need for proactive steps. Organizations must anticipate future ethical quandaries and develop strategies to meet them. This may entail conducting ethical impact assessments, consulting with stakeholders, and developing ethical guidelines for IIoT development and implementation.

Transparency is one of the key elements in dealing with ethical challenges. Organizations should be transparent about their decision-making processes and the potential impacts of their IIoT systems. Transparency builds trust with stakeholders and ensures that ethical considerations are not overlooked in the pursuit of technological advancement [12].

2.3 Ethical Audits and Continuous Assessment

Ethical audits and continuous assessments are the responsibilities of organizations in maintaining ethical considerations throughout the IIoT lifecycle. Ethical audits systematically evaluate an organization's IIoT practices, identify potential ethical risks, and ensure compliance with established guidelines.

Continuous assessment involves reviewing and updating ethics guidelines regularly to note new developments and challenges. This could mean monitoring the results of IIoT systems to stakeholders, conducting regular consultations with the different stakeholders, and updating ethics guidelines whenever necessary [13].

Organizations can ensure that their IIoT systems are aligned with societal values and ethical principles by implementing ethical audits and continuous assessments [14]. This will not only prevent ethical breaches but also help inculcate a sense of responsibility and accountability within organizations.

3. Embedding Sustainability in Design and Implementation

3.1 Environmental Impact of IIoT Systems

The environmental impact of IIoT systems is a major concern. As IIoT systems are more and more implemented, they contribute to increase energy consumption, electronic waste, and resource depletion. The manufacturing, deployment, and operation of IIoT devices require significant energy and raw materials, leading to a substantial environmental footprint.

One of the primary environmental concerns of IIoT is energy consumption. IIoT devices, such as sensors, actuators, and communication modules, consume continuous power to operate. With an increase in the number of connected devices, the demand for energy increases, straining existing energy infrastructure and contributing to greenhouse gas emissions.

Another major environmental concern is electronic waste, or e-waste. IIoT devices, as with all electronics, have a limited lifespan and eventually become outdated [15]. When disposed of, they are part of the growing e-waste problem that contains hazardous materials that can have adverse effects on the environment and human health.

Addressing the environmental impact of IIoT systems requires a concerted effort to embed sustainability into their design and implementation. This involves adopting sustainable design principles, improving energy efficiency, and developing strategies for managing e-waste.

3.2 Principles of Sustainable Design in IIoT

Sustainable design addresses the entire lifecycle of IIoT devices from raw material extraction to end-of-life disposal. Sustainable design allows organizations to reduce the environmental footprint of their IIoT systems, thus contributing to a more sustainable future.

Energy efficiency is one of the most basic principles of sustainable design. Energy-efficient designs reduce the energy required to run IIoT devices, which in turn reduces greenhouse gas emissions and limits the pressure on energy infrastructure [16]. This can be achieved through low power consuming components, optimized algorithms, and energy harvesting technologies.

The design of renewable materials is another important principle. Materials to be used in IIoT devices should be abundant, renewable, and with low environmental impact. Such materials include recycled material, biodegradable plastics, and sustainable metals. Thus, dependency on non-renewable resources may be reduced, and the organization decreases its environmental footprint from its IIoT systems.

Another important aspect is designing for recyclability. IIoT devices must be designed such that they are easily disassembled and recycled at the end of their life. This helps reduce e-waste and recovers valuable materials for use in new devices [17].

3.3 Case Studies of Sustainable IIoT Implementations

One of the best ways to understand the practical use of sustainable design principles is through case studies on successful implementations of IIoT, with a focus on sustainability. Such case studies show how one can minimize environmental impact in their IIoT system with a minimum decline in performance.

For instance, smart grids utilize energy-efficient IIoT systems. Smart grids deploy IIoT devices to monitor and optimize the distribution of electricity into and out of homes, with the result being reduced energy waste and maximum efficiency in the supply of energy [17]. Building with energy-efficient designs and taking note of renewable resources, smart grids can make a significant dent in greenhouse gases and hence improve the sustainability of energy sources.

Another example is incorporating sustainable materials in designing IIoT devices for farming. Agricultural devices, such as soil sensors, and weather stations, are likely to be harshly exposed to environmental conditions. Implementing durable, bio degradable, and recyclable materials can diminish the environmental consequences of these agricultural devices and reduce waste [15].

These case studies demonstrate the possibility of building and deploying highly efficient and sustainable IIoT systems. An emphasis on sustainability in the design and implementation of IIoT systems will therefore enable organizations to contribute to a more sustainable future while reaping the benefits of technological innovation.

4. Ensuring Security, Privacy, and Data Governance

4.1 Addressing Security Challenges in IIoT

Security is a paramount concern in the IIoT environment as devices and systems become increasingly interconnected with one another and vulnerable to cyberattacks and data breaches. With the growing trend of IIoT systems, security threats are also likely to increase, necessitating implementation of strong security at every stage of the IIoT lifecycle.

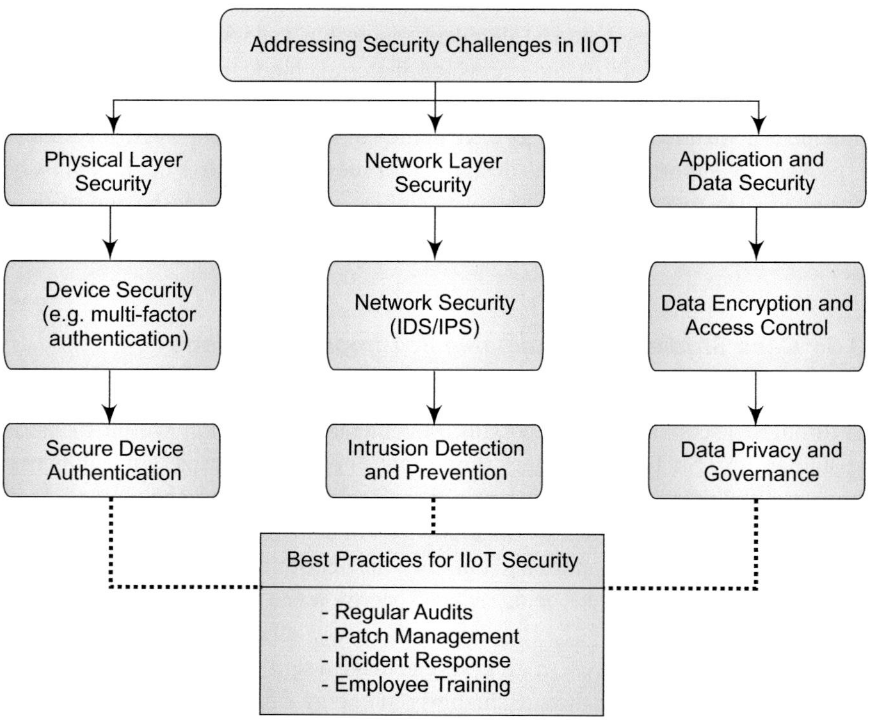

Fig. 1 Ensuring Security, Privacy, and Data Governance.

One of the primary security challenges with IIoT is the size of the systems. IIoT networks may comprise thousands of connected devices, or even millions, and represent a potential entry point for an attacker [18]. This creates a challenge in the securing of the entire system when devices are spread out in inaccessible or remote locations.

Another challenge is the diversity of IIoT devices, with varying security requirements and capabilities. Some devices might have limited processing power or memory, making advanced security features impossible to implement, while others might be designed for specific use cases that require unique security measures.

To overcome these challenges, organizations must adapt a multi-layered security approach. These include physical security, network security, and data security approaches [19]. The latter accounts for the implementation of encryption, authentication, and access control mechanisms. These ensure data protection at rest as well as during transit. Furthermore, organizations should frequently update their security protocols to ensure they address new threats that arise.

4.2 Privacy and Data Protection in IIoT

Privacy is another critical issue in IIoT environments, as huge amounts of data are being collected, processed, and analyzed. The data generated by the IIoT devices can be sensitive to people, including their location, behavior, and preferences. Thus, it is crucial to protect this data to maintain the trust of the users and meet the privacy regulations [20].

One of the challenges in protecting privacy in IIoT is that data flows are complex. There are usually several stakeholders involved in IIoT systems, each of whom has access to different types of data. This makes it difficult to track and control the flow of data, especially when it crosses different networks and jurisdictions [21].

Another challenge is the possibility of data breaches and unauthorized access. IIoT devices are vulnerable to hacking, data theft, and other forms of cybercrime. When data is compromised, it can have serious consequences for individuals and organizations, including financial loss, reputational damage, and legal liability.

To preserve privacy in IIoT, the organizations, therefore, should have strong data protection measures such as encryption, anonymization and access controls set up. Along with this, there should also be defined data governance policies to define who has access to the data, how it is used, and how it is stored. Organizations can ensure that the privacy of the individuals is protected once they take steps accordingly for making the IIoT systems compliant with relevant privacy regulations.

4.3 Building Trust through Transparent Data Governance

Trust is an important ingredient in successful IIoT implementation. If there is no trust, then the stakeholders are reluctant to embrace the IIoT technologies, data sharing, and interaction with IIoT systems. To create and sustain such trust, the data governance must be transparent with practices ensuring responsible and ethical treatment of data.

One of the essential elements of transparent data governance is clear data ownership policies. Stakeholders should know who owns the data produced by IIoT devices and how that data will be used. This includes the rights and responsibilities of data owners, users, and processors.

User consent is another very important aspect of transparent data governance [22]. Stakeholders need to be in control of their data usage and sharing. For this reason, organizations need to seek explicit user consent before collecting, processing, or sharing user data. Organizations also need to make users aware of the use and potential risks that their data might incur.

Regular audits and assessments are part of transparent data governance to ensure compliance with the data protection regulations and ethical guidelines [22]. Such audits allow organizations to recognize risks and act to correct the problems before they become serious. This protects the data and establishes trust with the stakeholders by demonstrating responsible data management.

5. Focusing on Human-Centric and Community-Oriented IIoT Solutions

5.1 The Significance of Human-Centric Design

Human-centric design is that approach to conceiving technologies that puts the needs, values, and experience of the people using technology first [23]. In the context of IIoT, human-centric design is therefore key towards ensuring that systems are accessible, usable, and aligned with the goals of the people whom they are intended to serve.

User-centered design is one of the major precepts of human-centric design. That means the participation of users should begin from the conceptual phase and involve an assessment of what their needs are and what preferences they may have in using an IIoT system. Engaging users early on and repeatedly creates IIoT systems that are intuitive, friendly to users, and responsive to user groups.

Usability is another great aspect of human-centric design [24]. Easy usability should be made available in these IIoT systems for people to use, considering the lack of technical knowledge at their end. It includes simplifying interfaces which are clear, consistent, with clear instructions on how to conduct operations and sufficient feedback to the user.

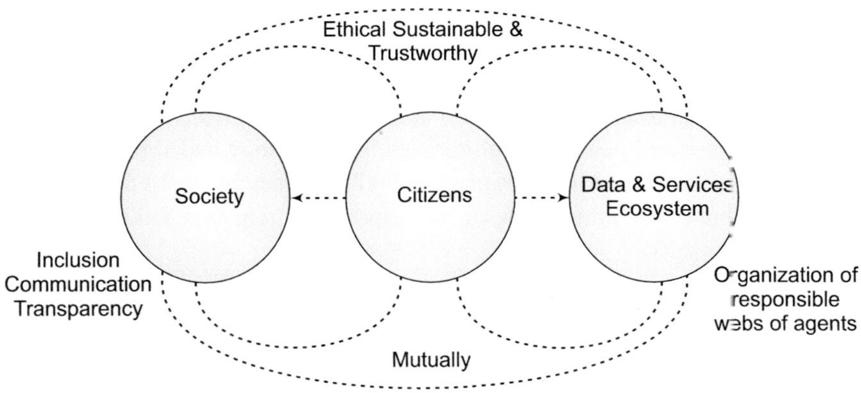

Fig. 2 Human-Centric and Community-Oriented IIoT Solutions

Human-centered design also encompasses the social and cultural environment in which IIoT systems are implemented [23]. This entails consideration of values, norms, and practices among various communities, ensuring that IIoT systems are culturally appropriate and responsive to the needs of diverse user groups.

6. Social Impact of IIoT on Communities

In contrast, with IIoT deployment, drastic change has swept into many corners around the globe that affect various community and work-related elements. With widespread applications of IIoT, a great future with its related boons and disadvantages can influence and change lifestyles for people's ways of working or living.

In Barcelona, Spain, the city has deployed an integrated IIoT system consisting of smart water, streetlight, remote control park irrigation, and smart parking meters. All these technological advancements have contributed much to improve urban living, enhancing the living quality for people. These technological developments have had to force most of the city's workers into retraining, thereby demonstrating that the human laborer should also undergo constant learning and development for survival [25].

Germany, known for its strong manufacturing sector, has seen significant job displacement due to IIoT automation. The introduction of IIoT technologies in manufacturing has automated many tasks, resulting in the loss of traditional jobs. In response, the German government and industries have heavily invested in retraining programs to help workers transition to new roles in the digital economy. This has been crucial in mitigating the effects of job displacement and preparing the workforce for challenges ahead.

Tata Steel has implemented IIoT technologies into its Indian operations to help make its operations more efficient and safer. This has led to loss of jobs from traditional work, but Tata Steel has invested in massive training programs, so its employees are ready for this shifting landscape [26].

The Indian government has also launched the Smart Cities Mission to make use of IIoT for improving urban infrastructure and services in different cities. This initiative has created new job opportunities in technology and urban planning. However, it has also underlined the need for digital literacy and skills development across different communities to ensure widespread participation and benefit from these advancements.

General Electric (GE) in the United States is one of the first organizations that have employed IIoT for predictive maintenance and enhancing operational efficiency. The IIoT has dislodged traditional manufacturing roles, but also introduced new data analytics and management jobs in the field of IIoT systems. GE is a classic example of the duality of impact by IIoT—eradicating jobs for traditional manufacturing, but providing opportunities for emerging careers.

In Detroit, Michigan, IIoT has drastically changed the automobile industry. There is now massive automation in the manufacturing processes that has eliminated many of the traditional jobs. However, the workers are being retrained for new jobs such as the management and maintenance of the IIoT systems, a fact that requires workforce adaptability in the face of technological change [27].

These examples depict the diverse social impacts of IIoT in different regions. Although IIoT technologies are bringing along the challenges of job displacement and increased inequality, proactive measures such as retraining programs and infrastructure investments can help to counter these problems. Organizations and governments must, therefore, take a proactive approach in considering the needs and concerns of the affected communities. This will include training and support for displaced workers through IIoT technologies, skills that are relevant to thriving in a changing job market, and investment in infrastructure and education for equitable access to the benefits of IIoT.

Overall, while IIoT technologies offer numerous benefits, it is crucial to manage their implementation carefully to minimize adverse social impacts and promote inclusive growth. By adopting a balanced approach that addresses both the opportunities and challenges presented by IIoT, societies can ensure that the benefits of these technologies are more evenly distributed.

6.1 Engaging Communities in IIoT Projects

Community engagement in planning and implementing IIoT projects is essential for the success and sustainability of such projects. Community engagement is about involving local stakeholders in decision-making processes, ensuring that their needs and concerns are addressed, and that they have a say in how IIoT technologies are deployed.

Community engagement will help build trust and buy-in from local stakeholders. When communities feel that they have a voice in the development of IIoT projects, they are more likely to support these projects and contribute to their success. This can lead to more sustainable and effective IIoT implementations that benefit all stakeholders.

Community engagement also helps in ensuring that the IIoT projects undertaken align with the local needs and priorities [28]. The organization can ensure that its IIoT projects are responding to the specific challenges and opportunities of the communities where it is deployed by involving the local stakeholders in the planning process. This will result in more relevant and impactful IIoT solutions for real-world problems.

To successfully engage communities in IIoT projects, organizations must adopt a participatory approach that involves stakeholders at every stage of the project lifecycle. This may include conducting consultations, focus groups, and surveys to gather input from community members, as well as establishing advisory boards or working groups to provide ongoing feedback and guidance.

7. Promoting Collaboration and Cross-Industry Synergies

7.1 The Importance of Collaborative Innovation

Collaborative innovation is a key driver of success in the IIoT space. Organizations from different industries can combine their expertise, resources, and perspectives to develop more robust and effective IIoT solutions by working together. Collaborative innovation can lead to new insights, breakthroughs, and synergies that would not be possible in isolation.

One of the main advantages of collaborative innovation is that it enables organizations to pool their resources and share the risks associated with IIoT development. Companies can gain access to new technologies, markets, and expertise by collaborating with other organizations, thus reducing the costs and risks of developing IIoT systems on their own.

Collaborative innovation also opens up avenues of idea and knowledge exchange between sectors. Cross-industry pollination of ideas and solutions may sometimes lead to more innovative approaches or solutions to everyday problems. For instance, lessons gleaned from healthcare might be brought into industrial fields to improve on safety and reliability, or techniques learned from automotive might be transposed to improving the efficiency of energy systems.

To encourage collaborative innovation, organizations should look for partnerships and collaborations with other companies, research institutions, and government agencies. These collaborations can take many forms, from joint research projects and technology transfer agreements to public-private partnerships and industry consortia. Through collaboration, organizations can achieve more than they could alone and contribute to developing a more responsible and sustainable IIoT ecosystem.

7.2 Insights from Cross-Industry Collaborations

Cross-industry collaborations have been instrumental in driving the development of IIoT technologies. These collaborations have enabled organizations to share knowledge, resources, and expertise, leading to the development of more effective and innovative IIoT solutions.

An example of cross-industry collaboration is the partnership between the automobile and telecommunications industries for the development of connected car technologies. Together, these industries developed new standards, protocols, and associated technologies that can be applied for communication between vehicles and with the infrastructure for safety, efficiency, and convenience.

Another example is the collaboration between the energy and manufacturing industries to develop smart grid technologies. By integrating IIoT devices into energy grids, these industries have optimized energy distribution, reduced waste, and improved energy systems' reliability [3].

The above examples illustrate how the value of collaboration across different industries helps stimulate innovation for IIoT. Organizations with various strengths and competencies, therefore, come together to provide effective and impactful solutions for the solution of similar issues and creating new ones.

7.3 Future Directions for Collaborative IIoT Development

Looking forward, the collaboration aspect would be another reason for developing IIoT in the future. IIoT will be so complicated and integrated in the near future that the problems and benefits could only be faced or seized together.

Future collaboration includes the development of new standards and protocols for IIoT. As more and more IIoT systems come to be used, there will be a greater demand for standardized approaches to data sharing, communication, and interoperability. Collaboration among organizations can lead to the development of compatible, secure, and scalable IIoT systems [29].

Another area of collaboration is the development of new business models for IIoT. With new forms of value creation enabled by IIoT technology, organizations will work together to develop new business models for capturing value and fairly redistributing it to stakeholders. For example, this could be in the form of new pricing models, revenue-sharing agreements, or collaborative platforms that allow organizations to share resources and benefits [6].

Lastly, collaboration is needed to be able to respond to the ethical and social implications of IIoT. With collective efforts, organizations can collectively build common frameworks for ethical decision-making, social impact assessment, and community engagement in the development and deployment of socially responsible IIoT technologies, thus ensuring that it aligns with the values of all stakeholders.

8. Upholding Transparency, Accountability, and Regulatory Compliance

8.1 The Role of Transparency in IIoT

Transparency is a critical component of responsible IIoT development [30]. Without transparency, stakeholders may be reluctant to trust IIoT technologies, share data, or engage with IIoT systems. Transparency ensures that stakeholders are informed about how IIoT systems operate, how data is collected and used, and how decisions are made.

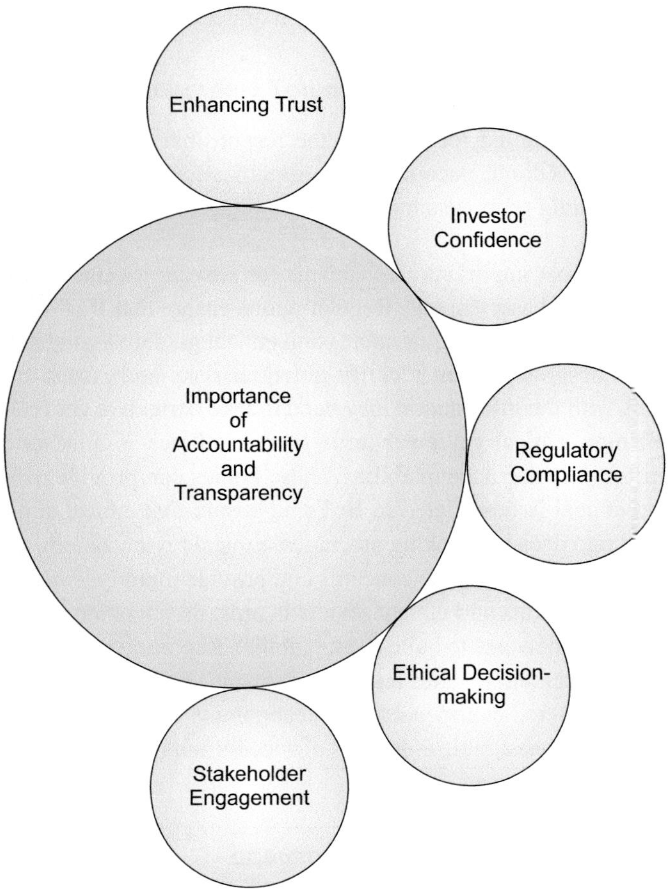

Fig 3 Role of Transparency and Accountability.

One of the transparency elements is the clear communication. Clear and accessible information about the organizations' IIoT systems needs to be communicated to stakeholders; this includes data collection, storage, and utilization.

This entails giving users clear explanations of the risks and benefits associated with IIoT technologies as well as details on how their data will be protected.

Another important aspect of transparency is accountability [30]. Organizations need to be held accountable for their actions and decisions and be prepared to explain and justify them to stakeholders. This includes being transparent about the potential risks and impacts of IIoT systems and the measures taken to mitigate such risks.

To enhance transparency in IIoT, organizations should embrace open and inclusive decision-making procedures. This may be by including stakeholders in the development of IIoT systems, holding regular consultations with stakeholders, and informing stakeholders about information or decision-making procedures.

8.2 Accountability Mechanisms in IIoT Operations

Accountability is required for ensuring the responsible and ethical functioning of IIoT systems. Without accountability, organizations might prioritize short-term gains over long-term sustainability or disregard the needs and concerns of stakeholders [31].

One of the most important mechanisms for ensuring accountability in IIoT is through audits and assessments. Regular audits ensure that IIoT systems are in compliance with regulatory requirements and ethical guidelines and are operating as intended. These audits can identify potential risks and issues and provide organizations with the information they need to take corrective action [30].

Establishing ethical review boards or committees is another important mechanism for ensuring accountability. These bodies can provide oversight and guidance on ethical issues related to IIoT and ensure that ethical considerations are integrated into decision-making processes. Ethical review boards also provide a forum for stakeholders to raise concerns and provide input on ethical issues.

In addition to audits and ethical review boards, organizations can also make use of reporting frameworks to build accountability. Reporting frameworks provide stakeholders with regular updates on the performance and impact of IIoT systems and help build trust and confidence in the technology. The reporting frameworks must contain clear, measurable indicators of success and information on how risks and issues may be addressed.

8.3 Navigating the Regulatory Landscape

The regulatory landscape for IIoT is complex and ever-changing. With the increasing proliferation of IIoT technologies, there is a need for clear and consistent regulations that govern the development and deployment of these technologies. These regulations must balance the need for innovation with the need to protect public safety, privacy, and the environment.

Another key challenge in the regulatory landscape of IIoT is the variability of regulatory frameworks across different industries and regions [30]. Each country and region has its unique regulations regarding data protection, cybersecurity, environmental protection, and more aspects of IIoT. This creates problems for organizations operating in multiple jurisdictions, as they have to abide by a patchwork of various regulations.

A proactive approach to regulatory compliance is one in which the organizations are constantly kept abreast of changes in regulations, have communication with the regulators and policymakers, and come up with strategies of compliance in conformity with the requirements of regulation.

Besides adhering to the extant regulations, organizations should be proactive in helping shape the regulatory landscape for IIoT [25]. This can be achieved by participating in industry associations, contributing to the development of standards and best practices, and advocating for policies that support responsible IIoT development. Through engagement with regulators and policymakers, organizations can ensure that the regulatory environment for IIoT is conducive to innovation and yet supportive of public interests.

9. Conclusion: A Call to Action

9.1 The Responsibility of all Stakeholders

As we close this book, it is evident that the responsibility for shaping a technological future that is ethical, sustainable, and equitable rests with all stakeholders. This includes developers, engineers, business leaders, policymakers, and end-users. Each of these groups has a role to play in ensuring that IIoT technologies are developed and deployed in ways that benefit society and protect the environment.

Therefore, integration of ethical concerns, sustainability, and security issues in the designing and implementation of the IIoT system will also be the implication for developers and engineers. The implication is acting proactively for the identification of potential risks to be addressed appropriately and being open about the trade-off decisions made.

For business leaders, this means prioritizing long-run sustainability over short-term profits. It means being responsible for the role of IIoT systems towards the stakeholders and ensuring that any ethical and social considerations be included in business strategies and decision-making processes.

For policymakers, it will mean developing regulations and policies that support responsible IIoT development. It means engaging with industry and other stakeholders to ensure that regulations are aligned with the needs and realities of the IIoT ecosystem and that they promote innovation while also protecting public interests.

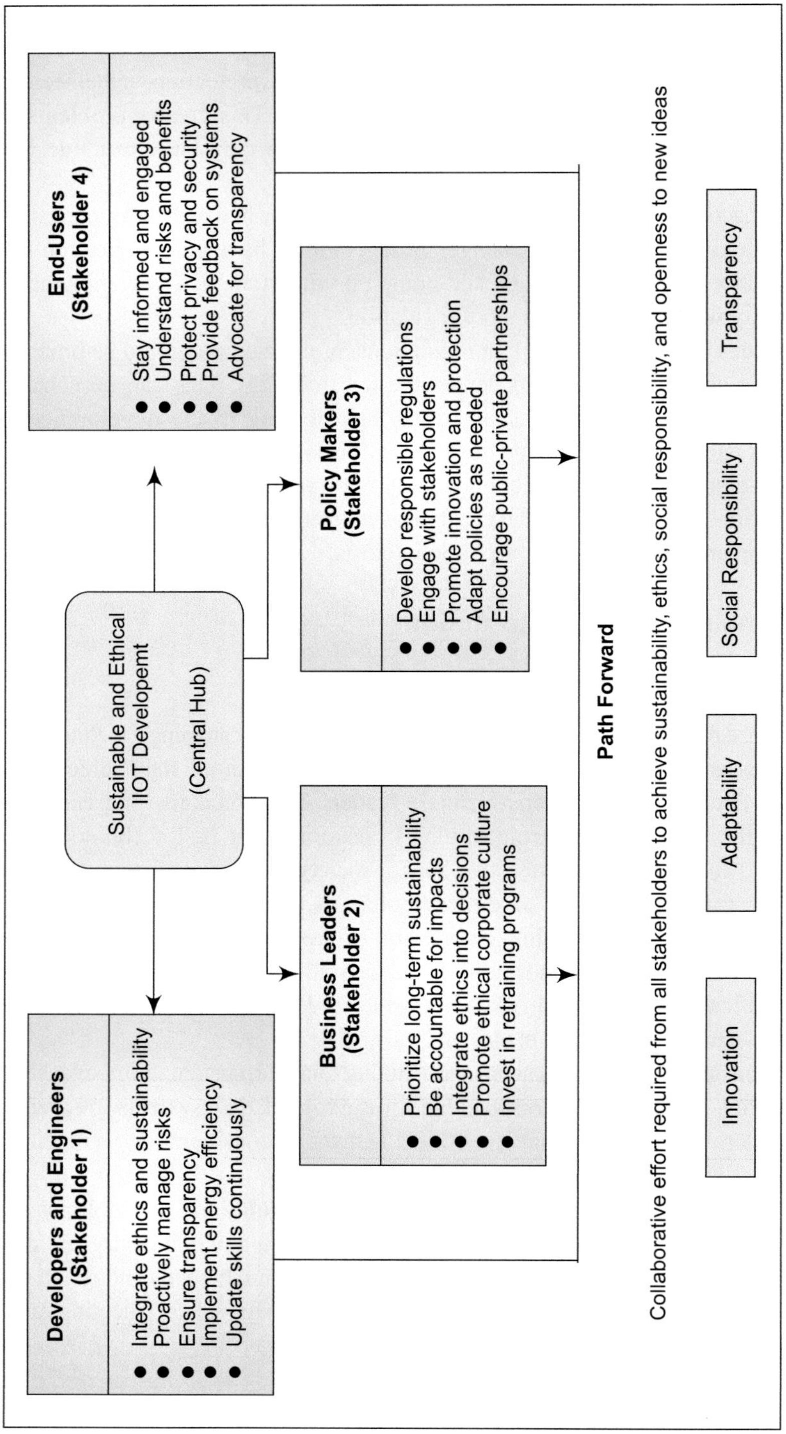

Fig 4 Stakeholder Responsibility Framework for Sustainable and Ethical IIoT Development.

To end-users, this means being well-informed and actively engaged. This is to understand potential risks and benefits of IIoT technologies, hence taking proactive actions in protecting one's privacy and security. Moreover, it entails being active participants in the development and deployment of IIoT systems, which involves providing feedback and input for these systems to be aligned with their needs and values.

9.2 Moving Towards a Sustainable and Ethical Future

The journey toward a sustainable and ethical future is in no way near the end. It will need constant watchfulness, creativity, and an association with the principles talked about in this book. It also needs the willingness to adapt and evolve in response to new challenges and opportunities.

With continued development in IIoT technologies, it is critical that all stakeholders remain committed to the principles of sustainability, ethics, and social responsibility. It means continued development and implementation of IIoT systems that are energy-efficient, environmentally friendly, and aligned with the needs of all stakeholders. It also involves being proactive in identifying and addressing potential ethical and social risks and being transparent and accountable in the decision-making process.

Moving forward, there will be a need to discover new approaches and solutions that can help overcome the challenges and opportunities associated with IIoT. This may include the development of new business models, the adoption of new technologies, or collaborating with other stakeholders to develop innovative solutions. We can ensure that IIoT technologies contribute to a more sustainable and ethical future by staying committed to these principles and being open to new ideas and approaches.

9.3 Final Reflections and the Path Forward

By thinking of the essential messages of this book, it is seen that the IIoT development and deployment represent both challenges and opportunities. For one, it will transform an industry, promote efficiency, and help open up innovation and growth areas. On the other hand, there are more ethical, social, and environmental concerns with IIoT technologies than before.

The path forward will require a collective effort from all stakeholders. It will require a commitment to the principles of sustainability, ethics, and social responsibility, as well as a willingness to engage with and learn from each other. It will also require a recognition that the development and deployment of IIoT technologies is not just a technical challenge, but a social and ethical one as well.

Moving forward in embracing IIoT as responsibility and with an added purpose means we all would work to build more innovative and sustainable, ethically aligned systems and infrastructures, where the IIoT meets diverse needs of society. In other words, developing and deploying it will enable creating a better and more sustainable as well as ethics-oriented future of all kinds, based on solid, technology-guaranteed ground that will never damage human relationships over time.

References

[1] Dangat, M.T., Chandgude, R., Kushwaha, P., Champeli, M. and Pardeshi, P. (2024). Industrial Internet of Things (IIOT). *International Journal for Research in Applied Science & Engineering Technology (IJRASET)*, *12*(3). https://doi.org/10.22214/ijraset.2024.59103.

[2] Khan, N., Solvang, W.D. and Yu, H. (2024). Industrial Internet of Things (IIoT) and Other Industry 4.0 Technologies in Spare Parts Warehousing in the Oil and Gas Industry: A Systematic Literature Review. *Logistics*, *8*(1), 16. https://doi.org/10.3390/logistics8010016.

[3] Javaid, M., Haleem, A., Singh, R.P., Suman, R. and Gonzalez, E.S. (2021). Understanding the adoption of Industry 4.0 technologies in improving environmental sustainability. *Sustainable Operations and Computers*, *3*, 203–217. https://doi.org/10.1016/j.susoc.2022.01.008.

[4] Shams Forruque Ahmed, Md. Sakib Bin Alam, Mahfara Hoque, Aiman Lameesa, Shaila Afrin, Tasfia Farah, Maliha Kabir, GM Shafiullah and S.M. Muyeen, "Industrial Internet of Things enabled technologies, challenges, and future directions", Computers and Electrical Engineering, Volume 110, 2023, 108847, ISSN 0045-7906, https://doi.org/10.1016/j.compeleceng.2023.108847.

[5] Cisco. (2019). *Cisco cybersecurity series 2019: Consumer Privacy Survey*. Cisco Systems. Retrieved from https://www.cisco.com/c/dam/en/us/products/collateral/security/cybersecurity-series-2019-cps.pdf.

[6] Bertino, E., Islam, N. (2017). Botnets and internet of things security. *Computer*, *50*, 76–79.

[7] Arampatzis, A. *Top 10 Vulnerabilities That Make IoT Devices Insecure*. Available online: https://venafi.com/blog/top-10-vulnerabilities-make-iot-devices-insecure/.

[8] Anand, P.. Singh, Y., Selwal, A., Alazab, M., Tanwar, S. and Kumar, N. (2020). IoT Vulnerability Assessment for Sustainable Computing: Threats, Current Solutions, and Open Challenges. *IEEE Access*, *8*, 168825–168853.

[9] Antonakakis, M., April, T., Bailey, M., Bernhard, M., Bursztein, E., Cochran, J., Durumeric, Z., Halderman, J.A., Invernizzi, L. and Kallitsis, M. (2017). Understanding the mirai botnet. In: *Proceedings of the 26th (USENIX) Security Symposium (USENIX) Security 17)*), Vancouver, BC, Canada, 16–18 August (pp. 1093–1110).

[10] Kolias, C., Kambourakis, G., Stavrou, A. and Voas, J. (2017). DDoS IoT: Mirai and other botnets. *Computer*, *50*, 80–84.

[11] Chen, K., Zhang, S., Li, Z., Zhang, Y., Deng, Q., Ray, S. and Jin, Y. (2018). Internet-of-Things security and vulnerabilities: Taxonomy, challenges, and practice. *J. Hardw. Syst. Secur.*, *2*, 97–110.

[12] Akinrinola, Olatunji, Okoye, Chinwe, Ofodile and Onyeka, Ugochukwu, Chinonye. (2024). Navigating and reviewing ethical dilemmas in AI development: Strategies for transparency, fairness, and accountability. *GSC Advanced Research and Reviews*, *18*, 050–058. doi: 10.30574/gscarr.2024.18.3.0088.

[13] Allioui, H. and Mourdi, Y. (2022). Exploring the Full Potentials of IoT for Better Financial Growth and Stability: A Comprehensive Survey. *Sensors*, *23*(19), 8015. https://doi.org/10.3390/s23198015.

[14] Nemade, B. and Shah, D. (2022). An efficient IoT based prediction system for classification of water using novel adaptive incremental learning framework. *Journal of King Saud University-Computer and Information Sciences*, *34*, (8), 5121–5131. [Online]. Available: https://doi.org/10.1016/j.jksuci.2022.01.009.

[15] Cairns, C.N., (2005). *E-waste and the Consumer: Improving Options to Reduce, Reuse, and Recycle*. 237–242. 10.1109/ISEE.2005.1437033.

[16] Bhandari, K.S. and Cho, G.H. (2020). An Energy Efficient Routing Approach for Cloud-Assisted Green Industrial IoT Networks. *Sustainability*. *12*(18), 7358. https://doi.org/10.3390/su12187358.

[17] Suiting Ding, Arnold Tukker, Hauke Ward. (2023). Opportunities and risks of internet of things (IoT) technologies for circular business models: A literature review. *Journal of Environmental Management*, *336*, 117662, ISSN 0301-4797, https://doi.org/10.1016/j.jenvman.2023.117662.

[18] Wang, M., Sun, Y., Sun, H. and Zhang, B. (2023). Security Issues on Industrial Internet of Things: Overview and Challenges. *Computers*, *12*(12), 256. https://doi.org/10.3390/computers12120256.

[19] Kumar, R., Kandpal, B. and Ahmad, V. (2023). Industrial IoT (IIOT): Security Threats and Countermeasures. In: *2023 International Conference on Innovative Data Communication Technologies and Application* (*ICIDCA),* Uttarakhand, India (pp. 829– 833). doi: 10.1109/ICIDCA56705.2023.10100145.

[20] Pant, A., Deshmukh, A.R., Kumar, Y. and Soi, A. (2023). Importance of data security and privacy compliance. *International Journal for Research in Applied Science & Engineering Technology* (*IJRASET*), *11*(8). https://doi.org/10.22214/ijraset.2023.56862.

[21] Maharana, K., Mondal, S. and Nemade, B.A review: Data pre-processing and data augmentation techniques. In: Global Transitions Proceedings, *3* (1), 91–99. [Online]. Available: https://doi.org/10.1016/j.gltp.2022.04.020.

[22] Schubert, K.D. and Barrett, D. (2024). Data Governance, Privacy and Ethics. In: Lacity, M.C. and Coon, L. (Eds.), *Human Privacy in Virtual and Physical Worlds: Technology, Work, and Globalization*. Palgrave Macmillan, Cham. https://doi.org/10.1007/978-3-031-51063-2_5.

[23] Madeleine Borthwick, Martin Tomitsch and Melinda Gaughwin. (2022). From human-centred to life-centred design: Considering environmental and ethical concerns in the design of interactive products. *Journal of Responsible Technology*, *10*, 100032. ISSN 2666–6596. https://doi.org/10.1016/j.jrt.2022.100032.

[24] Harte, R., Glynn, L., Rodríguez-Molinero, A., Baker, P.M.A., Scharf, T., Quinlan, L. R. and ÓLaighin, G. (2017). A human-centered design methodology to enhance the usability, human factors, and user experience of connected health systems: A three-phase methodology. *JMIR Factors Hum.*, *4*(1).

[25] World Economic Forum. (2018, January 11). *6 Ways the Internet of Things is Improving Our |Lives*. https://www.weforum.org/agenda/2018/01/6-ways-the-internet-of-things-is-improving-our-lives/.

[26] McKinsey Global Institute. (2015, June). *The Internet of Things. Mapping the Value Beyond the Hype*. https://www.mckinsey.com/~/media/mckinsey/industries/technology%20media%20and%20telecommunications/high%20tech/our%20insights/

the%20internet%20of%20things%20the%20value%20of%20digitizing%20the%20physical%20world/the-internet-of-things-mapping-the-value-beyond-the-hype.pdf.

[27] World Economic Forum. (2019, April). *A New Era of Manufacturing in the Fourth Industrial Revolution: $7 Billion of Possibilities Uncovered in Michigan.* https://www3.weforum.org/docs/WEF_The_New_Era_of_Manufacturing_in_the_Fourth_Industrial_Revolution.pdf.

[28] Anthony Jnr, Bokolo. (2023). The Role of Community Engagement in Urban Innovation Towards the Co-Creation of Smart Sustainable Cities. *Journal of the Knowledge Economy, 15*. doi: 10.1007/s13132-023-01176-1.

[29] Nemade, B. (2016). Automatic traffic surveillance using video tracking. *Procedia Computer Science, 79*, 402–409. [Online]. Available: https://doi.org/10.1016/j.procs.2016.03.052.

[30] Mohsen Soori, Fooad Karimi, Ghaleh Jough, Roza Dastres and Behrooz Arezoo. (2024). Blockchains for industrial Internet of Things in sustainable supply chain management of industry 4.0: A review. *Sustainable Manufacturing and Service Economics, 3*. 100026. ISSN 2667-3444, https://doi.org/10.1016/j.smse.2024.100026.

[31] Zvarivadza, T., Onifade, M., Dayo-Olupona, O., Said, K.O., Githiria, J.M., Genc, B. and Celik, T. (2024). On the impact of Industrial Internet of Things (IIoT): Mining sector perspectives. *International Journal of Mining, Reclamation, and Environment*, 1–39. https://doi.org/10.1080/17480930.2024.2347131.

Index

About the Editors

B.K. Mishra is a distinguished academician and administrator with over 33 years of experience in teaching, research, and educational leadership. Born in 1966, he holds a B.E. in Electronics from Amravati University (1988), an M.E. in Electronics and Communication Engineering (1992), and a Ph.D. in the same field from Birla Institute of Technology, Ranchi (1998). Currently serving as the Principal of Thakur College of Engineering & Technology (TCET) since June 2006, His expertise lies in Communication Engineering and Electronic Circuits, with over 400 publications in national and international journals and conference proceedings. He has authored six textbooks, edited over 20 conference proceedings, and contributed to many book chapters on emerging technologies. Till date Dr. Mishra has guided 20 Ph.D. and 21 postgraduate scholars. A pioneer in academic event organization, he has led over 70 faculty development programs, 49 conferences, and delivered more than 100 motivational talk. His commitment to quality education is evident in his leadership in securing multiple times NBA and NAAC accreditations for TCET, alongside driving its autonomous status.

Gaurav Dhiman is a highly accomplished academician and researcher with an impressive track record of excellence in computer science. He holds a Ph.D. in Computer Engineering from Thapar Institute of Engineering & Technology, Patiala, and has a Master's in Computer Applications from the same institute. Currently, he is an Assistant Professor in the School of Sciences and Emerging Technologies, Jagat Guru Nanak Dev Punjab State Open University, Patiala. His research has been recognized at the highest level, as he has been named one of the world's top researchers by Stanford University, USA list of world's top 2% scientists prepared by Elsevier. He is the senior member of IEEE. He has also received accolades for his outstanding reviewer work from Knowledge-Based Systems. He has authored over 300 peer-reviewed research papers, all of which are indexed in SCI-SCIE, and 10 international books.

Hemant P. Kasturiwale holds a B.E. degree (1995) and an M.E. in Electronics Engineering (2009) from Sant Gadge Baba Amravati University, India. He earned his Ph.D. in Electronics & Telecommunication Engineering from the same university. Currently, he serves as the Head of the Department and Professor in the Department of Electronics and Computer Science at Thakur College of

Engineering and Technology (TCET), Mumbai, Maharashtra, India. Dr. Hemant Kasturiwale has authored over 50 research papers in reputed international journals and conferences. His areas of research interest include Robotics and Automation, as well as Healthcare Devices and Applications using Machine Learning. He is an active member of the Indian Society for Technical Education (ISTE) & Fellow member of Institution of Electronics and Telecommunication Engineers (IETE) and many more reputed institutions.

Sujata Alegavi holds a B.E. degree (2006) and an M.E. in Electronics and Telecommunication Engineering (2010) from Thakur College of Engineering and Technology, University of Mumbai, India. She has secured a Ph.D. from Thakur College of Engineering and Technology, University of Mumbai, India. Currently she as the Head of the Department and Associate Professor in the Department of Internet of Things (IoT) at Thakur College of Engineering and Technology, Mumbai, India. Dr. Sujata Alegavi has authored over 30 research papers in reputed international journals and conferences. Her research interests includes Remote Sensing, Internet of Things (IoT), Image Processing, Artificial Intelligence and Deep Learning. She is a active member of the Indian Society for Technical Education (ISTE), Associate member of the Institution of Engineers – India (IEI) and Professional member of the Institute of Electrical and Electronics Engineers (IEEE) and many more reputed institutions.

Kusum Yadav holds the position of Associate Professor at the College of Computer Science and Engineering, University of Ha'il, Kingdom of Saudi Arabia. With an extensive academic background spanning over two decades, she has imparted knowledge in computer science at both graduate and postgraduate levels in India and the Kingdom of Saudi Arabia. Dr. Yadav's educational credentials include a Master's in Computer Applications (MCA), a Master of Technology (M.Tech), and a Doctorate (Ph.D.) in Computer Science.

Dr Yadav, an accomplished researcher, has more than 65 publications in esteemed international journals and conferences, including books, chapters, and papers. Her innovative pursuits extend to intellectual property rights, evident in her publication of 15 patents under IPR India, alongside three granted patents, two UK-granted patents, two Canadian patents, and 2 Australian patents. Demonstrating her expertise, Dr. Yadav has delivered numerous expert lectures, chaired technical sessions, and held board memberships in reputable international and Scopus-indexed journals. Active in cutting-edge fields like blockchain, artificial intelligence, the Internet of Things, big data, and cybersecurity, she is also involved in professional societies like IEEE, ACM, IAENG, and ISTE. Dr Yadav's multidisciplinary contributions have established her as a distinguished and influential figure in academics and research.

For Product Safety Concerns and Information please contact our
EU representative GPSR@taylorandfrancis.com Taylor & Francis
Verlag GmbH, Kaufingerstraße 24, 80331 München, Germany